GADGETS AND NECESSITIES

An Encyclopedia of
Household Innovations

GADGETS AND NECESSITIES

An Encyclopedia of Household Innovations

Pauline Webb and Mark Suggitt

ABC-CLIO
Santa Barbara, California
Denver, Colorado
Oxford, England

The publishers are grateful to the Museum of Science and Industry in Manchester for waiving reproduction charges for the images used in this book. The Museum welcomes inquiries about the use of images of its collections.

Library of Congress Cataloging-in-Publication Data
Webb, Pauline, 1955–
Gadgets and necessities : an encyclopedia of household innovations / Pauline Webb and Mark Suggitt.
p. cm.
Includes bibliographical references and index.
ISBN 1-57607-081-6 (hardcover) — ISBN 1-57607-383-1 (e-book)
1. Implements, utensils, etc. I. Suggitt, Mark, 1956– II. Title.
TX298.W43 2000
683'.8—dc21
00-011678

This book is also available on the World Wide Web as an e-book. Visit www.abc-clio.com for details.

06 05 04 03 02 01 00 10 9 8 7 6 5 4 3 2 1

ABC-CLIO, Inc.
130 Cremona Drive, P.O. Box 1911
Santa Barbara, California 93116-1911

This book is printed on acid-free paper ♾.

Manufactured in the United States of America

CONTENTS

GADGETS AND NECESSITIES
An Encyclopedia of Household Innovations

PREFACE

> Constant revolutionizing of production, uninterrupted disturbance of all social relations, everlasting uncertainty and agitation, distinguish the bourgeois epoch from all earlier times.
>
> Karl Marx, *Manifesto,* 1848

> Even the humblest material artefact, which is the product and symbol of a particular civilisation, is an emissary of the culture out of which it comes.
>
> T. S. Eliot, *Notes towards the Definition of Culture,* 1947

> Save 8 hours Every Week with a Hotpoint All-Electric Kitchen—gain extra time for all your extra duties.
>
> Advertisement in *Good Housekeeping Magazine,* 1951

The title of this book implies that it is about things—gadgets and necessities within the twentieth-century home. In a way it is, but we hope that it is a little more than that. It is about people and their ideas, about their relationships with things and the value they placed upon them. It is also about people as the products of an industrial society that required them to be both producers and consumers.

It is interesting to note that the word *gadget* appeared as a colloquial term for a small tool or mechanism in the 1880s. By this stage of the century most Victorians would have felt quite comfortable around machines of one sort or another. A gadget is something that already exists; it is tangible. But what of necessity? a need, the wanting of a thing. As humans we have basic natural needs that must be satisfied to keep us alive, but as the development of civilization has illustrated, what was a luxury to one generation becomes a necessity to the next. Why and when this change is decreed is far from obvious, but what is becoming clearer is that for the majority of people living in industrialized societies many of their needs are cultural rather than natural. So, if we are to look at the relationship between gadgets and necessities, we have to attempt a synthesis of approaches, to look at the material (the gadgets) and their necessity (the culture that informed their production and consumption). An alternative title to this book could therefore be *Electricity and Desire.*

The home has long been a microcosm of the larger society of which it is a part. The things that we choose to bring into our homes say a lot about us. Domestic objects have always been a rich source of information about the way people have lived and what they believed to be important. Everyday objects have long been valuable evidence for the archaeologist, even when written records are available. Nevertheless, the proliferation of written records has meant that the study of everyday things within the modern period has often been neglected. Historians have concentrated on documents rather than things. This is certainly true for the twentieth century, which has a greater weight of documents, photographs, films, and tapes than any other. Paradoxically, the museums that were largely creations of the industrial age have until recently concentrated on fine and decorative arts and antiquities. They have often ignored other avenues of inquiry, mistakenly letting artifacts try to speak for themselves.

The last thirty years have witnessed a change in both these areas, with historians of modern history taking a greater interest in the artifacts and their relationships with the culture that generated them. Museums have also become far more interested in the collection and study of post-1945 material. They have been influenced by the recognition of a broader, more inclusive definition of culture. Artifacts are no longer studied or displayed in isolation but placed in a context. The growth of "history from below," an approach informed by Marxism, feminism, and personal testimony, has led museum workers and historians to take a serious look at the ubiquitous objects and practices that constitute much of our domestic life and landscape. The late twentieth century has also seen a growth in researching local and family histories. In addition, there has been a recognition that design history, sociology, and semiotics have much to offer. Objects are no longer seen in aesthetic isolation from the societies from which they originated. They possess differing meanings, which can be intentional or appropriated; therefore they can be "read" in a variety of ways.

This book offers one such reading. It is an attempt to produce a useful, if not entirely comprehensive, survey of the key appliances that have been produced and used within the home during the twentieth century. It is a study of material culture, a term that links artifacts with the world that made and used them. As a result it looks at the social, economic, and cultural influences that effected their creation and consumption, mainly from the experience of the United Kingdom and the United States. There were important innovations within Europe, and these have been included. Nevertheless, the twentieth century began with the United Kingdom clinging to its industrial superpower status, which was being challenged by Germany and the United States. All three countries had produced pioneers in the generation of electricity and innovation in product design, but by 1945 it was clear that the United States was in a dominant position, one that it would maintain for the rest of the century.

Although this book concentrates on the twentieth century we have occasionally gone back to earlier periods in order to explain the development of a particular appliance or social phenomenon. This is because much of twentieth-century domestic history has been a working through of trends that were beginning in the late nineteenth century. Mass-produced gadgets and machines for specific tasks have proliferated since the 1850s. The majority of these, such as mincers, choppers, and grinders, were muscle powered, but they were responding to a form of domestic life that although different, would be recognizable to us today. The nature of the work that these appliances were designed to improve has changed little. The main tasks of what we call housework—childcare, cleaning, cooking, washing, and ironing—remain, along with the addition of other areas such as shopping and preserving food.

What has changed during the course of the twentieth century has been the growth rate of technology, communication, and urbanization. Weaving its way through this matrix has been a mixture of both gradual and explosive shifts in culture, economics, and politics. These have either driven technological change or followed in its wake. Humans have a remarkable capacity to deal with change, accept it, and then move on. It can be, and indeed is, all too easy to take so much for granted. It may sound ridiculous or pompous to include the objects of everyday use into such a world-view, but we would argue that if the history of a locality, the history on our doorsteps, can help define who we are and were we have come from, then so can the history of what is, quite literally, under our noses!

Mark Suggitt and Pauline Webb
March 2000

ACKNOWLEDGMENTS

The object and archive collections and library of the Museum of Science and Industry in Manchester were my starting point in researching this book. I hope that the book encourages more people to study and enjoy the domestic technology collections of this museum and others. As a British author struggling to write for a British *and* American readership, I hope that American readers will excuse any major omissions and cultural lapses or misinterpretations.

As this book would never have been finished if I spent the average amount of time on housework, I am grateful for the example of Quentin Crisp, my personal housework role model, who once remarked: "After the first four years the dirt doesn't get any worse." I would also like to thank my friends and colleagues for their tolerance and encouragement during the long gestation period.

Pauline Webb

Many people have contributed to the writing of this book, most of them unwittingly. They are the unselfish donors of the thousands of artifacts to museums throughout the United States and the United Kingdom. Without them we would not have had the opportunity to work with and study domestic appliances during our curatorial careers. In addition, there are the curators who have sought out and collected such everyday objects and the memories that are associated with them.

Personally my greatest thanks go to Helena, Niam, and Oliver Suggitt—my long suffering family, who put up with me hiding away in the study for many evenings and weekends. Special thanks are also due to Leanda van Noort, for U.S. information and trips to malls in Chicago. Also to Peter Jenkinson and Sue Runyard, for deciding on the right time for a study tour to California. Thanks also go to our patient editor, Bob Neville, for providing the opportunity to write this book and also for his ability to elongate deadlines.

Thanks are also due to Anthony Breen of Pifco, Richard Stansfield of York Castle Museum, and John Narborough of Amberley Museum. Finally, in a book on domestic appliances, it seems entirely appropriate to pay homage to my Dualit electric kettle and Bodum cafetiere, which together provided the gallons of coffee required!

Mark Suggitt

INTRODUCTION

The typical homes of 1900 and 2000 in the developed world were very different. This transformation was the product of a complex interaction between technological factors and social, economic, and cultural factors. In the nineteenth century, the first industrial revolution affected the home primarily by reducing its role as a workplace. In the twentieth century, the second industrial revolution brought the same technologies, essentially electrical and electronic, into the home as into the workplace.

Industrialization and the Home

The Industrial Revolution, which began in the British textile industry in the late eighteenth century, marked the shift in primacy from the domestic to the social mode of production. In preindustrial society, the household was the main unit of production and the household economy functioned by combining the production of goods for subsistence (that is, for consumption within the household) and goods for cash or exchange. From early times, the household family unit had featured the division of labor by gender—the male hunter and the female gatherer. In agricultural societies, this division persisted but men and women had roles both inside the house and outside in the fields. For instance, domestic textile production was characterized by the female spinner and the male weaver. Industrialization preserved this division of this labor, but it removed cash production from the home to the factory or workshop.

The Industrial Revolution also marked a shift from the use of tools to the use of machines. In anthropological terms, the distinction between a tool and a machine is that a tool relies on human muscle power and therefore operates primarily as an extension of the mechanical functions of the body, while a machine relies on nonhuman energy and the human operator is relegated to the role of attendant. The first phase of industrialization was based on the harnessing of waterpower to drive machinery, but waterpower was soon succeeded by steam power, which freed manufacturers from the locational restrictions of waterpower. Neither waterpower nor steam power was suited to the domestic context, as both required the installation and maintenance of a large plant.

With the removal of work as an economic activity from the household, there began to be a gradual change in the way that the home and the role of women within it were perceived. While women were employed outside the home from the onset of industrialization, the home became characterized as the female sphere or domain while the workplace was the male sphere or domain. The home became romanticized and expectations rose. Housework, a term only coined in the mid-nineteenth century, did not produce income, so it could not be "real work." Yet, washing, cleaning, and other household chores were laborious tasks. The solid fuels for heating, lighting, and cooking had to be carried about and produced dirty residues. The materials used in the home, such as cast iron for grates and ranges, were not easy to clean. Although bottled, canned, and other processed foods became commercially available, they were outside the budget of many households. Cotton clothes were easier to wash than their woolen predecessors were, but as standards of personal hygiene rose, clothes were washed more frequently. In large, wealthy households, housework was carried out by an army of servants with specialized functions. The possession of at least one servant was essential to any aspiring middle-class household. In the majority of homes, however, the "housewife" toiled alone without assistance from paid help and with few, if any, truly laborsaving devices.

The Seeds of Change

The typical urban houses of 1800 and 1900 were not greatly different, but the nineteenth century did bring two significant changes to the household: piped gas and piped water. Public gas and water-utility undertakings initially focused on meeting the demands of industry, but by 1900 they were equally committed to serving domestic consumers. The production of coal gas for commercial supply began in the early nineteenth century in both Britain and the United States. Coal gas was also known as town gas because gas works were located in towns, where demand was concentrated in a small area, enabling local gas suppliers to spread the infrastructure costs over more customers. At first, gas was conceived purely as a source of power for lighting. Gas stoves and heaters were available by the mid-nineteenth century but were seldom used in the home until after 1880, when electricity began to erode the dominance of gas lighting. Gas suppliers and appliance manufacturers responded to the competition from their electricity counterparts by pushing the use of gas for heating and cooking. Nevertheless, in1900, lighting still accounted for the bulk of gas consumption.

In Britain and the United States, urban water-supply and sewer systems expanded significantly from the middle of the nineteenth century on. Towns and cities began to build reservoirs so that they could cope with increasing water consumption. In Europe, a series of cholera epidemics had focused attention on the need for public-health improvements. In particular, it was noted that the life expectancy of urban dwellers was much lower than that of their rural counterparts, even though small country towns often had areas with appalling sanitary conditions. The provision of pure water supplies and better sanitation was seen as the key to eradicating certain infectious diseases and thereby lowering urban mortality rates. Water suppliers, however, were concerned about demand exceeding supply and controlled the supply to homes by providing piped water to communal pumps or standpipes. Households could pay to have water piped direct to the home, but this was only a realistic option for the privileged few. So, while the quality and quantity of the water supply improved, the onerous task of carrying water home remained for most households. It was not until the late nineteenth century that suppliers began to pipe water directly to houses as a standard service. Even then, most homes would have limited internal plumbing, in many cases restricted to a single cold-water tap in the kitchen and a plumbed water closet in the back yard.

While public health officers monitored the suppliers, their work was complemented by that of voluntary associations that fostered public awareness of hygiene issues. Fueled by the work of the French chemist and microbiologist Louis Pasteur and the English surgeon Joseph Lister on the germ theory of disease, the hygiene movement of the late nineteenth century charged women with the responsibility for family health and welfare. Women were expected to raise their standards of household cleanliness or be prepared to face the consequences. Hygiene campaigners saw the provision of piped water as an essential component in the fight against dirt and disease. Freed from the effort of obtaining water, women could put more time and energy into washing and cleaning.

Piped gas and water went some way toward making housework more efficient and the home more comfortable, but they produced little fundamental change in the processes of housework. The utility that would eventually revolutionize housework was electricity. In the late 1870s and 1880s, public electric utilities were set up in many towns and cities in Britain and the United States. The pattern of development of electricity supply was similar to that of gas. At first, the emphasis was on public lighting and industrial use. A range of electric domestic appliances—irons, toasters, heaters, stoves—were commercially available by 1900, but very few homes were wired for electricity.

Concurrent with the introduction of electrical technologies were the development and introduction of new communication technologies. In the 1870s, the telephone and phonograph appeared, followed in the 1890s by the radio, motion picture cameras and projectors, and the cathode ray tube (later to become a fundamental component of the television set). These new technologies did not make their impact on the home until the twentieth century.

There were also a number of changes in the late nineteenth century that collectively represented the growing democratization of society. These democratic forces also raised working-class expectations of pay and living standards. One strand of change was the spread of education. In the United States, the importance of education had long been recognized, and by 1850, the provision in all states of publicly funded schools, open and free to all, enshrined education as a right rather than a privilege. In the 1860s, the granting of federal land for states to establish colleges, particularly vocational colleges, extended access to higher education. In Britain, the 1870 Elementary Education Act set up local school boards to ensure that there was room in free schools for all children. New "redbrick" universities were set up in industrial towns and vocational education was provided through mechanics institutes and technical schools.

Workers sought to increase their political and economic influence through organization and demonstration. In Britain, the first meeting of the Trades Union Congress (TUC) took place in Manchester in 1868, attended by delegates from thirty-four trade unions. Three years later, the 1871 Trade Union Act gave legal recognition to trade unions. The success of the London dock workers' strike in 1889 encouraged unskilled workers to form unions, and the formation of the Independent Labour Party in 1893 provided a vehicle for working-class men to stand for parliamentary election. In the United States, there was even more reason for workers to be restive, as there was no legislation controlling work conditions until 1874. A national labor organization called the Noble Order of the Knights of Labor was set up in 1869, but was later superseded by the American Federation of Labor, formed in 1886. A number of violent labor conflicts occurred in the last quarter of the century, including the Great Rail Strike of 1877, the Carnegie strike of 1892, and the Pullman strike of 1894. Meanwhile, a period of hardship for farmers in the South and West resulted in the formation of Farmers' Alliances and the Populist Party, which contested the 1892 elections.

The New Century (1900–1918)

No sudden changes accompanied the dawning of the twentieth century. The electrification of the home proceeded slowly, but more rapidly in the United States than in Britain. In the United States, supply companies and manufacturers were quick to realize that standardization would lower prices and permit the faster growth of the electricity industry. As the key patents were held by a small number of companies, the process of negotiating joint standards and implementing them through cross licensing was simplified. The leading manufacturers, General Electric and Westinghouse, held stock in many of the electric utility companies. In 1885, the utility companies banded together to form the National Electric Light Association (NELA). Cooperation and discussion were also fostered through the Electric Club, founded in 1886 and based in New York. Accordingly, by 1910 the whole United States had a standard electricity supply in the form of alternating current delivered to the user at 120 volts. Standardization brought economies that made the price of electricity far lower in the United States than in other countries, encouraging its faster introduction to the home.

Even in the United States, however, manufacturers concentrated on meeting the needs of industry and regarded the domestic market as peripheral. Simple economics dictated that it was more cost-effective to tailor infrastructure to serve large industrial customers. Moreover, electrical manufacturers were used to customizing their plant and equipment for individual contracts, rather than producing many identical pieces of equipment. Until the 1920s, domestic appliance manufacture tended to be the preserve of smaller companies. By 1907, 8 percent of American homes were wired for electricity and this percentage had doubled by 1912. In Britain, where electricity supply had not been standardized and existed in both direct and alternating forms with a variety of voltages, only 2 percent of homes were wired for electricity by 1910. In both countries, most of these homes were wired for lighting alone, so there was only a minuscule market for electric appliances. Therefore, there was, as yet, little incentive for appliance manufacturers to adopt the mass-production techniques that were being espoused by other branches of manufacturing. In the United States, mass-production techniques based on the principle of standard, interchangeable components were well established, but there was still the potential to improve the flow of the manufacturing process. In 1901, Ransom Olds began manufacturing automobiles on the assembly-line principle. Two other Americans, Frederick Winslow Taylor and Henry Ford, took this production concept further.

Taylor was an engineer who moved into management consulting and in 1911 published his *Principles of Scientific Management,* in which he expounded his ideas for increasing industrial efficiency. Taylor was an advocate of step-by-step planning and the reduction of the manufacturing process to a series of simple tasks, each to be carried out by separate workers. He also believed in the use of time-and-motion studies to determine the optimum structure of the manufacturing process. In 1903, Henry Ford, who had trained as a machinist and later worked for the Edison Illuminating Company, formed the Ford Motor Company in Detroit, Michigan. Ford was committed to development of affordable automobiles for the mass market. His cheap, sturdy Model T was launched in 1908, and by 1913 he had found a way to cut production costs further by introducing a conveyor belt to create the moving assembly line.

Household hygiene continued to be a matter of great concern, so it is not surprising that the major domestic innovations in the early part of the twentieth century related to the functions of cleaning and washing. In 1901, the English engineer Hubert Cecil Booth invented the electric vacuum cleaner. Booth attracted a great deal of attention by holding demonstrations at which onlookers were fascinated and horrified to see how much dirt was collected by Booth's machine. Although it was many years before vacuum cleaning became a widespread household practice, in 1908 the American William Hoover became the first manufacturer to market a vacuum cleaner that was compact enough for normal household use. New products were commonly introduced to consumers through exhibitions and trade fairs, such as the world fairs and smaller local expositions. In Britain, a new series of annual exhibitions began in 1908, when the *Daily Mail* newspaper sponsored the first "Ideal Home" exhibition. The very fact that electric appliances were an expensive luxury conferred prestige and a sense of modernity on their owners. The scarcity of electric appliances in the 1912 Sears, Roebuck catalogue was an indication that retailers thought it futile to attempt to mass-market such products at this stage. Three more new electric appliances were launched in the United States before the outbreak of World War I: the washing machine (1908), the dishwasher (1910), and the refrigerator (1913), but there was little opportunity for manufacturers, particularly in Britain, to exploit these new inventions before the war required industry to concentrate on the production of munitions.

The development of practical electric appliances for household use was stimulated by improvements in electric motors. At first, the main emphasis was on performance in terms of operating industrial machinery. All early models were direct current motors. Their main advantage was that their speed could be easily regulated. However, direct current could not be transmitted over long distances without significant energy loss. Alternating current was more efficient because its voltage could be easily "stepped" (raised or lowered) by a transformer so that the high voltage current suitable for long distance transmission was delivered to the end user as safer, lower voltage current. The first practical alternating-current induction motor was invented by the United States–based Croatian engineer, Nikola Tesla, in 1888. Westinghouse, which acquired Tesla's patents, manufactured the first appliance to incorporate a compact alternating-current motor, an electric fan, in 1889. The next significant development for the domestic appliance industry was the fractional-horsepower universal motor, prototyped in 1905 by Chester Beach, an employee of the Arnold Electric Company in Racine, Wisconsin. Beach's motor was lightweight, high-speed, and could be driven by direct or alternating current, hence the "universal" tag. Beach and coworker L. H. Hamilton formed the Hamilton Beach Manufacturing Company in 1911 to exploit the universal motor by producing a range of small appliances.

The invention of laborsaving domestic appliances during this period coincided with social changes affecting the role of women in society. On the one hand, women were being instructed to devote more time and attention to the home and family welfare; on the other hand, women were finding new opportunities for paid employment and were demanding equal political rights. The role of women, particularly mothers, as guardians of family welfare was perhaps most famously expounded by the German emperor, Kaiser Wilhelm II, who directed women to devote themselves to "Kinder, Küche, Kirche" (children, cooking, church), a slogan later adopted by the Nazi party. Manuals of household management had been popular in middle-class homes since the mid-nineteenth century when writers such as Catherine Beecher of the United States and Mrs. (Isabella Mary Mayson) Beeton of Britain produced comprehensive treatises. Frederick Taylor's scientific management techniques opened up a new approach to household management. In 1915, two U.S. authors, Christine Frederick and Mary Pattison, produced books (*Household Engineering* and *The Business of Home Management*) that applied Taylorism to the home. The basic argument was that by analyzing household tasks and organizing work accordingly, women could do housework more efficiently. This approach became increasingly influential from the 1920s on as ownership of laborsaving appliances increased.

Meanwhile, women were seeking greater involvement in the world outside the home. In Britain and the United States, women had been actively campaigning for the right to vote since the mid-nineteenth century. Britain's National Society for Women's Suffrage was founded in 1867, and the American equivalent, the National Woman Suffrage Association, was founded in 1869. By 1900, the American campaigners had enjoyed some success as, for example, the territory of

Wyoming and the state of Colorado granted women the vote in local elections. In Britain, a number of bills promoting female suffrage failed to gain full parliamentary approval. In 1903, dissatisfied with these failures, Emmeline Pankhurst formed the Women's Social and Political Union, which began to mount an increasingly militant campaign. A year later, Susan Brownell Anthony, cofounder of the National Woman Suffrage Association, sought to bring women suffragists together globally by founding the International Woman Suffrage Alliance. Both the British and U.S. campaigns bore fruit only after World War I.

World War I opened up new employment opportunities for women. Prior to the war, the most significant change in the range of paid work available to women had arisen as a result of the adoption of the typewriter as a standard piece of office equipment in the 1880s. Previously, clerical work had been almost exclusively the province of men. The typewriter created a new job, that of typist, for which women with their smaller fingers, were deemed to be suitable. The war introduced women to other male work strongholds, such as engineering. Women were needed to fill the vacancies created by the conscription of male workers into the armed forces. Because many factories were by then electrically powered, women war workers had the opportunity to experience at first hand the new form of power that could also transform their homes.

The gradual transformation of the home was an aesthetic process as well as a technological process. In the late nineteenth century, new design movements rebelled against high Victorian ornamental clutter. These movements prevailed until the Modern Movement and art deco took hold in the 1920s. The main design movements were, in Britain, the Arts and Crafts Movement, which espoused functionality, simple lines, and individual craft values, that is, a single craftsperson producing unique items by hand; and in continental Europe, Art Nouveau (or Jugendstil), which also celebrated craft values and was characterized by sweeping organic shapes. In the United States, the Shakers, who for many years had been producing simple, handcrafted furniture for their own religious communities, established a factory at Mount Lebanon, New York, in 1871. Other American advocates of craft values were Gustav Stickley, Louis Comfort Tiffany, and Frank Lloyd Wright. Stickley, who was heavily influenced by the Arts and Crafts Movement, produced both wooden furniture and metalware. Tiffany, whose father was a leading New York jeweler and silversmith, developed a form of iridescent glassware, called Favrile, in flowing Art Nouveau shapes. Although Wright was principally an architect, his holistic approach to architecture led him to design furniture and fittings for his buildings. The geometric simplicity and, where appropriate, asymmetry of Wright's designs served the dictum "form follows function," coined by American architect Louis Sullivan, Wright's early mentor.

The emphasis of these design movements on the individually crafted item meant that their influence was restricted to wealthy households. However, new materials that made their first appearance in the early twentieth century were later to provide the means of translating these design principles for a mass market. For example, the United States–based Belgian chemist, Leo Baekeland, patented Bakelite, the first wholly synthetic plastic, in 1907. Initially, it was valued principally for its insulating properties and used for the casing of electric fittings such as sockets, plugs, and switches. Bakelite is a thermosetting plastic, which means that it cannot be remelted. Just before World War I, two developments occurred that paved the way for later thermoplastics, which are remeltable and recyclable: the recognition of the role of plasticizers and the thermal cracking of petroleum. New metal alloys also made their appearance at this time. In 1906, Albert L. Marsh developed Nichrome, an alloy of nickel and chromium that was a more durable and efficient material for use as resistance wire in electric heating elements than iron or platinum. In 1913, stainless steel, an alloy of iron, chromium, and nickel, was first manufactured in Britain. Stainless steel has two properties that were to make it valuable in the production of kitchen appliances: it has high tensile strength (so it could withstand being bent sharply) and it is resistant to corrosion.

Homes Fit for Heroes: The Interwar Period (1919–1939)

Historians have come to regard World War I as a political and social watershed and in many ways the true juncture between the nineteenth and twentieth centuries. The Great War earned its name because of its sheer scale, in both geographic and human terms. About 6 million British men, more than a quarter of the male population, served in the war, of whom 750,000 died, 1.5 million sustained serious injuries, and at least another 1 million suffered lesser injuries or shell-shock. Although the United States only took part in the last eighteen months of the war, by the end it had deployed more than 1,750,000 soldiers in France, as well as providing naval support. The number of American fatalities exceeded 115,000 and the number of casualties was just over 200,000.

Although World War I was industrialized slaughter, it was also a watershed for imperialism and served to accelerate the processes of political and social democratization. On the front, men of different social backgrounds had come into contact to an intimate degree that would have been inconceivable in their previous civilian lives. This had a more profound effect in Britain, where class was based primarily on inherited social station, than in the United States, where class was more fluid, being more merit-based. There was a strong feeling that returning soldiers had earned the right to expect, at the very least, a decent home life. Prime Minister David Lloyd George of Britain summed this up in 1918 when he pledged himself to the task of making Britain "a fit country for heroes to live in." The consequent homes-fit-for-heroes schemes entailed slum clearance and the creation of suburban council estates and new towns. Women, too, were to be rewarded for their efforts on the home front. In Britain, the 1918 Representation of the People Act extended the vote to all men over the age of 21 and women householders over the age of 30. It was another ten years before women gained equal voting rights. In the United States, the 19th Amendment, which gave women the franchise on the same basis as men (at the age of 21), became law in 1920.

The decade of the 1920s began with great optimism and ended with the Wall Street Crash and the onset of economic depression. Radio communication technology, which played a vital role in military communication, had benefited from the war, and once public radio broadcasting was launched in both Britain and the United States in the early 1920s, ownership of radio receivers grew rapidly. Ownership was not restricted to homes that were electrically wired, because radio receivers could be battery-operated. The immediate popularity of radio proved that there was a market for new household appliances, provided that they offered genuinely new or improved service and were affordable. The radio began to take the focus of the room away from the hearth, where it had been for centuries. Outside the home, the cinema and the dance hall became major leisure attractions. Cinema was a globalizing force and influenced tastes in clothing, hairstyles, home decor, and other aspects of lifestyle. In the United States, these trends led to the labeling of the 1920s as the Roaring Twenties and the Jazz Age.

Domestic appliance manufacturers and electric utilities began to woo the domestic consumer more assiduously. In both Britain and the United States, the electric utilities faced the same problem: their success in promoting the industrial use of electricity had created an imbalance in the power load, with a heavy load demand during the day and light demand at night. As generating capacity had to be capable of achieving the peak load, capacity was wasted during the off-peak period. By balancing the load, power stations would become more efficient and unit prices would fall. Increasing domestic consumption of electricity was the obvious solution. In 1920, only 6 percent of British homes were electrically wired, compared with 35 percent of American homes. In Britain, the major obstacle to domestic growth was the lack of standardization of electricity supply. The 1919 Electricity Supply Act created the Electricity Commissioners to coordinate the British supply industry. In the same year, an independent initiative, forged through the alliance of supply engineers and appliance manufacturers, set up the British Electrical Development Association (BEDA) to generate promotional material of common value to both wings of the industry. Standardization came seven years later with the formation of a Central Electricity Board to rationalize power stations and connect them through a national grid. The electricity grid was fully operational by 1933.

In the United States, the already standardized supply industry began to mount nationwide campaigns targeting the domestic consumer through NELA. Meanwhile, the big two American electrical equipment manufacturers, General Electric and Westinghouse, began to acquire small appliance companies to increase their stake in the household market. They also recognized the potential benefits of forming a new alliance to complement NELA. In 1926, the National Electrical Manufacturers Association (NEMA) was formed to develop technical standards by defining products and processes, thereby encouraging fair competition and consumer confidence.

Although Britain lagged behind the United States in terms of market penetration, the electricity industries in both countries used the same approaches to stimulate domestic demand. While the wiring of new homes became common practice, incentive schemes were offered to encourage the wiring of older homes, for example, by deferring payment through monthly installments. In Britain, where electric domestic appliances were significantly more expensive than in the United States, the electric utilities introduced low-cost appliance rental schemes. As the main objective was to increase electricity consumption, renting appliances was regarded as a loss leader.

Gas was still a cheaper option, and for some functions, particularly heating and cooking, it was also a more efficient and reliable option. The electricity companies therefore concentrated their promotional messages on aspects of electricity that constituted tangible benefits, such as its cleanliness and laborsaving potential. The latter was especially resonant at a time when an increase in the availability of better-paid unskilled work in factories made domestic service a less attractive proposition than in the past. In both Britain and the United States, the existence of a "servant problem," a shortage of supply, was recognized even before World War I. In the United States, domestic service had long been seen as a last resort, as it held connotations of dependence and deference. Statistics on domestic servants are notoriously unreliable, but the significant change after World War I seems to have been a decrease in multiple-servant households and live-in servants. There was, however, a concurrent increase in the number of households with one servant, often a "daily help," reflecting the growth of the middle class.

Because employment of a servant was an important status symbol—even though it was a hopeless aspiration for many—the electricity industry sought to fill the gap by promoting electricity as the "silent servant." Indeed, the electrical servant was described as superior to the human servant—always available, inexhaustible, and infinitely dependable. It would be demeaning for the middle-class housewife to be engaged in household drudgery, so advertisements gave the exaggerated impression that electric appliances did all the work, with the operator acting merely as an attendant. In reality, it was undoubtedly less time-consuming and less onerous to wash clothes in an electric washing machine than by hand, for example, but the washing machine did not load or unload the clothes. Nevertheless, it was not just the (predominantly male) copywriters who bought into this myth of the self-sustaining electrical home. The liberating potential of electricity filled some women with positively evangelical zeal. This can be exemplified by comparing the statements, two decades apart, of an American male inventor and entrepreneur and a British female engineer and electricity advocate:

> The Woman of the Future will be rather a domestic engineer than a domestic laborer, with the greatest of all handmaidens, electricity, at her service (Thomas Alva Edison, 1913).

> We are coming to an Age when the spiritual and higher side of life will have freer development and this is only possible when women are liberated from soul-destroying drudgery (Caroline Haslett, 1934).

Caroline Haslett was the first director of the Electrical Association of Women (EAW), an organization launched at a meeting of the Women's Engineering Society in 1924. The EAW was independent of the electricity industry, and its objectives were to inform women about "electrical apparatus" and to provide a platform for women's views on all matters relating to electricity, but it's wide-ranging educational program— encompassing courses tailored to specific categories of women (teachers and demonstrators, as well as housewives), handbooks, and other publications—was particularly useful to the industry. Education was vital because although the industry regarded the invisibility of electricity as an asset and safety as self-evident, many women needed to be convinced of these assertions. The U.S. equivalent of the EAW was the NELA Women's Committee, which had a similar mission but was not independent of the industry. Meanwhile, the industry itself also sponsored electrical homemaking courses and employed trained female demonstrators in showrooms and at exhibitions.

By 1930, it was evident that the industry's marketing strategies were paying off—the proportion of homes wired for electricity had risen to more than two-thirds in the United States and to one-third in Britain. Appliance ownership remained low, however, with only the electric iron and the vacuum cleaner well established in the household. Both of these appliances were clearly technically superior to the alternatives and consumed relatively little electricity. Door-to-door selling, as practiced by the Hoover Company, helped to personalize the transaction and allowed women to see appliances demonstrated at leisure and ask questions. After-sales service was another persuasive factor. Manufacturers and utilities were still keen to push larger appliances, such as stoves and washing machines, that would yield greater profits, but the Wall Street Crash of 1929 and the ensuing worldwide economic depression meant that the industry had to reassess its approach. In the United States, President Franklin Delano Roosevelt's New Deal of 1933 ushered in a series of "welfare state" measures. In Britain, similar measures had been introduced from 1908 onwards. The U.S. government also broke with its tradition of leaving the private sector to regulate itself and, through the National Power Policy Committee, began to scrutinize the electric utilities and their pricing policies. NEMA took heed and set up its own National Recovery Administration code to monitor production and prices.

Aside from pricing, the area where electricity had most room for improvement in relation to household usage was appliance design. While the industry had vociferously proclaimed the futuristic qualities of electricity during the 1920s, this image was not apparent in appliance design. Most electric appliances of the 1920s, excluding those that had no direct predecessors, were distinctly traditional in appearance and construction. Washing machines had evolved from wooden tubs to enameled steel tubs and cast-iron stove bodies had also gained enamel finishes, so that their surfaces were easier to clean, but little attention had been given to shape. Even new products, such as the radio receiver, were at best anonymous in appearance.

In order to stimulate interest, appliance manufacturers began to adopt construction techniques used in industries such as automobile production and to employ designers to style products in line with prevailing aesthetic trends. Cast and pressed steel superseded heavier cast iron, with cast aluminum and magnesium being used to a lesser extent where their greater lightness of weight was valuable. The use of pressed steel for the casings of large kitchen appliances allowed designers to break away from box-like shapes with sharp edges and corners. Pressed steel also introduced the categorization of stoves, washing machines and refrigerators as "white goods," referring to the characteristic white or cream enamel finishes that took over from the mottled grays, greens, and blues previously favored. White was expressive of the lighter weight of pressed steel and also symbolized hygiene. In the case of smaller appliances, such as radios and hair dryers, which did not need to be particularly robust, the use of Bakelite and similar plastics granted designers considerable freedom of expression within the dictates of functional requirements. New plastics developed in the 1930s included polymethyl methacrylate (known as Perspex in the U.K. and Lucite in the U.S.) and polyethylene (or polythene). Plywood, a material associated particularly with Scandinavian designers, offered molding benefits similar to those of plastics: it could be heated and bent.

The three dominant design influences of the 1930s were modernism, art deco, and streamlining. The Modern Movement had its origins in Germany before World War I and is most closely associated with the Bauhaus, the German art-and-design school founded in Weimar in 1919. Modernism was primarily influential in the fields of architecture and furniture design and had relatively little impact on appliance design. One exception was the commissioning in the 1930s by the British company Ekco of the Russian designer/architects Serge Chermayeff and Misha Black and the Canadian designer and architect Wells Coates to produce designs for Bakelite radio cabinets. Coates's circular radio cabinets may be considered as much a product of art deco influences as of modernism. Art deco is the name derived from a Paris exposition of 1925 and given to designs that feature elements such as geometric shapes, bold use of color, and Egyptian influences. It was particularly suited to radio cabinets and loudspeakers. A spectacular example was the skyscraper-shaped radio designed by the American designers Harold van Doren and J. G. Rideout for Air-King Products in 1931. Van Doren was one of many American designers who became exponents of streamlining, which described the wrapping of the product mechanism in a sleek, rounded casing. Streamlining derived its basic characteristics from the application of aerodynamic principles to the design of land- and air-transport vehicles. Aerodynamics, as such, had no practical value in the design of domestic appliances, but the streamlined form when applied to white goods brought a pleasing unified look to the kitchen and suggested modernity. Streamlining and styling were, like many of the rising industrial designers, products of the depression. Linked to annual model changes, they helped products to sell. Influential streamlined appliances included Raymond Loewy's 1934 Coldspot refrigerator (for Sears, Roebuck) and Norman Bel Geddes's 1932 Oriole stove (for Standard Gas Equipment Corporation). Streamlining continued to influence appliance design throughout the 1940s and 1950s.

World War II and Aftermath (1939–1960)

By the outbreak of World War II in 1939, 70 percent of British homes and 80 percent of

American homes were wired for electricity, indicating that the wiring of urban homes was virtually complete. Rural homes were more problematic as the infrastructure costs were so much higher in relation to demand. In the United States, ownership of larger domestic appliances had risen sharply over the past decade and about half of U.S. households had refrigerators and washing machines. British homes were less well endowed. The war seriously disrupted appliance production: factories were required to turn their attention to munitions manufacture, and the use of scarce materials for nonessential production was strictly controlled. In Britain, the Board of Trade's Design Panel was responsible for approving consumer production and issued its own utility standard designs, which economized on materials. New ventures, including early experiments in television broadcasting in both countries, were suspended.

The aftermath of World War II was in many ways similar to the aftermath of World War I. While the loss of life and casualties were much lower, the financial costs were higher. This reflected the more technological nature of the second war, particularly the role of aerial combat. Britain sustained heavy damages to property through enemy bombing raids. This included the destruction of almost 0.5 million houses and damage to a further 3.5 million, together representing a third of the housing stock. The consequence was that Britain ended the war heavily in debt and with a massive reconstruction program required. Austerity measures, including the rationing of food and clothes, remained in force into the early 1950s. One of the economic measures taken by the Labour government elected in 1945 was its promised program of nationalization of key industries. The 1947 Electricity Act nationalized the electricity-generation and -supply industry. The welfare state was strengthened by reforms, based on the 1942 Beveridge Report, including the provision of free health care funded by national insurance contributions. In contrast, the United States was in a stronger financial position and normal civilian life resumed almost immediately.

Among the wartime technological developments that soon found civil applications were aluminum fabrication, microwave technology, and computer electronics. Global aluminum production rose from just over 0.5 million tons at the beginning of the war to 2 million tons during the war, owing principally to the demands of the burgeoning aircraft industry. At the end of the war, stocks of aluminum remained and became available at cut rates, encouraging the wider use of aluminum in products such as refrigerators and freezers. Wartime research and development into radar technology in Britain and the United States improved the performance of the magnetron, a special thermionic valve (or vacuum tube) that generates high frequency oscillations. The cavity magnetron, first developed in Britain in 1939 and refined in the United States at the Radiation Laboratory, emitted very short waves in the centimeter/millimeter wavelength band. These microwaves proved to have a domestic application, microwave cooking, but in the immediate postwar years microwave ovens only became available as commercial catering models. Wartime developments in computing had been motivated by the need to crack German communication codes. Within five years of the end of the war, computers became commercially available, but these machines were only reduced to a domestic scale in the late 1970s.

For women the end of war marked a return to the home. As had World War I, World War II engaged the services of women in manufacturing industry to fill the jobs vacated by men on military service. This time, however, there was no postwar reward. Women had already gained equal political rights, and there was no desire to give them equal economic rights. The war had solved the unemployment problems that had persisted throughout the 1930s, and governments were keen to avoid a recurrence through more women competing in the job market. In order to keep married women in the home, the role of housewife was elevated and glorified. The ideas of rational household management, ergonomics, and household technological expertise were marshaled and packaged to present housework and family care as the equivalent of an outside job, albeit unpaid.

Suppliers seduced women with the lure of new appliances—freezers, dishwashers, and clothes dryers. Rising wages in the 1950s and the growing availability of installment plans, or hire purchase, and other credit schemes assisted acquisition. Encouraging household consumption of appliances and other finished goods fulfilled a dual function: it boosted production and hence employment and kept women locked into a cycle of ever-rising domestic standards. The tyranny of drudgery was replaced by the tyranny of perfectionism. This process is summed up by the words of a motivational research director, quoted by the American psychologist and social researcher Betty Friedan in *The Feminine Mystique:* "In a free enterprise economy we have to develop the need for new products. And to do that we have to liberate women to desire these new products. We help them rediscover that homemaking is more creative than to compete with men." Resentment of this kind of manipulation was one of the motivating factors behind the rise of the women's movement in the 1960s.

Immediately after the war, television broadcasting resumed in Britain and the United States. In the United States, the consumer boom of the 1950s saw home television ownership rise from 9 percent to 87 percent. In Britain, where the consumer boom was delayed by the prioritization of core industrial reconstruction and higher prices, television ownership grew more slowly, with a miniboom in 1953 caused by the televising of the coronation of Elizabeth II. The 1950s was the period when the electronics industry began to establish itself as a major economic force. Miniature thermionic valves, or vacuum tubes, had become available in the late 1930s, but it was the invention of the transistor in 1947 that permitted true miniaturization of electronic consumer goods, such as radios and televisions. Three American scientists working at Bell Laboratories, William Shockley, John Bardeen, and Walter Brattain, who jointly received the Nobel Prize for physics in 1956, developed the transistor. The transistor is a solid-state electronic component, based on the semiconducting properties of certain natural and synthetic crystals. Transistors began to replace vacuum tubes in radios in the mid-1950s and in televisions in the late 1950s. They also began to be used in power amplifiers, paving the way for the high-fidelity sound systems that became standard in the 1960s.

The transistor became closely associated with two postwar phenomena, one sociocultural and the other economic: the emergence of youth culture and the rise of Japan as a global manufacturing leader, particularly in the electronics industry. The buoyant economy of the 1950s created a new consumer group composed of teenagers and young adults with disposable income. Young people sought to set themselves apart from their parents' generation by adopting new tastes in music, clothing, and lifestyle. Youthful rebellion and disaffection was embodied in a new form of popular music, rock and roll, and reflected in "teen movies," such as *Rebel without a Cause* and *The Wild One.* Despite beginning to look and think differently, however, most young people continued to live in the parental home, hence the popularity of portable transistor radios and record players that could be taken to meeting places. In the United States, where car ownership was much higher, youth lifestyle was highly mobile and encouraged the proliferation of coffee bars, hamburger joints, and other fast-food outlets, as well as drive-in movie theaters.

The Japanese economy had undergone a process of modernization beginning in the late nineteenth century, after two centuries of self-imposed isolation. Licensing agreements with manufacturers in the West assisted Japanese industrial growth. Japan had begun to develop an electronics industry through independent research and development before World War II. In the 1950s, Japanese companies, such as Sony, Toshiba, and Matsushita, eagerly embraced the new transistor technology, which was in harmony with the Japanese love of miniaturization. Japan was rapidly able to assume a strong position in the global electronics market through the cheapness of its goods, which reflected low labor costs and a business culture that emphasized company loyalty and quelled labor unrest.

Postwar Italian industry also recovered thanks to United States aid and a rising home market. During the 1950s, Italian companies laid the foundation of quality appliance production based on innovative design.

One of the consequences of the 1950s consumer boom was that the producers of cheap, mass-produced consumer goods cut costs by ignoring the precepts of good taste, as defined by the design movements of previous decades. The term "kitsch," derived from the German verb "*verkitschen,*" meaning to cheapen, was applied to such products. Mass-market goods were often based, albeit crudely, on more expensive, "high design" originals. The importance of the automobile in American life led U.S. appliance designers to favor the bulbous lines of car bonnets (hoods) and the application of chrome details. Biomorphism, exemplified by the exaggerated kidney shape, was another popular element of kitsch. While streamlining remained popular in the United States, the International Style became increasingly influential. This name was coined for an exhibition about 1920s architecture held at the Museum of Modern Art, New York, in 1932. The International Style was essentially a branch of modernism that became an accepted design style largely through the arrival of European designers, including Ludwig Mies van der Rohe and Eero Saarinen, in the United States in the 1930s, and became widespread later in the 1930s.

One of the purest exponents of the International Style was the German designer Dieter Rams who worked for Braun from 1955, designing audio equipment and small domestic appliances. Their clean lines and the dominance of a single color, often white or black, characterized Rams's designs. In Britain, the Council of Industrial Design (later, the Design Council), formed in 1944 to promote "good design," tried to educate public taste through exhibitions, such as the 1951 Festival of Britain and its Design Centre. The Council's aspirations were reflected in the work of British designers such as Kenneth Grange, who designed the Kodak 44a Brownie camera in 1959, but the public delighted in the more colorful kitsch designs.

The Global Village (1960–1969)

In the 1960s, the acceleration of the space race, a tangible expression of the cold war between the United States and the Soviet Union, brought about a revolution in global communications. Telstar, the first active communications satellite, was launched by Bell Laboratories of the United States in 1962 and was used to demonstrate the transatlantic transmission of television signals. The U.S. Congress immediately set up the Communications Satellite Corporation (Comsat) to take sole responsibility for satellite telecommunications. Two years later, more than 100 countries joined together to form the International Telecommunications Satellite Consortium (Intelsat). Initially, communications satellites were used for long-distance telephone, television, and radio transmissions, but later became used for fax, electronic mail, video-conferencing, and data transmissions. Each satellite carries a number of transponders, each of which has a dedicated role.

In 1962, the Canadian communications theorist Marshall McLuhan characterized the new world united by electronic communications as "the global village." He later coined the phrase "the medium is the message" to describe his belief that the substance of any communication is shaped by the nature of the medium. While McLuhan viewed electronic communication predominantly as a positive force for social change, because it "encourages unification and involvement," he also recognized its potential negative side effects—information overload, infringement of personal privacy, and technological anxiety. The same hopes and fears accompanied the dramatic upsurge of computerized communications a quarter of a century later.

Social change in Britain and the United States in the 1960s gained more impetus from the actions of citizens than from the initiatives of government. In 1960, the Democratic candidate, John Fitzgerald Kennedy, won the U.S. presidential election on the basis of a platform of proposed social reforms. However, he was unable to get a majority in Congress to back increased funding for public education or the provision of medical care for the elderly and only succeeded in raising the minimum wage slightly. After Kennedy's assassination, the Democrats' social program came to fruition under the Johnson administration. The American civil rights movement had gained prominence beginning in the mid-1950s under the leadership of Martin Luther King, Jr., and the movement that had most effect on the home, the women's movement, emerged in a new form in the early 1960s. After the major battle for equal political rights had been won, feminism lost its high public profile. The publication in 1949 of *The Second Sex,* by the French socialist and feminist writer Simone de Beauvoir reawakened interest in women's politics. De Beauvoir asserted that the subordinate role of women was the result of calculated repression by a male-dominated society.

In the United States, the publication of Betty Friedan's *The Feminine Mystique* in 1963 dissected the myth of the "happy housewife heroine." In the 1920s and 1930s, female advocates of domestic electrification had genuinely believed that widespread ownership of electric appliances would release women from domestic thralldom. They calculated the number of hours that could be saved by electrifying the execution of household tasks and concluded that time spent on housework could be reduced by as much as three-quarters. That was the theory. Betty Friedan concluded that in practice "housewifery expands to fill the time available." In 1964, the U.S. Civil Rights Act introduced a government agency, the Equal Employment Opportunities Commission, to police discriminatory employment practices. This was only one area, however, where women were seeking liberation and, in 1966, Friedan cofounded the National Organization for Women (NOW) and became its first president. NOW's agenda covered issues such as portrayal of women by the media and women's reproductive rights.

In the mid-1950s, the introduction of an improved intrauterine device (the coil, or IUD) and successful trials of a combined estrogen-gestagen contraceptive pill widened women's contraceptive choices. The Pill became generally available in the 1960s. Effective contraception made it possible for women to plan a family to suit their career needs, rather than the reverse. In Britain, the introduction of the 1967 Abortion Act, which made legal terminations easier to obtain, further increased women's control of their reproductivity. American women gained more limited abortion rights as a result of the Supreme Court ruling in the test *Roe v. Wade* case in 1973. The women's movement also campaigned for wages for housework and childcare, but that cause was predictably doomed to failure.

Meanwhile, the availability of appliances that speeded up housework continued to grow, with the introduction of the refrigerator-freezer and domestic microwave oven. For working women, these two appliances in tandem made it possible to serve an evening meal within a short time of returning home. Both of these products, however, only began to sell extensively in the mid-1970s. The consumer boom of the 1950s meant that ownership of longer-established appliances was approaching saturation point. In order to maintain sales levels, manufacturers tried to guarantee themselves a continuing market by following a strategy of planned obsolescence. This trend had begun in the United States in the depression years of the 1930s. The strategy entailed the use of cheaper, less durable materials and the constant restyling of products. The theory was that consumers would either be forced to repurchase appliances regularly because of mechanical failures or cosmetic breakages or would want to buy the latest model in order to keep up with changing tastes. The ultimate in disposability was the fashion for paper clothes, which proved to be as ephemeral as the clothes themselves. In contrast, some European manufacturers, particularly in Germany and Scandinavia, disdained this approach and based their marketing on intrinsic product quality. Manufacturers also continued to court the youth market by developing new products such as the audiocassette recorder.

A consequence of the manufacturers' hard-sell tactics was that consumers became more militant about their rights. The American writer Vance Packard raised concerns about the insidiousness of product advertising (in *The Hidden Persuaders*) and the wastefulness of built-in obsolescence (in *The Waste Makers*). In 1962, the Kennedy administration defined the four basic rights of consumers: safety, information, choice, and to be heard. In spite of the existence in the United States, since 1899, of a National Consumer League, consumer protection became a hot issue. Ralph Nader, an American lawyer, emerged as a forceful consumer advocate in 1965 when his book *Unsafe at Any Speed* led to the introduction of car-safety legislation. He went on to lobby on many other consumer issues and founded the consumer association, Public Citizen, in 1971. In Britain, while the British Standards Institute had set and monitored product standards since 1901, the creation of the Consumers' Association in 1957 provided the public with useful information in the form of comparative product test reports published in a monthly magazine, *Which?*. Consumer rights were also protected through the 1968 Trade Descriptions Act and the 1972 Fair Trading Act.

The 1960s was a decade of eclecticism in design. The reduction in the life expectancy of products freed designers to adopt an aesthetic that valued irreverence, fun, and color. The Pop Art movement that began in London in the early 1950s and was popularized by American artists, including Andy Warhol and Roy Lichtenstein, influenced designs aimed at the youth market. Pop design drew on many influences, but its expression in domestic appliances was primarily futuristic. One of the classic pop appliances was the 1970 spherical Nivico television made by the Japanese company JVC. Even at the high end of design, many designers began to reject the tasteful austerity of modernism and the International Style. Italy had begun to build a reputation for thoughtful and distinctive design in the 1950s. Italian designers evolved an aesthetic approach that favored organic rather than geometric shapes. This evolved into the radical design and antidesign movements in the late 1960s. Good examples of 1960s Italian products include the range of typewriters designed by Ettore Sottsass for Olivetti and the televisions designed by Marco Zanuso and Richard Sapper (a German designer) for BrionVega. Polypropylene, a thermosetting plastic developed in Italy in the 1950s, became an influential material because it could be injection-molded, permitting the creation of an almost infinite range of shapes.

The consumerism that characterized much of 1950s and early 1960s America was roundly criticized by the "alternative society" of the late 1960s. Politicized by its opposition to the Vietnam War, the hippie world—an amalgam of beats, folkies, surfers, and rock and rollers—promoted an abandonment of conventional suburban life but, with its echoes of a preindustrial age, often left women with the washing, the cooking, and the children. Nevertheless, many of the fringe concerns of the 1960s, such as concern for the environment, would surface in the next decade.

Environmentalism (1970–1979)

In the early 1970s, there was a backlash against the conspicuous consumption and waste of the previous two decades. In particular, a growing environmentalist movement questioned global energy policies, conservation practices, and emission of pollutants. In the United States, the voluntary organization, Friends of the Earth (FoE), was formed in 1969. The government response was to set up the Environmental Protection Agency in 1970. The Friends of the Earth concept attracted international support, and FoE groups were founded in other countries, including Britain in 1971. A more radical international pressure group, Greenpeace, also emerged in 1971. German economist Ernst Friedrich Schumacher highlighted environmental and social concerns in his influential 1973 book *Small Is Beautiful*. Schumacher argued that the rise of gigantic multinational corporations and the pursuit of incessant economic growth created intolerable pressures on society and the planet. Sustainability, through the conservation of energy and resources, became a watchword for these campaigners.

Major targets of the environmentalists during the 1970s were the oil and electricity industries and the nuclear-processing industry, which reprocesses nuclear waste. Oil production had rocketed during the 1960s. Demand was high as a result of increasing car ownership and usage and the growing use in a wide range of products of plastics and synthetic fibers derived from petrochemicals. On the supply side, the oil-producing countries of the Middle East had acquired a virtual stranglehold on the world petroleum-export market. Environmentalists were concerned about both oil-related air pollution (through the release of carbon monoxide and carbon dioxide) and marine pollution. In 1967, the *Torrey Canyon* oil tanker had capsized off the southwest coast of England, spilling over 100,000 tons of crude oil. This disaster established the principle that oil companies were responsible for the costs of cleaning up in the wake of such accidents. At the end of the 1970s, three major oil spills intensified concern: the *Amoco Cadiz* tanker ran aground off the French coast in 1978, spilling 225,000 tons of oil; in 1979, the *Intoc 1* oil rig accidentally released more than 0.5 million tons of oil in the Gulf of Mexico; and also in 1979, two tankers collided in the Caribbean Sea near Tobago, spilling 230,000 tons of oil.

In the case of the oil industry, environmental pressures to cut consumption coalesced with economic pressures. In 1973, the involvement of the United States in the Arab-Israeli war led Saudi Arabia to impose an embargo on oil exports to the United States. Capitalizing on this situation, the other members of the Organization of Petroleum-Exporting Countries (OPEC) announced the quadrupling of their prices. The major oil-consuming countries in the West resisted the price increase, and the ensuing oil shortages caused an economic slump that lasted beyond the resolution of the crisis in 1974. In the past, the multinational oil companies had largely controlled supply, but the oil-producing countries began to assert their rights to safeguard their resources and gain maximum benefit. In 1979, OPEC introduced another price hike. This time the result was that the oil-consuming countries began to exploit other oil sources, such as oil deposits below the North Sea, and where possible, to seek alternatives to oil.

The consumption of oil and other fossil fuels in electric-power plants was attracting scrutiny on both economic and financial grounds. Industrialized countries had been exploiting their coal resources for much longer than they had been using petroleum. Consequently, coal deposits were being rapidly depleted and mining costs were rising. Solid-fuel-power plants were therefore becoming expensive producers of electricity. As emitters of sulfur dioxide and carbon dioxide, they were also identified as major culprits in the incidence of acid rain and global warming. The absorption of sulfur dioxide by rainwater to form dilute sulfuric acid (acid rain), was linked to the destruction of forests and damage to buildings in northwest Europe and North America. Climatic records revealed a significant rise in global temperatures over the twentieth century. This was attributed to the build-up of so-called greenhouse gases in the upper atmosphere. Greenhouse gases, including carbon dioxide and water vapor, absorb reflected solar radiation and prevent its escape to space, trapping heat in the atmosphere. Global warming, if it continues, will have many environmental consequences, including drought and a rise in sea level caused by the melting of polar ice caps.

To the electricity producers, the obvious solution was to increase nuclear generation of electricity. Nuclear electricity was touted as cheap and resource-efficient because only tiny amounts of radioactive materials are consumed in relation to the amount of energy released. It was also presented as clean and environmentally friendly because it did not involve the emission of carbon dioxide and sulfur dioxide. Civil nuclear applications had been heavily promoted since the mid-1950s, partly to justify and offset the enormous costs of military nuclear research and development. The world's first commercial nuclear power plant opened at Calder Hall in Britain in 1957. Antinuclear pressure groups had formed in the 1950s, in the wake of the 1945 Hiroshima and Nagasaki nuclear bombings, primarily to resist the proliferation of atomic weapons. Campaigners turned their attention to civil nuclear use as fears about the safety of nuclear power plants grew. In 1979, there was a major accident at an U.S. nuclear power plant at Three Mile Island in Pennsylvania when a pressurized water reactor leaked radioactive material. This provoked a public outcry and the plant was closed, only reopening in 1985.

Thus, it appeared that there were no easy solutions to the challenge of safer and more environmentally responsible electricity generation. While there was public concern about the consequences of ever-increasing electricity consumption, there was no sign of demand declining. People were loath to give up or reduce the use of the electric appliances that had made their lives easier and more comfortable. The energy crisis of the 1970s forced governments to become more active in promoting energy conservation. They began to provide advice on saving energy and to encourage passive measures such as better insulation of buildings. Manufacturers were slow to redesign appliances to be more energy-efficient. The only significant development came in the field of electric lighting. The Dutch company Philips, the world's leading manufacturer of electric light bulbs, introduced new energy-saving light bulbs, based on the fluorescent technology that had come to dominate industrial and commercial lighting.

Young people continued to be a major target audience for manufacturers, particularly in the field of audio entertainment. The growing role and influence of youth in society was recognized by the reduction in the voting age. The voting age was lowered to 18 in Britain in 1970 and in the United States, through the 26th Amendment, in 1971.

The 1970s saw the birth of high-tech consumer products, due primarily to the accelerating development of the electronics industry, with space technology providing a subsidiary influence. The depressed economy of the 1970s caused Americans to become more critical of the vast expenditure on space exploration. Therefore, the manufacturers that were profiting from the space program were keen to demonstrate that space technology had spin-off benefits. One example was Black and Decker's application of the cordless technology that it had developed for the Apollo program to consumer products, including hand-held vacuum cleaners and power tools. Teflon, the trade name for polytetrafluoroethylene (PTFE), had first been synthesized in 1938, but it proved difficult and expensive to produce in quantity. Teflon was first used as a coating for nonstick pans in the mid-1950s but didn't become widespread until the 1970s. Its use in space technology (resistance to heat and chemical agents make it an ideal substance for coating spacecraft) had led to improvements in Teflon manufacturing, reduced the price, and enhanced its performance and durability.

Japanese companies consolidated their dominance of the home entertainment sector of the consumer electronics market through innovation and cutthroat competition. Video was the new buzzword. The major new product of the 1970s was the videocassette recorder. Although the Dutch company Philips and the American company RCA took an early interest in this technology, by the late 1970s the commercialization of the videocassette recorder had developed into a head-to-head battle between two Japanese companies, Sony and Matsushita (and its subsidiary JVC). Matsushita was the bigger company and used its financial clout to ensure that its VHS system became the standard, rather than Sony's technically superior Betamax system. Video also flourished in the form of the video games and videotex. In the 1970s, video games were the preserve of American companies, particularly Atari and Mattel, although Japanese companies moved in with predictable results in the 1980s. The world's first teletext systems (like videotex, systems for transmitting information as text displayed on television), Ceefax and Oracle, were developed by the British broadcasting groups BBC and ITV and launched in 1976. One failure in the video world was the videodisc player, launched in the late 1970s and virtually defunct by the mid-1980s.

In the electronics industry, the microchip or microprocessor superseded the integrated circuit of the 1960s. Two American scientists perfected the integrated circuit in the late 1950s. First, Jack Kilby at Texas Instruments created a simple integrated circuit by assembling five components on one silicon chip. Next, Robert Noyce at Fairchild Semiconductor created the planar transistor, a silicon chip coated and etched to embed electronic instructions. The planar process was repeatable, enabling a complex circuit to be built up in layers. In 1969, Marcian Edward (Ted) Hoff took this process a stage further when he began work at Intel on a microchip containing thousands of components that would act as the entire central processing unit of a computer. The outcome was the Intel 4004 microprocessor, which became commercially available in 1971 and immediately resulted in the launch of the first pocket calculators. Over the course of the 1970s, more and more components were packed into microprocessors, increasing their power and speed of processing. By 1975, it was possible to make a computer small enough to sit on a desktop and therefore suitable for home use. A number of home or personal computers, including the Apple II and the Commodore PET 2001, hit the market in the 1970s, but because they were too expensive and lacked useful software, they had little impact in the home.

Design in the 1970s acted as bridge between 1960s eclecticism and 1980s niche designerism. The radical design and antidesign movements evolved into postmodernism. Intellectually, postmodernism was an overt reversal of modernist theory. Postmodernists argued that the functionalist geometric and monochromatic severity of the modernist aesthetic was not user-friendly. However, while postmodernism provided the theoretical umbrella to unite a range of styles, these styles seemed completely disparate to the uneducated consumer. Postmodernism allowed designers the freedom to plunder a range of existing styles while maintaining industrial and modernist methods of production. Whether applied to a radio or a building, postmodernism was essentially an ironic envelope for existing technologies. At the kitsch end of design, ornamental novelty phones and lamps featured popular cartoon characters, while "high-tech" design was similar to modernism in its functionalism and simplicity of color (silver and black were basic components of the palette). At the end of the 1970s, the concept of lifestyle design and marketing emerged. One of the first products conceived in terms of lifestyle was the 1979 Sony Walkman personal cassette player.

Lifestyle and "Designerism" (1980–1989)

Increasing global competition, particularly from the spectacular growth economies of Southeast Asia, made it difficult for Western economies to recover after the 1970s oil crises. The continuation of the economic depression saw the climate of optimism and idealism of the 1960s and early 1970s give way to a climate of conservatism and self-interest. Environmental interests were marginalized, although the German Green Party continued to be relatively successful. This was the "Me" decade, where society became more polarized than it had been for years between the "haves" and "have nots." At the one extreme were the yuppies, the young upwardly mobile professionals who worked long hours to further their careers and spent their income on a conspicuously affluent lifestyle. At the other extreme were the long-time unemployed, who felt disempowered and resentful and had ample time to develop a sense of relative deprivation. Television programs such as the U.S. soap operas *Dallas* and *Dynasty* flaunted glamorous lifestyles. It is, therefore, unsurprising that both Britain and the United States experienced a sharp rise in both violent and nonviolent property crimes, such as burglary, car theft, and mugging, in the 1980s. The 1980s has been termed the "Designer Decade," largely due to the rise of clothes designers such as Giorgio Armani, Calvin Klein, Donna Karan, Paul Smith, and Gianni Versace. The allure of labels and the cult of the designer spread into other areas, including the home; if you had the right clothes, then you also needed the right coffeemaker!

Conservative governments in Britain and America espoused the doctrines of free market and individual choice, at the expense of social welfare. One of the industries most radically affected by the glorification of the free enterprise economy was the communications industry. In the United States, American Telephone and Telegraph (AT&T) had a monopoly of the telephone network and services until the late 1960s, when the government granted limited licenses to two other companies. In 1974, the Justice Department indicted AT&T on antitrust grounds, apparently with the intention of separating its manufacturing and network service functions. The legal suit only reached the trial stage in 1981. In a surprise move in 1982, AT&T agreed to a negotiated settlement, ratified by the courts in 1983. Under the deal, AT&T kept its manufacturing division (Western Electric) and retained the provision of long-distance telephone services, but lost the provision of regional services. The latter was restructured to create seven independent regional telephone-operating companies, nicknamed the "Baby Bells." Residential customers were not convinced that deregulation had been to their immediate benefit, but the changes did result in the faster upgrading of the infrastructure technology.

In Britain, telephone services were the monopoly of the state-owned General Post Office until 1980. In 1980, the Post Office lost its exclusive role as a supplier of consumer telecommunications equipment before being split into two independent divisions. The telecommunications division was labeled British Telecom and was privatized in 1984. Meanwhile, a second telephone service supplier, Mercury Communications, was granted a license. As in the United States, British residential customers felt that competition and revised charges had done more to benefit business customers than themselves.

On the technological front, the two big innovations in the telecommunications industries in the 1980s were the widespread adoption of optical-fiber (or fiber-optic) cables and the introduction of cellular telephone networks. Optical-fiber technology has benefited both the telephone and television industries. Although the optical properties of glass fibers were already recognized, it was only in the 1960s that scientists in both Britain and the United States began working on their application for voice transmission. It was the British researchers who first successfully demonstrated voice transmission by optical fiber, but it was the American glass company Corning that in 1970 invented the process of vapor deposition whereby it was possible to guarantee uniform quality along the fibers. In the late 1970s, Japanese optical-fiber manufacturers refined the production process to enable true mass production. Optical-fiber cable outperforms copper cable for voice, data, and video transmission in terms of both its capacity and its speed. It is also based on a cheap and plentiful material—sand. Since 1980, optical-fiber cable has become widely used for long-distance telecommunications and data communications, including the submarine intercontinental cables that form part of the Internet.

The introduction of cellular telephone networks in the 1980s was not so much a technological innovation as a business innovation. Cellular phone technology is simply an application of radio technology, tailored to the particular set of operational needs. Its introduction came about in response to the recognition of specific lifestyle demands, initially more to do with business lifestyle than personal lifestyle. Mobile telephone systems had existed since the 1940s, but their structural design was such that they had very limited capacity and were therefore expensive as well as poor in terms of service quality. The cellular system is based on the division of a service area into small cells, each served by a low-power radio transmitter. Because the user picks up signals from the nearest of many transmitters, a limited range of frequencies can carry many calls without interference.

Modern mobile phone services got under way in the United States in 1978 when AT&T was granted a trial license, seven years after its initial proposal to the Federal Communications Commission (FCC). The first trial, in Chicago, was repeated elsewhere, and the success of these pilot schemes led the FCC to announce a system for licensing mobile phone services in 1982. Britain lagged a few years behind the United States in adopting cellular phone technology. In 1982, the British government decided in favor of granting licenses to two consortia, one led by British Telecom and the other led by Racal Electronics (later Vodafone). Both groups had set up services by 1985.

Optical technology also made its mark on the home with the launch, in 1982, of the prerecorded compact disc and compact disc player, jointly developed by Philips and Sony. The compact disc player introduced digital audio and proved to be an instant success, becoming the fastest-selling consumer electronics product to that time. Choices in home entertainment were further widened by the proliferation of cable television operators in the United States and the introduction of satellite television in Britain. Video games became more sophisticated as the Japanese companies Nintendo and Sega vied to capture the lion's share of the market by developing ever more complex game scenarios. The use of the personal computer grew dramatically in the business world following the introduction of IBM's PC format in 1981, which led to fierce competition between manufacturers of IBM PC clones and, consequently, a sharp decline in prices. The computer, however, remained a minority product in the home until the mid-1990s.

Lifestyle marketing and design is based both on understanding what the consumer wants and persuading him or her that you are what you own. The yuppie mode of power dressing was complemented in the styling of power products, particularly on high-tech lines. Mobile lifestyles were matched by mobile products—not only mobile phones, but also portable personal televisions and compact disc players, as popularized by Sony.

High-tech, a style targeted primarily at men, was not only applied to home entertainment products, but also made an appearance in the kitchen. Manufacturers recognized that the microwave oven was as much of a boon to single men as to working women, and so the microwave shed its image as a white good and emerged as a black good too. Smoked glass, which was associated with audio cabinets, was adopted for oven doors. In 1981, the Italian designer Ettore Sottsass founded the Memphis design studio in Milan with the intention of promoting postmodernism to a wider audience. However, a good example of how postmodernism failed to translate to a mass audience is the failure of the Zanussi Wizard refrigerator, designed by Roberto Pezzetta. The Italian company Zanussi had earned a reputation for technical quality and good design, but the Wizard was too whimsical for its market.

The rise of lifestyle products and niche marketing brought significant changes to the retail map. In the United States, rising car ownership in the 1950s had encouraged the development of out-of-town shopping centers. The number of such centers rose from only 8 in 1945 to almost 4,000 in 1960. At first, traditional department stores, such as Sears, Roebuck, were the lynchpins of the shopping mall and prospered accordingly. By 1980, consumers were more impressed by the value-for-money, no frills approach of discount stores, such as Wal-Mart and Kmart. These stores kept prices down by locating on cheap suburban land and wasting no effort or expense on presentation. Both Wal-Mart and Kmart overtook Sears, whose retail dominance had long seemed unshakeable. In the 1980s, the Swedish retailer IKEA became an international success by adapting discount store principles to the home furnishing market. IKEA offered the essentials of good Scandinavian design at low prices by selling furniture in a flat-pack format for assembly at home. This was in tune with the fashion for do-it-yourself home improvement that also boosted sales of electric power tools.

However, while discount stores continued to generate the biggest sales in the 1980s, niche marketing further undermined the role of the department store. Up-market city-center shopping malls had to offer more distinctive purchasing opportunities. In the United States, specialty stores occupied particular niches. For example, Crate&Barrel began as a single store in Chicago selling cookware and housewares and by 1990 had become a small chain with a wide geographic spread within the United States. The British retailer Next began as a clothing chain catering to middle-income working women and broadened its lifestyle base by moving into men's clothing and housewares. The American chain Urban Outfitters offers a similar range for a younger audience.

The Microprocessor Meets the Millennium (1990–1999)

Toward the end of the 1980s, there was a revolt against the worst excesses of designerism and lifestyle marketing. 1980s values were characterized as superficial and uncaring. Moreover, the economies of the West were recovering, having undergone a shift from manufacturing to service industries. The change in values was epitomized by the emergence of the New Age movement, an antimaterialistic counterculture loosely based on the hippie ideals of the 1960s. The more mystical elements of New Age beliefs, such as crystal and pyramid therapies, had a very marginal impact, but the renewed advocacy of environmentalism was more widely shared.

Perhaps the most significant event of the 1980s in terms of reawakening environmentalism was the 1986 nuclear accident at the Chernobyl power station in the Ukraine. Two explosions destroyed a reactor and released 9 tons of radioactive material, 90 times the amount released by the Hiroshima bomb. A cloud of radioactive isotopes spread as far west as Ireland and as far south as the Mediterranean, exposing 5 million people to radiation; 31 people in the immediate vicinity died. In spite of increased concerns about nuclear safety, the need to slow down global warming by reducing greenhouse gas emissions made it difficult for countries to abandon nuclear electricity. In the wake of the 1970s energy crisis, electric utilities did explore renewable energy alternatives to fossil fuels. Solar energy has limited potential in the temperate belt of Europe and North America, but two solar power stations were built in California's Mojave Desert in the 1990s. A disincentive even where long hours of sunshine do exist is the high cost of solar-power installations. Hydroelectric power was already well established in North America, but in Britain it is only feasible in parts of Scotland and Wales, and only France has attempted to harness tidal power. There has been a steady growth in electricity production from wind power, but by the mid-1990s the total global production was only equivalent to the output of a single nuclear power station. Moreover, some environmentalists have opposed the construction of wind farms on the grounds that the tall, highly aerodynamic turbines are both a visual and, potentially, an ecological blight in areas of great natural beauty, where the noise they generate is thought to disturb wild animals.

At the 1992 United Nations Conference on Environment and Development in Rio de Janeiro, also known as the first Earth Summit, commitments were made to reduce greenhouse gas emissions. By the second Earth Summit, held in New York in 1997, it was clear that the targets were not being met. Britain and Germany were among the few nations to be on schedule. The 1997 United Nations Kyoto Protocol set the new target for industrialized countries of a 5 percent reduction in emissions by 2012, but the required number of countries have not ratified the protocol. The pressures to combat global warming have, therefore, sanitized nuclear electricity. However, the economic arguments in favor of nuclear electricity have been dented. As part of the process of preparing for the privatization of the British electricity industry in the early 1990s, nuclear power operations were defined as a separate division. A financial prospectus had to be prepared for potential future shareholders. This revealed that, when the costs of waste disposal and the potential costs of decommissioning older nuclear power stations were taken into account, nuclear electricity was not an attractive investment. As a result, the nuclear power stations were withdrawn from the privatization package and remained in state ownership. In 1992, a sixth of the world's electricity came from nuclear power. In Britain and the United States, the proportion was just over a fifth, whereas in France and Sweden, it was more than a half. Nuclear electricity's contribution rose to more than a third in Britain after the opening of Sizewell B, the world's most advanced pressurized-water reactor power station, in 1995. Both Britain and the United States have expended considerable efforts in research and development into nuclear fusion during the 1990s in hopes that this will be a safer form of nuclear energy production.

The other big environmental challenge for industry in the 1990s has been the issue of ozone depletion, blamed largely on the release of chlorofluorocarbons (CFCs) into the atmosphere. A class of CFCs known as Freons was first synthesized in the United States in 1930. Apparently harmless—inert, nontoxic, inflammable, and odorless—they were adopted for use as refrigerants and aerosol propellants. In 1965, the British scientist James Lovelock became the first person to detect the presence of CFCs in the upper atmosphere, but he did not fully appreciate their effects. It was another nine years before scientists recognized that CFCs were depleting the ozone layer, which acts as a filter of harmful ultraviolet radiation. During the 1980s, aerosol manufacturers responded to public pressure and replaced CFCs with less harmful, although not harmless, propellants. Refrigerator manufacturers were slower to respond on the grounds that their products were not inherently disposable. In 1985, the United Nations Environment Programme held the Convention on the Protection of the Ozone Layer in Vienna. This led to the Montreal Protocol on ozone depleters, which has since been amended several times. Ozone depletion was also on the agenda of the first Earth Summit. As a result, in the early 1990s, refrigerator manufacturers began to replace CFCs with HFCs (hydrofluorocarbons), which are not ozone depleters. However, HFCs are not ideal because, like CFCs, they are greenhouse gases. In the late 1990s, Greenpeace has endorsed an environmentally friendly Greenfreeze technology.

Manufacturers of white goods, which had the greatest potential for improved energy efficiency, began to produce energy-saving models, generally more expensive than the standard models. The washing machine was particularly ripe for remodeling along eco-friendly lines, the key lying in load sensitivity. By 1990, it was standard for washing machines to feature a range of wash programs, primarily according to fabric type. The availability of fuzzy logic microchips meant that the washing machine could be designed to program itself so that it would regulate water temperature and cycle length to minimize energy use and ration water and detergent/softener intake in line with load size. The Italian company Zanussi was a leader in eco-friendly washing technology, having developed an original technique of water-efficient spray-washing in the mid-1980s. The microwave oven was another appliance that became capable of performing "intelligently" through the wonders of the microchip.

In the 1990s, it became feasible to transform the entire household through the use of digital technology. The core digital device, the computer, finally achieved the status of the universal machine in the home as well as the office. The growing popularity of computers, which accelerated as the decade progressed, owed much to their increased scope for providing entertainment, through computer games and accessing the Internet. This did, however, provoke concerns that computers were becoming an addiction for the current generation of children and young adults, concerns similar to those raised twenty years earlier about the role of television. Indeed, the introduction of digital television broadcasting and the digital videodisc player in the late 1990s added to the competition for home leisure time. Debates about the social and cultural effects of the computer seem certain to continue. As the millennium drew to a close, the way forward seemed to lie in the merging of digital technologies. The latest mobile phones combined their primary function with text messaging, personal organizer facilities, and Internet access. In the 1960s, portrayals of ideal homes of the future typically envisaged the completely automated home controlled by a central computer. In 1999, this vision was close to becoming an affordable reality, as evidenced when the Swedish company Electrolux demonstrated its Screenfridge, not just a refrigerator but also a household communications and control system.

So, what do these changes mean for women? At the end of the century, almost as many women as men in Britain and the United States are in paid employment. In spite of equal-opportunities legislation, however, women are significantly less well paid. In Britain, the 1970 Equal Pay Act and the 1975 Sex Discrimination Act addressed some of the demands of the women's movement. Yet, in 1993, when women made up 49 percent of the workforce, the average woman's wage was 40 percent lower than the average man's wage. This was partly because many more women than men were in part-time work, but also because women were underrepresented in better paid jobs, such as management, and in certain professions, including the law and politics. Women have the vote on equal terms, but make up only about a tenth of elected members in both the British and United States legislatures. In the United States, the proposed Equal Rights Amendment that would have outlawed sex discrimination was passed by Congress in 1972 but then failed by a short margin to achieve ratification by the necessary three-quarters of the

50 states by 1982. In 1994, American women made up almost half of the workforce and three-quarters of women in the 25–54 age group were in paid employment. The growth of women's employment has made little difference to women's role in the home. Studies have consistently shown that in the majority of cases where both female and male household partners work, women still do most of the housework. A plethora of household appliances may make that housework less strenuous, but it still consumes time that for men can be given to leisure pursuits.

Environmentalism also influenced product design and the choice of materials. The first Earth Summit advocated policies of sustainability, thereafter referred to as Agenda 21. As the opposite of planned obsolescence, sustainability requires products to be designed and made in a way that helps to extend their life through repair or reuse or that permits recycling and minimizes waste in the production process. One of the most high-profile exponents of "green design" is the French designer Philippe Starck. His early work was more flamboyant and "designerist," but Starck developed a more holistic consciousness in the early 1990s, exemplified by his Jim Nature television for the French-owned manufacturer Saba. The television casing was made of high-density chipboard, a material made from waste wood chips. An equivalent mass-market example is the Recyclone vacuum cleaner made by the British company Dyson Appliances. Plastic waste recovered from the manufacturing process is recycled to make the Recyclone's shell.

Green design looks to the future, whereas another 1990s design style, retro, draws on the past. Retro design was popular in the 1960s when it typically drew on Art Nouveau and Arts and Crafts influences. In the 1990s, retro was mainly inspired by the 1940s and 1950s and was particularly prevalent in kitchen appliances such as toasters, food mixers, kettles, and refrigerators. In some cases, this merely involved manufacturers reproducing their own earlier designs, labeled as "classics." The British company Russell Hobbs and the U.S. company Sunbeam both fall into this category, whereas the Italian company Smeg produced refrigerators that were based on the typical 1950s American fridge. 1990s retro was the reflection of a sense of nostalgia as the century drew to a close and the conscious selection of symbolic products.

It is interesting to witness electrical nostalgia at the end of the electric century. Two female novelists, one British and one American, chose electricity as a theme for works of historical fiction. Published in the last five years of the century, *Electricity: A Novel* by Victoria Glendinning and *City of Light* by Lauren Belfer both treat electricity as a metaphor for change, an interpretation for which this book provides evidence.

GADGETS AND NECESSITIES

An Encyclopedia of Household Innovations

A

AEG

Allgemeine Elektrizitats Gesellschaft was founded in Berlin by Emil Rathenau in 1883. AEG was originally known as DEG (the German Edison Company for Applied Electricity), as Rathenau had been impressed by Edison's light bulb at Exposition Internationale d'Electricité in Paris in 1881.

AEG grew with the newly unified Germany and by 1900 it was one of the country's major companies. By 1907 it had become the biggest manufacturer in the world of generators, cables, transformers, motors, light bulbs, and arc lamps. In that year it appointed an architect, Peter Behrens, as chief designer. He not only designed products such as kettles, fans, and telephones but also undertook factory buildings. Behrens was a pioneer of corporate identity, giving an elegant and uniform look to AEG posters, trade literature, and advertisements. He left the company in 1914.

Behrens was a member of the Deutsche Werkbund, a body devoted to the reconciliation of art, craft, industry, and trade. Germany's adoption of this philosophy at the beginning of the century led to a greater commitment to research and design, resulting in products that were often superior to those of Britain and the United States. Behrens kettles illustrate this point. There were three basic forms, round, oval, and octagonal, set on a molded base. Through the use of standardized parts AEG was able to offer a range of thirty types.

AEG continued to produce a wide range of electrical products that have a restrained, elegant appearance. Since 1996 the operative parts of AEG have become independent companies utilizing the AEG trademark under license from its legal successor, EHG Elektroholding GmbH, a DaimlerChrysler company.

See also Industrial Design and Designers; Kettles; The Modern Movement and Appliance Design

Reference:

Bayley, Stephen, ed. *The Conran Directory of Design.* London: Octopus Conran, 1985.

Aga Cookers

The Aga cooker (or stove, in American usage) was invented by the blind Swedish physicist Gustav Dalen in 1924. Dalen was born in Stenstorp in 1869 and became a successful physicist. He initially made his name in lighthouse technology, for which he was awarded the Nobel Prize in physics in 1912, the year in which he was blinded in an experiment involving gas cylinders. While convalescing he realized the problems that his wife encountered with a traditional cast-iron, coal-fired closed range. The food needed constant attention, and the stove used fuel inefficiently. He set about redesigning the range in his own kitchen. The result was a continuous-coal-burning stove that was clean, efficient, and controllable. It was made of cast iron and insulated with kieselguhr, a heat-resistant padding used in the manufacture of explosives. It featured simmering and roasting ovens, two hotplates, and a water-heating capacity. The oven temperature was controlled by a thermostat, which opened or closed an air damper.

Launched in 1929 and named after the manufacturing company Svenska Akyiebolaget Gasacumulator (Swedish Gas Accumulator), the Aga was successful because it filled the gap between outdated kitchen ranges and the increasingly efficient gas cookers. They were manufactured under license in Britain and became popular in country areas, especially those without a gas supply. Later models, designed to run on wood, oil, and gas, provided a constant source of warmth and hot water. Hand built and installed, Agas were always expensive and beyond the means of most households in urban and suburban areas.

The Aga changed little over the twentieth century and despite its Swedish origins is seen as an icon of English country living. (Some novels set in the country have been called "Aga-sagas," and there is an Aga magazine for owners.) They have achieved cult status and are now popular in both town and country. Writing in 1985, Deyan Sudjic called the Aga "the earth goddess of suburbia, the last vestige of the hearth at the center of the home." The original cream enamel finish has been augmented by a range of colors. They have become symbols of rustic domesticity and urban aspiration. This duality has ensured their popularity, despite the fact that modern gas and electric cookers are now more practical and efficient, and they are distributed worldwide.

Reference:

Sudjic, Deyan. *Cult Objects.* London: Paladin/Granada, 1985.

Air-Conditioning

Air-conditioning is an integrated, automated system for controlling the temperature, humidity, and cleanliness of air in a building. The concept arose from the known sensitivity of certain industrial processes to air temperature and humidity. In nineteenth-century textile mills, the use of water sprays to cool and humidify the air was a primitive attempt at air-conditioning. True air-conditioning was the invention of an American mechanical engineer, Willis Haviland Carrier. In 1902, only a year after graduating from Cornell University, Carrier installed a system for controlling air temperature and humidity in a printing plant in Brooklyn, New York. He patented his "apparatus for treating air" in 1906. Early customers for Carrier's system included textile mills, where dry air could cause fibers to become unmanageable owing to the effects of static electricity. Interest was not restricted to cotton mills in the American South; the first foreign customer was a silk mill in Yokohama, Japan. In 1911, Carrier made public the formulae for his air-conditioning calculations, which still form the basis of air-conditioning technology today.

In order to capitalize on his invention, Carrier formed the Carrier Engineering Corporation with six partners in 1915 and began manufacturing air-conditioning units in 1922. By then, Carrier had invented a new machine—the centrifugal chiller. This refrigeration device provided the first practical solution to the problem of cooling very large spaces. As air-conditioning became well established in the industrial context, operators of other types of large buildings, such as theaters and hotels, became aware of the more general benefits of air-conditioning as a means of improving human comfort levels. The first public building to feature a Carrier centrifugal chiller was the J. L. Hudson Department Store in Detroit, Michigan, where three chillers were installed in 1924. Four years later, the Carrier company developed the "Weathermaker," a small air-conditioning unit specifically designed for household use.

The economic depression of the 1930s killed the market for residential air-conditioning. While interest revived after World War II in the United States, home air-conditioning has never become widespread elsewhere. In temperate climatic zones, including much of Europe, the benefits of air-conditioning are greatly outweighed by the cost, complexity, and intrusiveness of the hardware.

The simplest part of the air-conditioning apparatus is the heating plant, which corresponds to conventional central heating. A disadvantage of central heating is that it tends to dry air below the optimum level of humidity for human comfort. Air-conditioning regulates humidity by passing air through water sprays or over water reservoirs to increase moisture content. Air cooling is achieved by passing air through a refrigeration unit consisting of a compressor and a condenser. Because relative humidity increases as temperature falls, chilled air requires dehumidification. This is achieved through extracting excess moisture by passing air across cool metal plates, which causes water vapor to condense. The third element of air-conditioning, cleansing, is achieved by passing through filters and over activated charcoal to remove dust and odors. The conditioned air is circulated by fans.

See also Heating

References:

www.carrier.com

Alessi

Giovanni Alessi Anghini established a plate-turning workshop at Bagnella, Omegna, Italy, and founded the Alessi Company in 1921. It initially worked in nickel silver and brass and later electroplated with nickel, chrome, and silver. The first articles produced were coffeepots, trays, and table accessories. In 1928 the company moved to Crusinallo in order to utilize hydroelectric power and began to shift from the traditional turned products to pressed ones in stainless steel.

Alessi always produced stylish products, including Carlo Alessi Anghini's Bombe coffee set of 1945. Ettore Alessi, the technical director, opened the company up to collaboration with external designers in 1955. Working with architects, the company produced stylish objects such as the stainless-steel-wire Citrus basket that is still in production. When Alberto Alessi took over the running of the company he began to use star designers, and the company increased its reputation for very stylish objects during the 1970s, working with Ettore Sottsass from 1972. This trend continued into the 1980s and 1990s with commissions from well-known designers and architects for kettles, coffee sets, and table accessories. These included Michael Graves, Philippe Starck, Aldo Rossi, Richard Sapper, Robert Venturi, and Frank Gehry. Although not always the most functional of objects, they were seen as status symbols for the style conscious. Produced for the top end of the market, these and in-house Alessi designs have influenced the look of many mainstream products. In the 1990s Alessi collaborated with Philips on kettles and toasters.

Alessi is an unusual company with a mission to act as a patron for designers; according to Alberto Alessi, it is "not a normal factory—it is closer to being an applied art research laboratory." It produces three ranges, Alessi (mass-produced stainless steels and plastics), Officina Alessi (small or middle-series production, including reproductions of outstanding late-nineteenth- and early-twentieth-century designs by the likes of Christopher Dresser and Marianne Brandt), and Tendentse (porcelain). Still a family-owned company, it continues to celebrate both tradition and innovation.

See also The Modern Movement and Appliance Design; Philips Electronics

References:

Alessi, Alberto. *The Dream Factory: Alessi since 1921.* Cologne: Konemann, 1998.

Sweet, Fay. *Alessi: Art and Poetry.* London: Thames and Hudson, 1998.

Answering Machine

An answering machine is a device that allows incoming phone calls to be received and messages recorded when the phone user is unavailable. It may be a separate machine that is connected to a telephone or combined with a telephone to form an integrated telephone answering machine. Today's answering machines may be either analog or digital. Analog answering machines use conventional audiotape recording, whereas digital answering machines use memory chips. The modern tape format is the continuous-loop microcassette, identical in size to the double-sided microcassettes used in dictating machines. Most machines take one cassette, which holds both the user's message and incoming messages, but some machines take two cassettes so that the user has two prerecorded messages available. When analog answering machines first became available in the late 1940s, they were marketed as business machines. The use of the name "electronic secretary" conveyed their main selling point—as a secretarial substitute. In the 1980s, as answering machines became cheaper and more compact, they became standard items for the home as well as the office.

See also Tape Recorders; Telephones

References:

Goodman, Robert. *How Electronic Things Work . . . and What to Do When They Don't.* New York: McGraw-Hill, 1999.

Apple Computer, Inc.

Steven Jobs and Stephen Wozniak founded Apple in 1976. In 1975, Jobs and Wozniak were both working in Silicon Valley, California, for Atari and Hewlett Packard, respectively. They were also members of the Home Brew Computer Club, a group of computer enthusiasts. Inspired by recent microprocessor developments, they built their own microcomputer, the Apple I, in Jobs's garage. They received orders from a local computer shop and began small-scale production. Encouraged by this, they looked for financial backing. Mike Markkula became the third partner, taking over the financial and administrative side. The company was incorporated in 1977.

Steve Wozniak with an Apple II, May 1986 (Roger Ressmeyer / Corbis)

The world's first commercial personal computer, the Apple II, was displayed at a San Francisco computer fair in 1977 and was an immediate success. Wozniak was injured in a plane crash in 1981, and although he returned briefly after recovering, he subsequently quit Apple. Apple became the first personal computer company to achieve annual sales of $1 billion and the fastest growing American corporation ever, but by 1983 it needed to counter the challenge of the fast-selling IBM PC. Its response was the Apple Lisa, a computer featuring an innovative, user-friendly interface. Program options were displayed in the form of graphic icons, pull-down menus, and windows, with easy navigation via a mouse. However, the Lisa was too expensive for the home computer market. Its basic features were incorporated in the Apple Macintosh, styled by the Californian branch of the German design consultancy, Frogdesign. The Mac received a high profile launch in January 1984 with the showing of an unusually long (60 seconds) TV commercial, directed by acclaimed British film director Ridley Scott. In addition to the graphical user interface (GUI), the Apple Mac had the advantage of compactness. With the monitor, processor unit, and disk drive contained in one streamlined shell, the Mac became a style icon.

Disagreements with John Sculley, the company's president and chief executive officer, led Jobs to resign in May 1985, and Apple faltered as IBM-compatible personal computers began to flood the market. A dispute with Microsoft over the alleged infringement of Apple's design rights by Windows 1.0 was settled in a way that left Microsoft free to mimic the Apple GUI in later versions of Windows. In 1986, Apple bounced back by making desktop publishing affordable with the launch of PageMaker software for the Mac and the LaserWriter printer. A portable Apple Mac was introduced in 1989. The introduction of Microsoft Windows 3.0 in 1990 made Apple increasingly vulnerable to competition from PC manufacturers and allied software companies. Apple initially sustained its market share thanks to its loyal existing customer base and then bolstered its position with the introduction in 1991 of the Macintosh PowerBook, a notebook computer with networking and multimedia capability.

A joint venture with the Japanese company Sharp led in 1993 to the Apple Newton. This hand-held computer featured built-in optical character recognition software, allowing input to be handwritten onto the liquid crystal display with a plastic stylus. In 1994, the technical performance of the Mac was boosted by the adoption of the PowerPC chip, a fast microprocessor. In spite of such innovations, Apple's market share declined steeply, prompting major corporate changes. The decision to license the Mac operating system came too late to be effective. In December 1996, Steven Jobs returned when he agreed to Apple's acquisition of his NeXT software company. Jobs soon regained control of Apple as acting chief executive officer and in 1997 negotiated a five-year software development deal with Microsoft whereby Microsoft also invested $150 million in Apple.

While Apple continued to improve technical performance through the G3 (and later G4) versions of the PowerMac and PowerBook, it also developed a more competitively priced computer, the iMac. Like the first Mac, the 1998 iMac has a single-shell monitor-cum-processor-and-disk drive. Its distinctiveness lies in its rounded lines and availability in a range of five strong colors combined with translucent white. Back on a profitable footing, Apple consolidated its position in 1999 with the introduction of the iBook notebook computer.

See also Computers; IBM (International Business Machines); Microsoft Corporation

References:

Butcher, Lee. *Accidental Millionaire: The Rise and Fall of Steve Jobs at Apple Computer.* New York: Paragon, 1989.

www.apple-history.com/history.html

Asahi Optical Corporation

The Japanese company Asahi Optical Corporation produces cameras under the Pentax brand name. It came to international prominence as a camera manufacturer in the 1950s, when Japanese companies began to challenge the dominance of European manufacturers in the 35mm rangefinder and single-lens reflex (SLR) camera sectors. Asahi developed a reputation for innovation and was responsible for a number of camera industry firsts. As well as producing Pentax cameras, lenses, and accessories, Asahi also manufactures eyeglass lenses, binoculars, printers, scanners, and endoscopes.

The company was formed in 1919 as the Asahi Optical Joint Stock Company and began to manufacture lenses for use in cameras, binoculars, and other optical instruments. In 1939, the company moved into the manufacture of aerial cameras for military use. After World War II, Asahi began to develop cameras for the consumer market. In 1952, Asahi launched the Asahiflex I camera, the first Japanese 35mm SLR camera. At this time, most SLR cameras suffered from a problem known as "mirror blackout:" the angled mirror behind the lens, which reflects the image to the viewfinder, is slow to retract when the shutter opens, thus blocking the path between the lens and the film. In 1954, Asahi solved this problem by fitting the Asahiflex II with an instant-return mirror. The first camera produced under the Pentax brand name appeared in 1957. The name was derived from the elision of the words "pentaprism," or five-sided prism, and "reflex," because the innovative feature of the Pentax camera was the incorporation of the pentaprism in the viewfinder, which made it possible for the viewfinder to be set vertically, providing a more natural viewing position.

Mass production of Pentax cameras began in 1959, reflecting the growing popularity of Japanese SLR cameras. The next major advance came with the introduction of the Pentax Spotmatic camera in 1964. This featured a photoelectric cell positioned behind the lens, which meant that the light reading was as accurate as possible. This arrangement, known as "through the lens" (TTL) metering, became standard in SLR cameras. TTL metering was taken a step further in the Pentax ES SLR camera of 1971, which introduced automatic exposure control by incorporating an electronic shutter that was programmed to select the exposure time according to the light reading. Accelerating sales meant that by 1971, Asahi had sold a total of three million SLR cameras since 1952, with a third of total sales coming in the last two years.

Asahi turned its attention to developing smaller, lightweight cameras. The Pentax MX SLR camera of 1976 was the world's smallest and lightest SLR camera to date. Another variant of the MX model, the ME, was the first camera that operated wholly by automatic exposure, with no manual override. Even at its most compact, the 35mm SLR camera was still heavy and bulky in comparison to the pocket-size Instamatic cameras, pioneered by Eastman Kodak in the 1960s. Asahi's solution was to create an SLR camera that used the same compact 110 film cartridges as the Instamatics. The Pentax System 10 SLR camera, launched in 1978, was compact and convenient, but also had the superior functionality provided by interchangeable lenses and numerous accessories. Meanwhile other companies were working on another lightweight format—the non-SLR 35mm compact camera. Asahi entered this field in 1982, when it introduced the Pentax Sport 35 camera. This camera also featured the innovative auto-focus lens technology introduced in the Pentax ME-F SLR camera of 1980. Asahi also became the first camera company to achieve total sales of 10 million SLR cameras in 1980.

Asahi continued to be a pioneer in the automation of camera functions. The Pentax Super Program SLR camera, introduced in 1983, offered the user a choice of six types of exposure control, including the use of auto flash. Two years later, the Pentax A3000 SLR camera provided fully automatic operation, with the addition of automatic film loading and winding and film-speed sensing. In 1986, Asahi improved the flexibility of the fixed-lens 35mm compact camera by marketing the world's first compact 35mm camera with a zoom lens, the Pentax IQZoom camera. Since then, Asahi has extended the range of the compact zoom lens, culminating in 1998 with the launch of the Pentax IQZoom 200 camera, with a 48 mm to 200 mm zoom lens, which is still the longest zoom lens on a compact camera. Meanwhile, the first digital Pentax camera, the EI-C90, came on the market in 1996, followed in 1997 by the first Pentax APS (Advanced Photographic System) camera, the efina. As Asahi was not a partner in the consortium that developed APS, it played to its strengths by concentrating on applying its compact zoom lens technology to APS.

See also Cameras

References:
Hick, Roger. *A History of the 35 mm Still Camera.* London and Boston, MA: Focal Press, 1984.
www.pentax.com

AT&T (American Telephone and Telegraph)

American Telephone and Telegraph (AT&T) had its roots in the company formed to exploit Alexander Graham Bell's telephone patents. For a century, it had a virtual monopoly of U.S. telephone services. Since the divestiture of regional telephone services in 1984, the company is still the provider of long-distance telephone services and a major manufacturer of telephone equipment.

Alexander Graham Bell

Alexander Graham Bell was born in Edinburgh, Scotland, in 1847. His father and grandfather were well established in the field of elocution and speech therapy, particularly in the teaching of deaf people. Even as a boy, he showed an interest in the family business by making a working model of the human speech organs out of a bellows, rubber, and cotton cloth. He studied at the universities of Edinburgh and London and became interested in scientific theories and inventions that offered new possibilities in terms of sound transmission. Bell was particularly interested in the acoustic experiments that the German scientist Hermann Ludwig Ferdinand von Helmholtz conducted, using resonance spheres and tuning forks. In 1870, the Bell family emigrated to Canada, where they stayed for two years before moving to Boston, Massachusetts.

In Boston, Bell continued to teach deaf people and to investigate the potential for sound transmission by wire. He attracted financial backing from Gardiner Greene Hubbard and Thomas Sanders, the wealthy fathers of two of his deaf pupils. Hubbard's daughter, Mabel, later became Bell's wife. With the assistance of Thomas Watson, a skilled mechanic, he began to make prototypes of what would become the first telephone. In February 1876, Bell filed a patent application only hours before that of a rival inventor, Elisha Gray. A month later, he succeeded in transmitting speech from a transmitter to a receiver in a separate room. Although Bell became a partner in the National Bell Telephone Company, he took little part in the running of the company and set up laboratories in Nova Scotia, Canada, and Washington, D.C., where he could work on other inventions.

From National Bell to AT&T

The National Bell Telephone Company was formed in Boston in 1877. Hubbard, Sanders, Bell, and Watson were the founding partners. The telegraph company, Western Union, also entered the telephone business, having acquired the rights to Gray's patent and the U.S. rights to Thomas Alva Edison's improved telephone mouthpiece. While a series of telephone patent disputes were not finally settled in Bell's favor until 1893, Western Union decided to sign over its patent rights to National Bell in 1879, in return for a share of profits. National Bell was renamed the American Bell Telephone Company and gained a monopoly on telephone services until the expiry of Bell's patents in 1894. In Britain, where Edison held the rights to his improved mouthpiece, American Bell and Edison joined forces to form the United Telephone Company.

Further restructuring followed as American Bell sought to optimize the management of its expanding business. In 1882, the Western Electric Manufacturing Company was set up in Boston to manufacture telephones for American Bell. The American Telephone and Telegraph Company was registered in New York as a subsidiary of American Bell in 1885. At first, AT&T was solely responsible for American Bell's long-distance telephone services, complementing the Bell Telephone System's control of local telephone services, but it assumed the role of the parent company in 1899. This move gave the company

Alexander Graham Bell at the opening of the long-distance line from New York to Chicago, 1892 (Library of Congress)

more flexibility in dealing with the renewed competition that followed the expiry of the original patents, as New York's corporate laws were more permissive than those of Massachusetts were.

The Monopoly Issue

The legitimacy of AT&T's monopoly was an issue that recurred periodically. In 1907, AT&T's president, Theodore Vail, asserted that service efficiency dictated that telephone services should be controlled by a single provider, with government regulation supplying restraints in lieu of market forces. Many other countries accepted the notion of telephone services as a natural monopoly and created state-owned telephone utilities to prevent the potential abuses that might arise from private ownership. While state ownership did not accord with American free enterprise ideals, concerns about monopolies had resulted in the Sherman Antitrust Act of 1890, which was intended to prevent monopolistic practices. In practice, the Sherman Act was applied only selectively, and AT&T built up its business in the early twentieth century with little opposition.

One acquisition that did attract government scrutiny was AT&T's purchase of 30 percent of Western Union's stock in 1909. Threatened with the possibility of antitrust action, AT&T sold off the stock only four years later. Meanwhile, the company was expanding internationally and had Western Electric manufacturing plants in Europe, the Far East, Australasia, and South America. For a short period during World War I, the government took over the running of a number of key industries, including telephone and rail services. In the postwar boom, the government was keen to stimulate the economy and remove restrictions. One consequence of this was the Graham Act, which exempted telephone services from antitrust law. With its national role thus strengthened, AT&T decided to divest itself of its foreign interests, except those in Canada. In 1925, AT&T sold the International Western Electric Company to the new International Telephone and Telegraph Company. Four years later, AT&T became the first American company to reach the $1 billion mark in annual revenues.

Improving Telephone Services

Despite the divestiture of most of its foreign interests, AT&T was active in improving services to its American customers by developing international telephone services. Bell System engineers had been experimenting with transatlantic voice transmission by radio since 1915. The research and development activities of AT&T were boosted in 1925 by the establishment of Bell Laboratories as the company's dedicated research department. In 1927, AT&T launched a commercial telephone service between New York and London, using two-way radio. While international radiotelephone services expanded over the next ten years, capacity was limited and call quality was adversely affected by signal interference. These problems were removed only in the 1950s and 1960s when transoceanic telephone cables were installed. AT&T installed a transatlantic cable in 1956. Meanwhile, AT&T expanded the capacity for long-distance calls within the United States by adopting microwave technology. The first long-distance microwave telephone relay linked New York and Chicago in 1950. A year later, AT&T introduced direct long-distance dialing, which was available to 90 percent of American telephone subscribers by 1964. Another method of international and long-distance telephone transmission became available from 1962, when AT&T launched Telstar, the world's first commercial communications satellite.

AT&T also played a pioneering role in the development of mobile telephony. AT&T engineers recognized early on that limited capacity was a major impediment to the expansion of mobile phone use. They proposed a cellular solution in 1947, thirty years before such systems actually came into being. AT&T revived this proposal in the early 1970s when the Federal Communications Commission (FCC) invited proposals for a new mobile phone service to operate in an ultra–high frequency band around 800 MHz. The FCC's decision was delayed by the instigation of antitrust proceedings against AT&T in 1974, but AT&T was granted a trial license in 1977 and began its first public trial, in Chicago, in 1978. Commercial licenses were awarded in 1982 when AT&T was in the process of being restructured.

Research and Innovation

From the 1920s, AT&T became renowned for invention and innovation, not just in the field of telephony, but also in the fields of sound recording and electronic engineering. In 1924, it pioneered commercial facsimile services by developing the technology for "telephotography," the transmission of photographs by telephone line. Photographs were

The AT&T Global Network Operations Center, which was unveiled by the company at their headquarters in Bedminster, New Jersey, 10 February 2000 (AFP/Corbis)

successfully sent from Chicago and Cleveland to New York. This system became widely used by the newspaper industry. The next step after the transmission of still images by phone was the transmission of moving images. In 1927, before television broadcasting existed, Bell System demonstrated long-distance television transmission by sending live images of Herbert Hoover, the secretary of commerce, from Washington, D.C., to New

York. Bell System had introduced the concept of electrical sound recording in 1915, by demonstrating that the carbon microphone used as the telephone mouthpiece was equally suitable for capturing sound for phonographic reproduction. In 1931, engineers at Bell Laboratories developed an improved technique for "cutting" gramophone records, whereby the stylus vibrated up and down rather than from side to side. They went on to develop a method for stereophonic recording in 1933 that eventually became standard in the 1940s, although the first stereo experiments had been carried out in Britain two years earlier.

Perhaps the most significant single invention ever to emerge from Bell Laboratories was the transistor, which began the microelectronics revolution. This stemmed from wartime research into the properties of semiconducting crystals, such as silicon and germanium, in relation to radar. In December 1947, three Bell research engineers—John Bardeen, William Shockley, and Walter Brattain—developed the n-type semiconductor diode. The three received the 1956 Nobel Prize in physics for their work on the transistor. Further research into silicon revealed that sunlight caused a release of energy from silicon that could be converted to electric current. The first solar cell, or battery, was created at Bell Laboratories in 1954. The physicist Arthur Leonard Schawlow joined the research team at Bell Laboratories in 1951. Schawlow developed existing ideas on the laser (light amplification by stimulated emission of radiation), an optical version of the maser (microwave amplification by stimulated emission of radiation). In 1961, the first continuous-beam laser was made at Bell Laboratories. It was the combination of the laser and optical fiber cables that enabled telephone service capacity to be increased in the 1980s.

AT&T became involved in research into computer operating systems and computer languages from its perspective as a user of computers for the control of switching systems and call routing. The nature of its computer use meant that it took an early interest in the development of systems that were interoperable and accommodated multiple users. In 1969, Bell Laboratories developed the UNIX operating system. In the early 1970s, the Bell researchers refined the language on which UNIX was based to create C, a high-level, general-purpose computer language. This made UNIX compatible with virtually any of the existing minicomputers. UNIX became widely used on networked computers. By the early 1980s, C was becoming restrictive for more demanding computer applications. In 1983, a Bell Laboratories researcher, Bjarne Stroustrup, added the principles of object-oriented programming to C to create C^{++}, which has become one of the most widely used programming languages.

Divestiture

Despite periodic government restraints, AT&T survived intact until 1984. In 1969, the recently licensed Microwave Communications Incorporated (MCI) obtained FCC approval to connect its microwave long-distance service to the local Bell networks. With major changes in the offing, including the expansion of mobile telephone services, the FCC decided to challenge AT&T's monopoly in a landmark 1974 antitrust suit, with the initial intention of separating its manufacturing and service functions. After prolonged pretrial hearings, the case finally reached the trial stage in 1981. Only a year later, AT&T surprisingly agreed to a negotiated settlement whereby the company was dismantled to create eight separate companies. The settlement was approved by the courts in 1983 and took effect in January 1984.

Prior to divestiture, AT&T was the world's largest private company by such a large margin that the downsized AT&T was still in the world's top three. AT&T retained its long-distance telephone network, its international telephone services, its manufacturing function (Western Electric), and its research and development function (Bell Laboratories). It also gained the right to branch out into data communications, an activity previously prohibited by the FCC. AT&T lost the twenty-two Bell System local networks, which were restructured to form seven regional operating companies, nicknamed the "Baby Bells"—Nynex, Bell Atlantic, Bell South, Southwestern Bell, Ameritech, U.S. West, and Pacific Telesis. Deregulation did not end there, as the Baby Bells began to test the regulatory limits by applying to expand both geographically and functionally. Moreover, in 1995, AT&T announced a voluntary demerger, whereby it would split into three independent companies. In October 1996, AT&T's manufacturing and research business was reconstituted as Lucent Technologies. Two months later, its computer business, the NCR Corporation followed. (NCR had been acquired after the original divestiture.) This left AT&T, now redesignated the AT&T Corporation, with the long-distance telephone network, cellular phone services, a business-communications consultancy, a credit facility, and Internet services.

See also Computers; Mobile Phones; Record Players; Telephones

References:

Bruce, Robert V. *Alexander Graham Bell and the Conquest of Solitude.* Ithaca, NY: Cornell University Press, 1990.

Forester, Tom. *High-Tech Society: The Story of the Information Technology Revolution.* Cambridge, MA: MIT Press, 1987.

Gardner, Robert, and Dennis Shortelle. *From Talking Drums to the Internet: An Encyclopedia of Communications Technology.* Santa Barbara, CA: ABC-CLIO, 1997.

Tunstall, Jeremy. *Communications Deregulation: The Unleashing of America's Communications Industry.* Oxford, UK, and New York: Basil Blackwell, 1986.

www.att.com

www.research.att.com

Atari, Inc.

Atari Inc., the world's first electronic games company, was founded by the American engineer and entrepreneur Nolan Bushnell in 1972. Bushnell was previously employed by the Ampex Corporation, manufacturers of audio and videotape recorders. The name Atari describes a move in the Japanese game "Go." The company was based in Sunnyvale, California. Bushnell's first electronic game was Pong, an electronic version of tennis, originally developed as a coin-operated machine for use in bars and amusement arcades. Atari introduced a home version of Pong two years later, but by then, another American company, Magnavox, had launched its Odyssey electronic ball game. While Atari was doing business with its coin-operated games machines, it lacked the capital to compete with the larger companies that were entering the home-entertainment market. Therefore, in 1976, Bushnell decided to sell Atari to Warner Communications, although he continued to work for Atari until 1978.

In 1977, Atari introduced a games console called the Video Computer System (VCS) that took interchangeable game cartridges. By 1979, there were more than twenty VCS games available, including Space Invaders, which had been developed as an arcade game by the Japanese company, Taito. The popularity of the home version of Space Invaders persuaded Atari to develop VCS versions of its own arcade games, which, by 1982, included Asteroids, Battle Zone, Missile Command, and Pac-Man.

Meanwhile, the company had also turned its attention to the growing home-computer market. Steve Jobs, cofounder of Apple, was employed by Atari in 1976 when he and Stephen Wozniak developed the Apple I microcomputer. While Atari rejected an offer to acquire the rights to the Apple I at a time when the market was untested, two years later it introduced the Atari 400 and 800 home computers. The Atari home computers were technically sound and were well supported with peripherals, but their commercial success suffered because of Atari's divided commitments. By the early 1980s,

Atari was facing stiff competition on all fronts. The launch of the IBM PC in 1981 was followed by a savage price war in 1982, as companies such as Commodore and Texas Instruments cut prices in an effort to maintain sales. While other companies were continuously updating their home computers, Atari was tied up in the rather lengthy development of a new line of home computers, the XL series. The introduction of the XL computers in 1983 did little to offset the financial problems caused by the slump in Atari's games sales.

In 1984, Warner Communications was glad to offload Atari by selling it to Jack Tramiel, who had founded Commodore in 1958. Tramiel streamlined Atari by reducing the workforce and suspending existing projects, and he also developed new products. Atari's prospects began to look healthier when the launch of the 5200ST home computer was favorably received. The 5200ST was nicknamed the "Jackintosh" because of its similarity to the Apple Macintosh. Aimed at and priced for the home market, the 5200ST's disadvantage was a limited range of software. To capitalize on the burgeoning range of PC software, Atari launched its first PC-compatible computer, the i8088 PC-1, in 1987. Its other strategy was to compete for Apple's share of the non-PC business-computer market by developing a more powerful successor to the 5200ST. Like Apple, Atari used Motorola microprocessors. The Atari TT computer, based on the Motorola 68030 chip, was launched in Europe in late 1989, a year before its American launch. Europe, where Apple had a lower market share than in the United States, had proved a more receptive market for the predecessor ST computer. The Atari TT was cheaper than its Apple equivalent but software was again a stumbling block.

Meanwhile, Atari had not entirely given up on the games market, but it was under increasing pressure from the Japanese companies Nintendo and Sega. While Atari was largely relying on its back catalogue of games, Nintendo and Sega were developing new, inventive games. The Atari 7800, 2800Jr., and XEGS games consoles were primitive technologically compared to the Japanese rival systems, but Atari showed that it was still capable of innovation in the games field when it unveiled its Lynx portable games console with a full color LCD display in 1989. However, Nintendo's cheaper Game Boy, also launched in 1989, was more commercially successful.

With its computer sales stagnating, Atari pinned its hopes on overtaking its Japanese rivals in the games market by developing an advanced console. The Atari Jaguar console, launched in December 1993, was the world's first 64-bit games console. It featured high quality sound and color rendering. A contract securing IBM manufacture of the hardware and the commitment of numerous software developers to produce games for the Jaguar were promising signs. However, Atari made the fatal mistake of underestimating the time required for software development after the release of the programming code. Consumers lost interest in the superior Jaguar hardware when a suite of games was slow to appear. Atari was unable to recoup sufficient lost ground before the arrival of the Sony PlayStation and Nintendo 64 consoles in 1996.

The commercial failure of the Jaguar was the last straw for Atari. No longer viable as an independent company, it merged with JTS, a disk-drive manufacturer. This proved to be a short-lived reprieve as JTS folded in 1998. The American toys and games company Hasbro purchased the rights to all Atari's games. This was just one of a series of strategic acquisitions in the 1980s and 1990s that broadened Hasbro's product range. In 1995, Hasbro had set up an "interactive" division to develop games on CD-ROM for the personal computer and the Sony PlayStation. Old favorite Atari games, such as Centipede and Frogger, are now available in CD-ROM format. In Europe, Atari's main computer market, computers are still being made to the specifications of Atari's architecture and operating systems.

See also Computers; Electronic Games

References:

Herz, J. C. *Joystick Nation*. London: Abacus, 1997.
http://homepage.tinet.ie/~morrikar/
www.atari-history.com
www.hasbrointeractive.com/atari/

B

Baby Monitors

Baby monitors became popular in the 1980s thanks to the availability of simple localized radio communication. Either battery and/or plug-in (mains-operated) standard models consist of a "baby's" monitor, which is placed in the child's room, and a "parent's" unit for the room where the parent or caregiver is. Some baby's units have incorporated night-lights and room temperature displays. The baby's unit has a microphone and a transmitter that will alert the parent or caregiver if the baby cries or requires attention. More sophisticated models vibrate like a telephone pager. They have a range of between 50 and 100 meters.

One of the largest manufacturers is the Japanese Tomy Corporation, which was founded in Tokyo in 1927. Although its main business is toys for young children it also produces a range of baby monitors.

Such appliances reflect changing social attitudes about caring for small children. Many parents no longer think their children should cry themselves to sleep. Also, recent publicity given to "crib death" (SIDS) and asthma has made such devices almost essential for concerned parents. These appliances take advantage of the developments in communications technology; the Tomy Baby Watch can transmit live images of the sleeping infant onto the family television screen.

Bagel Slicer

See Slicers

Baird Television Company

John Logie Baird, the first person to transmit television pictures, was born in Helensburgh, Dumbartonshire, Scotland, in 1888. He studied electrical engineering at the Royal Technical College, Glasgow, and began a degree at Glasgow University that was suspended by the outbreak of World War I. Ill health, which was a recurrent feature of his life, ruled him out of military service. Instead, he became superintendent engineer of the Clyde Valley Electrical Power Company. After the war, Baird did not resume his degree. He set up a successful business, marketing a range of goods including soap and patent socks.

In 1922, Baird suffered a serious physical and nervous breakdown, which made him unable to continue working. He began to experiment with television after moving to Hastings on the English south coast. Baird developed a mechanical scanning system, based on a design patented by the German engineer

Baird televisor—one of 1,000 made by Plessey in 1932 (The Museum of Science and Industry in Manchester)

Paul Nipkow in 1884. At this stage, Baird's experiments were a hobby with no immediate business prospects, so he was forced to improvise by using cheap or waste materials, such as biscuit (cookie) tins and bicycle lamp lenses. In early 1924, he succeeded in transmitting a still image of a Maltese cross to a receiver in the same room. Convinced of the potential of his invention, he moved to London and was hired to give television demonstrations in Selfridge's department store. With family financial backing, he set up Television Ltd. and refined his basic technology to improve the quality of the picture. By October 1925, he was able to transmit the live image of a person. He repeated this demonstration for members of the Royal Society in January 1926. Baird then applied for a license to transmit television signals and began trials over a distance of 10 miles. In 1927, he made the first long-distance telecast from London to Glasgow. The next milestone for Baird came in 1928 with the first transatlantic television broadcast from London to a radio station in Hartsdale, New York.

With new financial backing, Baird formed the Baird Television Development Company in 1927 and set up a studio near the Crystal Palace headquarters of the British Broadcasting Corporation (BBC) in 1928. He negotiated a contract with the BBC to provide trial television broadcasts, initially twice weekly for half an hour, using its Crystal Palace transmitter. Baird's receivers, known as "televisors," cost the equivalent of three month's average wages. Not surprisingly, fewer than a thousand homes in London invested in this novelty. In 1932, the BBC decided to take control of Baird's broadcasts.

More ominous for the long-term prospects of the Baird system was the launch of a powerful television consortium. EMI and Marconi, aware of American experiments with electronic television that promised picture delivery superior to Baird's 32-line picture at 12.5 frames per second, had been conducting their own research and development. In 1934, they formed the Marconi-EMI Television Company. A parliamentary committee, the Selsdon Committee, was set up in 1934 to investigate the existing systems and recommend standards of service. Baird decided to improve the performance of his system by making a licensing agreement with the American inventor, Philo Taylor Farnsworth, for use of his Image Dissector. In 1935, the BBC was given responsibility for television broadcasting and invited the Baird Television Company and Marconi-EMI to carry out trial broadcasts at "high definition" picture quality, defined as at least 240 lines. After four months of trials, in February 1937, the BBC decided in favor of the Marconi-EMI 405-line system.

The Baird system was rendered redundant, but Baird himself received some consolation when his pioneering work was rewarded with the gift of the Gold Medal of the International Faculty of Science, which had never previously been awarded to a Briton. Baird's company continued to manufacture televisions that met the Marconi-EMI standard, while Baird himself pursued a new challenge—color televison. In 1928, he had demonstrated color television using mechanical scanning, and he now returned to the development of color television. He experimented with a mixture of electronic and mechanical techniques that yielded 600-line color television pictures by late 1940. Earlier in 1940, the Rank Organisation had taken control of the Baird Television Company, which became Rank Cintel Ltd., leaving Baird free to pursue his color television interests. By 1944, he had developed the Telechrome tube, a two-color system that used two electron guns whose beams converged on a translucent screen that was coated on one side with blue-green phosphors and on the other side with red-orange phosphors. He found another financial backer in British music hall star and actor Jack Buchanan and set up John Logie Baird Ltd. Unfortunately, this new venture proved to be short-lived as Baird died in 1946.

See also Television

References:

Bennett-Levy, Michael. *Historic Televisions and Video Recorders.* Midlothian, UK: MBL Publications, 1994.

Sinclair, Ian. *Birth of the Box: The Story of Television.* Wilmslow, UK: Sigma Press, 1995.

Bang & Olufsen

The Danish company Bang & Olufsen is a leading manufacturer of top quality audio equipment and televisions. Bang & Olufsen products are renowned for their blend of high-tech performance and elegant, minimalist styling, summed up in the slogan it registered in 1931, "B&O—the Danish Quality Brand."

Two young Danish engineers, Peter Bang and Sven Olufsen, who had met while studying at the School of Engineering in Århus, founded the company in 1925. They were fortunate to have families wealthy enough to back them financially, and their first workshop was in the attic of the Olufsen family's country manor near Struer. Their first product was a mains radio receiver, that is, one that was powered by a wired electricity supply—unusual at a time when most radios were battery-powered. However, the company's first commercial success was not the mains radio itself, but its eliminator, the device that rectified the incoming alternating current to produce direct current. B&O began to manufacture the eliminator as a separate device that enabled any battery-powered radio to be run off mains electricity. Expanding production led Bang & Olufsen to open its first factory in 1927 in the town of Gimsing. In 1929, the company returned to producing mains radios with the launch of a five-valve radio that delivered high output.

In the 1930s, Bang & Olufsen diversified into the production of a range of audio equipment, including gramophones, amplifiers, and loudspeakers. The company's products and advertising graphics were heavily influenced by the design aesthetics of the Bauhaus school. The key design characteristics were simple, geometric lines and detailing that emphasized the function of the product and an absence of ornament for purely decorative effect. B&O was a pioneer of the radiogram, a radio receiver and record player combined in one cabinet. The first B&O radiogram, the Hyperbo, was launched in 1934. The tubular steel frame of the Hyperbo was influenced by the chair designs of the German Bauhaus designer Marcel Breuer. Bang & Olufsen's first radio with a Bakelite cabinet, the Beolit, was introduced in 1939. From the mid-1960s, the prefix "Beo" was incorporated in all B&O model names. In the same year, B&O's Master de Luxe radiogram incorporated a feature that became very popular—push-button radio-station selection. The radio was pretuned to 16 radio stations.

Bang & Olufsen went through a quiescent period during World War II because it refused to cooperate with the occupying German forces. Worse still, after liberation from German occupation in 1945, the factory was bombed by Danish Nazi sympathizers. After the rebuilding of the factory, Bang & Olufsen entered the field of television manufacture. In the 1950s, B&O commissioned a number of Danish architects, including Poul Henningsen and Ib Fabiansen, to design the cabinets for its audio and television equipment. It was keen to produce cabinets that were lighter and easier to move around. In 1962, B&O introduced the Horizon TV, its first television to be mounted on a four-wheeled metal stand.

The transistorization of audio equipment and televisions paved the way for compact, modern product designs. The Beomaster 900K, designed by the Danish architect Henning Moldenhawer, was the world's first low-line radio cabinet, a forerunner of the stereo receivers that formed part of the popular modular hi-fi systems of the late 1960s and 1970s. The designer who did most to establish a distinctive B&O style of audio equipment was Jakob Jensen. His designs, beginning with the Beolab 5000 music system of 1965, were expressive of the technical sophistication of B&O's products. This system introduced user-friendly sliding controls. The Beolab system was accompanied by cube stereo loudspeakers, with the angular speaker cone mounted on thin stems with a circular base. However, Jensen's most famous design for B&O was the Beogram 4000 stereo turntable of 1972, because this introduced the world's first tangential pickup arm. The straight double tone arm was electronically controlled by a spot of light, and its tangential path eliminated the wandering in the groove that curved arms were prone to.

Recognizing that its products were never going to achieve the mass-market penetration of rival Japanese electronics products because high quality meant high prices, B&O concentrated on lifestyle marketing and design. It targeted a wealthy international clientele for whom style and quality were the tantamount product characteristics. B&O's continuing commitment to functionality and ease of use was exemplified in the controls of the 1976 Beomaster 1900 receiver. The most frequently used controls were mounted visibly at the front for easy access, while the secondary controls were behind, concealed beneath a hinged lid. Similarly concealed controls became standard on televisions in the 1980s. The other innovative feature of the Beomaster 1900 controls was that the buttons were touch-sensitive electronic buttons, not mechanical push buttons. The Beosystem 5000 modular hi-fi system of 1983 eliminated controls from the hi-fi units in favor of a unified remote-control panel. This concept was taken a step further in 1984 with the introduction of the Beolink 1000 remote-control unit that incorporated television as well as audio controls.

In the 1990s, B&O broke away from stacking, modular hi-fi design in order to distinguish its products from those intended for the mainstream mass market. The Beosystem 2500 of 1991 was an integrated unit with the decks mounted vertically and therefore more visibly. The Beosystem 2500 and its successor, the BeoSound Century, also echoed the slim verticality of B&O's televisions. Introduced in 1984, the BeoVision MX 2000 television was the first of B&O's slim televisions. Its shallow cabinet and the minimal frame around the screen emphasized the picture, the core function. Audio and television were brought together in the BeoCenter AV5 of 1997, a complete home-entertainment system. As the twentieth century ended, Bang & Olufsen's final contribution to user convenience was the development of the BeoVision 1 television, which incorporates an intelligent automatic program selection function, whereby the user selects the preferred types of program and the television matches the selection to the programs available.

See also Radio; Record Players; Television

References:

www.bang-olufsen.com

Barbecues

Barbecues, or open-air meals, date back to large social events such as ox or hog roastings. Such events were communal affairs; today the barbecue is seen as a more private affair conducted in a suburban garden (yard). They still maintain their social functions, as they often double as parties. Barbecues became popular in the United States in the 1960s and spread to Northern Europe in the 1970s. The large barbecues offered as part of Mediterranean package holidays were another stimulus.

The garden barbecue grills or spit roasts food with the heat supplied by hot charcoal or compressed hardwood briquettes. There are many different shapes and sizes, but they are all used in much the same way.

The simplest type is based on the Japanese hibachi, or fire bowl, a simple rectangular container with a grilling rack. Larger models stand on legs and can also be circular. They usually have a windshield with slots to accommodate different grilling positions and spits. Rounded kettle barbecues are more sophisticated, with domed hoods and vents. When closed the hood allows it to act more like an oven, capable of broiling joints or fowl.

The most difficult part of barbecue cooking is to get the charcoal to light. This has been assisted by the use of solid and liquid firelighters. An easier way is to light the fuel by liquid petroleum gas. More sophisticated models also have small gas rings. Funnel barbecues use lightly folded newspaper that is set alight to deliver rapid intense heat to the fuel.

The appeal of barbecue cooking is that it can make simple sausages and burgers taste better. One interesting sociological factor is that, although women still do most of the cooking, the control of the barbecue is often a male preserve.

See also Broilers

Batteries

Many small electrical appliances rely on batteries for their power source. They can be for hand-held games such as the Nintendo Game Boy, portable radios, and systems like the Sony Walkman or for remote controls for videos, televisions, and hi-fi equipment.

The development of batteries began with the experiments of Alessandro Volta (1745–1827) and John Frederic Daniell (1790–1845). Volta discovered that when two different metals are in contact with moisture an electrical current is produced. His first "wet cell" battery used alternating zinc and silver discs separated by cloth moistened with a salt solution. Daniell improved on this by using zinc and copper electrodes, resulting in a more practical battery. The "dry cell" battery was developed in the 1860s. This led to the ubiquitous zinc-carbon battery that was to be in use for most of the twentieth century. The so-called dry cell has a moist paste electrode inside a zinc container. The positive electrode is a carbon rod in the center of the cell.

These batteries developed into two basic types, the small cylindrical battery for flashlights (torches), and so on, and the larger, rectangular power pack. The main manufacturers in the United States and the United Kingdom were both called Ever-Ready.

Batteries were important because they made electricity portable. The small inexpensive battery-operated flashlight soon became a household staple. Other appliances (such as radios) had to wait until the associated technological and social conditions allowed them to become smaller. The radios that followed the "cat's whisker" sets required heavy "wet-cell" batteries. The development of valves allowed the use of lighter "dry-cell" batteries and the portable radio. These models, such as the Pye Type 25 of 1928 (which featured the first of its famous "sunrise" speaker grilles) were heavy due to a combination of batteries and wooden cases. Although a little lighter, even late 1940s and early 1950s models were cumbersome. The British Ever-Ready battery company also produced its own portables.

The development of transistors and radios such as the Sony TR-55 (1955) and the UK Pam (1956) and increasing personal mobility led manufacturers to produce smaller products. The rise of the portable transistor radio was also aided by the growth of rock and roll and a youth culture fuelled by a generation of teenagers with more money and time.

The development of transistors and the culturally driven desire for music and communications on the move has led to a migration of the products from the domestic and office environment into the public realm, as exemplified by the Sony Walkman and the mobile telephone. To keep this revolution going, manufacturers have relied on increasingly efficient lightweight long life batteries.

This trend was exploited by the Duracell Company, which pioneered the marketing of longer-lasting alkaline batteries in the late 1970s and 1980s. It caught the manufacturers of zinc-carbon batteries by surprise, as they were unprepared for the competition. The UK Ever-Ready Company folded in 1981, only to be bought up by an American company, Ralston Purina. In less than ten years alkaline batteries accounted for over 50 percent of U.S. battery sales. Duracell is a division of the Gillette Company and trades in over fifteen countries, employing 4,500 people in the United States, Belgium, China, and India. Although 80 percent of the world market is still zinc carbon, the alkaline battery dominates the Western consumer goods market. Recent trends have been the introduction of power indicators on the sides of the batteries and longer-lasting rechargeable batteries.

See also Radio; Sony Corporation; Television

References:

Hill, Jonathan. *The Cat's Whisker: Fifty Years of Wireless Design*. London: Oresko, 1978.

Kochan, Nicholas, ed. *The World's Greatest Brands*. London: Macmillan/Interbrand, 1996.

Bed Warmers

Climbing into a cold bed has never been a pleasant experience. The traditional method of dealing with this problem was the warming pan or the hot water bottle. Nineteenth- and early twentieth-century hot water bottles were made of either copper, stoneware, or (later) rubber.

The rise of electricity use in the 1920s prompted manufacturers to experiment with the humble hot water bottle. The Supreme Miracle was a 12-inch tubular element that screwed into an ordinary rubber hot water bottle and heated the water inside. This idea was developed by F. S. Spooner Wates, who also patented an electric bed heater encased in an asbestos tube. His patent was taken up by the British company Rothermel who produced an electric bed warmer with a brown Bakelite case in the shape of a rubber hot water bottle. The flex entered where the stopper would have been and the on/off switch was placed at the neck. Such products were probably best used to heat the bed before getting into it, as they were not grounded.

The first electric blanket appeared in the United Kingdom in 1927. This was the small Thermega heating pad, which had flexible electric heating elements within woolen fabric. Sunbeam was a major manufacturer in the United States. These were relatively expensive items, and electric blankets only became popular during the 1950s and 1960s, thanks to more reliable insulation, thermostatic controls, and developments in flameproof materials. The British Burco Company, which began by making gas water heaters, started manufacturing electric blankets in the 1950s. It continues to produce them under the Cozee Cumfort brand. Most of these electric blankets were designed to go under the sheets and warm the bed before anyone slept in it, nevertheless the story of the electric blanket setting both bed and sleeper alight did enter popular folklore. The 1960s saw the introduction of electric over blankets designed to stay on all night. They could be washed in an electric washing machine, and the double models featured separate controls for each side of the bed, allowing sleeping partners to choose their own temperature. Although electric blankets are still in production, their popularity has declined due to the rise of central heating and warm duvets.

Reference:

de Haan, David. *Antique Household Gadgets and Appliances, 1860–1930.* Poole, UK: Blandford Press, 1977.

Belling Company

Charles R. Belling founded the Belling Company in 1912. Belling had previously worked at the British electrical companies of Crompton and Co. and Ediswan, where he had gained experience making electric heaters. He wanted to develop heaters further and set up in a small shed in Enfield, London, with £150. Two partners, C. L. Arnold and H. E. How, contributed £150 each.

Belling invented the "firebar," a resistance wire wound around a fireclay former. Like gas radiants, the firebar could be raised almost to incandescence. The result was the Standard, an electric heater that went into production in the same year. It featured six horizontal radiants set in an enameled cast-iron body. It had a hook at the top from which a kettle could be suspended, a fold-down rack for keeping a pot warm and an attachment for toasting bread. Belling acted as a traveling salesman for the the Standard, which was an immediate success, and the company dropped the idea of producing other lines, such as an immersion heater and a kettle, in order to concentrate on its money earner. By 1914 £11,000 worth of heaters had been sold, delivering a profit of £3,500.

The company went into wartime production during 1914–1918 and emerged in a strong position to develop in the 1920s. Belling expanded their range with hotplates, immersion heaters, water urns, irons, steamers, grills, and the No. 7 boiler, a water boiler. They also introduced their first cooker (stove), the Modernette, one of the first lightweight models featuring a sheet steel body, in 1919. The firebar was improved with the Multi-Parabola firebar of 1921. This was a block of fireclay into which grooves comprising parabolic wells had been made. The heating element was formed in a continuous coil and laid in the grooves. This type of element was to be used in millions of electric heaters worldwide. Belling also introduced an imitation coal-fire model in 1921 in response to the market's continuing love of an "open fire." Although much derided as a piece of phony design, they proved popular for many years.

Other innovations followed with the introduction of a lightweight horizontal cooker with a waist level oven in 1926 and the compact Baby Belling in 1929. The Baby Belling was similar to the oven unit of the 1926 cooker but with a hotplate on the top. This model remained in production throughout the century. It was ideal for small flats and offices. During the 1930s the company continued to expand its range of heaters, including the Solray reflector heater of 1934, and to produce innovative cookers with glass oven doors.

During World War II, Belling moved to wartime work, producing its only nonelectrical item, an incendiary bomb snuffer. As if by fate, when peace came in 1945 it was Belling who manufactured the Vee cooker for the prefabricated houses designed to help alleviate the severe housing shortage caused by bomb damage. Following the war it produced the 47AB cooker, which was one of the most successful British models of the 1950s. The company expanded during the 1950s and 1960s as postwar consumption grew.

Belling was acquired by the Glen Dimplex group in 1992 and now concentrates on the manufacture of free-standing cookers.

See also Cookers; Glen Dimplex; Heating

Reference:

Belling Company. *The Story of Belling,* 1962.

Bidets

A bidet is a low, narrow basin intended for washing the anal and genital areas after using the water closet (toilet), although it may also be used as a footbath. The name is derived from the French word for a little pony, referring to the action of stepping astride it. Portable bidet pans were in use in France from the early eighteenth century. The Marquise de Pompadour, the mistress of the Louis XV, had two bidets, one with a rosewood surround and the other in walnut. Fixed pedestal bidets became available in the late nineteenth century when indoor plumbing and water closets became more widespread. For example, W. R. Maguire patented a combined water closet and bidet in 1888. After World War II, bidets began to lose their luxury status, but they are still uncommon outside France and other parts of continental Europe. Today, in addition to the basic type of bidet filled by hot and cold taps above the rim, there are bidets that feature an internal rising spray.

See also Water Closets

References:

Wright, Lawrence. *Clean and Decent: The Fascinating History of the Bathroom and the Water Closet.* London: Routledge, 1980.

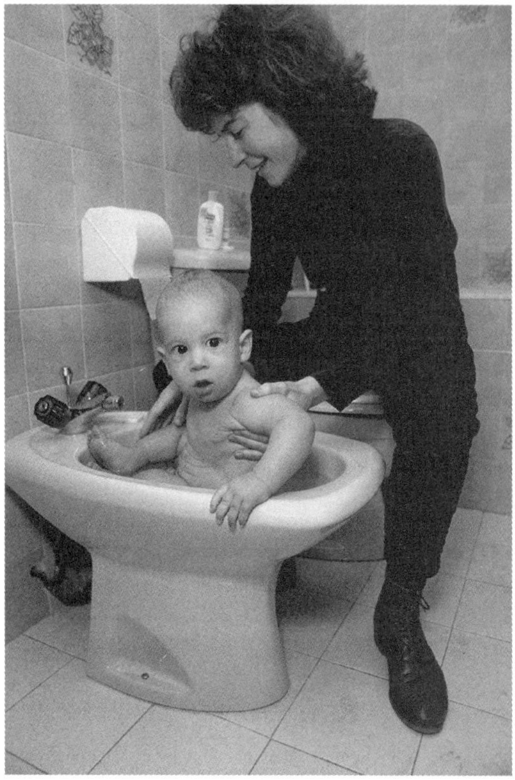

A baby gets a bath in a hotel room bidet in France, c. 1989 (Philip Gould/Corbis)

Black & Decker

The Black & Decker Company was founded in 1910 by S. Duncan Black and Alonzo G. Decker as a small machine shop in Baltimore, Maryland. They produced their first electric drill in 1915, followed by patented pistol grip and trigger switch. These innovations became standards within the electrical tool industry.

The company has developed over the post-1945 period to become a multinational corporation manufacturing power tools, domestic appliances, hardware, and building products. It has a number of subsidiaries making Emhart fastenings, DeWalt industrial tools, Kwikset home security, and Price Pfister plumbing products. Black & Decker acquired the electrical domestic appliances section of the U.S. General Electric Company in 1984. It is the world's largest producer of power tools and residential security systems and the third largest faucet manufacturer in the United States. Black & Decker products are produced in fourteen countries and marketed in over a hundred. It has a large power tool plant at Spennymoor within the United Kingdom. It also produced tools for the NASA space program in the 1960s and 1970s.

The company has continued to innovate in product development and can claim to have introduced genuinely new tools and appliances. Most notable are the DustBuster vacuum cleaner of 1979. This small rechargeable cordless hand-held cleaner was a logical development of the older Hoover Dustette. It remains popular for cleaning up small areas and spilt crumbs. Another innovation is the Workmate designed by English engineer Ron Hickman in 1968. Black & Decker was not convinced that the Workmate would be successful and had initially refused it. However, Black & Decker introduced it in the United Kingdom in 1972 and in 1975 in the United States. The Mouse (1998) is a small hand-held sander and polishing tool that produces 11,000 orbits a minute. It comes with twenty-three accessories.

More recently the corporation has invested in more sustainable methods of production using recyclable coppers and irons in its motors and identifiable plastics for future recycling. In the United States, it has a national disposition center in Nashville, established to deal with damaged or worn-out products.

See also Breadmakers; Can Openers; Do-It-Yourself; Electric Drills, Workmates

References:

Heskett, John. *Industrial Design.* New York: Oxford University Press, 1980.

Julier, Guy. *Dictionary of Twentieth-Century Design and Designers.* London: Thames and Hudson, 1993.

Kochan, Nicholas, ed. *The World's Greatest Brands.* London: Macmillan/Interbrand, 1996.

www.blackanddecker.com

Blenders/Juicers

Extracting juice from fruits has long been necessary either for producing drinks or for cooking. Juicers are a good example of the continuing use of traditional models alongside more sophisticated electric versions that have been introduced in response to the changing tastes of modern consumers.

Traditional citrus juicers of beech wood that are simply reamers to be placed into the fruit and twisted are still sold today, as are juicers with dome-shape reamers over which the halved fruit is twisted. Initially made of glass, the dome-shaped juicers were also made of aluminum and are now produced in plastic. One of the most stylish of these is the Italian Kartell lemon squeezer designed by Gino Colombini in 1958. Produced in low-density polyethylene it features a sharply fretted pivot inside a container onto which the halved lemon is placed. This is covered by a ribbed cap that is turned to pulp the lemon juice down channels in the pivot into the container. Electric versions operate in similar ways. Wooden or cast-iron hinged presses for lemon and limes were also popular for the

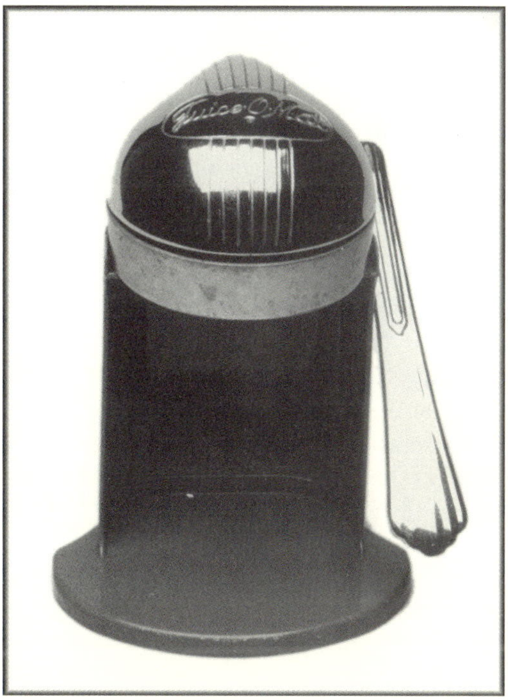

Juice-O-Mat manual juicer made by the Rival Company, Toledo, Ohio (The Museum of Science and Industry in Manchester)

first half of the century. Wooden models, usually of beech, often had glazed ceramic bowls. There are also aluminum hinged presses with inverted domed reamers.

Lever-operated juicers and presses were developed in the United States for professional use in restaurants, bars, and diners. The domestic versions of the 1950s, such as the Juice-O-Mat by the Rival Company, were available in sprayed aluminum or stainless steel.

Electric blenders were developed by the American Stephen Poplawski who had patented a commercial drink mixer in 1916. The Greene Manufacturing Company produced blenders based on this in 1932. Two successful models from the 1940s are the Osterizer by Oster Manufacturing Company (now part of Sunbeam) and the Blendor by Waring Corporation. These blenders (known in the United Kingdom as liquidisers) had a glass jug with a lid and cutting blades in the base, operated by an electric motor. Blenders are essentially simple appliances, and the main changes in them have been the addition of extra speeds. The arrival of the food processor meant that the blender became an inexpensive commodity. Recent changes have been the introduction of high-powered "pro-style" models capable of chipping ice without liquid and more dominant bases with electronic touch pad controls. The main manufacturers are Cuisinart, Hamilton Beach, Kenwood, KitchenAid, Krups, Moulinex, Oster, Philips, and Sunbeam.

The 1980s saw the introduction of smaller hand-held blenders capable of blending foods and liquids in a cup or small container. One of the earliest lines was introduced by Braun, which remains a leader in this market. Hand blenders are especially popular for making baby food.

Cultural trends in healthy eating have led to an increase in fresh fruit consumption and the promotion of premium juices produced only from fresh fruit. Manufacturers such as Kenwood, Braun, Sunbeam, Hamilton Beach/Proctor-Silex, and Breville have responded with improved electric countertop juice extractors designed to produce fruit juices directly into an internal jug in seconds for the domestic market. A simpler appliance designed to produce just enough juice for flavoring is the Lemon Mate, a small plastic device that turns a single lemon into a small jug by topping and tailing it with a screw-in reamer and a base to keep it upright.

Despite their efficiency, the most influential juicer of the late twentieth century was a reworking of the traditional hand-held juicer by the French designer Philippe Starck. His cast aluminum Juicy Salif lemon squeezer of 1990 for Alessi placed the reamer on three legs, which allowed it to be placed over a glass. The lemon juice flowed down the reamer to a point and then into the glass below. Available in either a polished or PTFE (polytetrafluoroethylene) nonstick coating, it quickly became a "must-have" item for the fashionable kitchen.

See also Alessi; Braun; Food Mixers; Food Processors; Hamilton Beach; Kenwood; KitchenAid; Proctor-Silex; Sunbeam

Bodum

The Danish Bodum company was founded in 1944 by Peter Bodum, importing glassware from Europe into Denmark. It produced its first coffeemaker, the Santos, in 1958. This was designed by the architect Kaas Klaeson and is still in production. In 1974, the company began the production of its highly successful cafetieres with the Bistro model that sat on a distinctive cork base.

The Bodum company has always invested in design, reflecting a humane Scandinavian modernism. It has expanded its range of glassware products and has also moved into plastics and a limited number of related appliances, such as the Osiris (1981) stove kettle and the Ibis (1998) electric kettle.

The company remains in family ownership; the present manager is Peter Bodum's son Jorgen. The company has a number of shops throughout Europe and a design company, Pi Design, based in Switzerland.

See also Coffeemakers; Kettles

Braun

The Braun company was founded in Frankfurt in 1921 by Max Braun (1890–1951), an engineer from East Prussia. It originally produced connectors for machine belts and later moved into components for radios and gramophones in 1923. By 1925 the company was producing many of its own plastic components, and by 1929 it had begun to make complete sets. Braun became one of Germany's largest radio manufacturers. It began to innovate during the 1930s, introducing a combined radio and phonograph in 1932 and a battery powered portable radio in 1936. By 1938 its modern Frankfurt factory employed 1,000 people.

During the postwar reconstruction it added domestic appliances and electric razors to its range of products. The Braun S50 shaver and the Multimix appeared in 1950. In 1954 Braun struck a deal with the Ronson Company, who were licensed to manufacture Braun shavers in the United States.

Max Braun was succeeded in the early 1950s by his sons Artur and Erwin, who were interested in design and brought in a range of talented designers to work on their products. Dieter Rams joined the company in 1955, along with Hans Gugelot, Otl Aicher, and Gerd Alfred Müller. The following year it set up its own design department, which Rams headed from 1960.

The result of this corporate approach was a unified range of products that possessed a sculptural simplicity. The electronics of razors, food mixers, and heaters were enveloped in white metal or plastic covers with minimal, easy-to-use controls. The KM 321 Kitchen Machine of 1957, a food mixer, exemplified this approach. This "neo-functionalist" approach could also be seen in the audio products such as the Phonosuper of 1956, nicknamed "Snow White's Coffin" because of its rectangular shape, white body, and clear Perspex lid. Braun set a standard that influenced other companies to take design more seriously. Its products were selected by the New York's Museum of Modern Art and praised at the 1958 Brussels

The Braun KM31 food mixer came with a range of attachments including a liquidizer jug, shown here. (The Museum of Science and Industry in Manchester)

World Fair as "outstanding examples of German manufacturing."

The aesthetic merit of Rams's designs was reflected in the work of U.K. "Pop" artist Richard Hamilton in his *Toaster* screen print and collage of 1967. He stated, "My admiration for the work of Dieter Rams is intense and I have for years been uniquely attracted towards his design sensibility; so much so that his consumer products have come to occupy a place in my heart and consciousness that the Mont Sainte-Victoire did in Cézanne's."

The controlling interest in Braun was bought by the U.S. Gillette Company in 1967. Since then its style has become a little diluted but the ET22 calculator and the Micron shaver have ensured that Braun products remain distinctive. Braun has, more than probably any other company, managed to successfully marry modernist principles to industrial production. The results have largely been just what Erwin Braun wished them to be: "honest, unobtrusive, and practical devices."

See also Calculators; Coffeemakers; Food Mixers; Hair dryers; Razors

References:

Bayley, Stephen. *In Good Shape: Style in Industrial Products, 1900–1960.* London: Design Council, 1979.

Sparke, Penny, *A Century of Design: Design Pioneers of the Twentieth Century.* London: Mitchell Beazley, 1998.

Sudjic, Deyan. *Cult Objects.* London: Paladin/Granada, 1985.

www.braun.com

Breadmakers

The baking of bread at home declined during the twentieth century due to the rise of industrialized baking and retailing. By 1950 most people bought their bread from small local bakeries, which in turn were overtaken by the large supermarket chains.

In the United Kingdom, the late 1970s saw a reaction, led by food writers, to the rather bland industrially produced bread and a demand for greater choice. This feeling was amplified when more people took holidays in France, where the tradition of the small local bakery has remained intact, even in large cities. Supermarkets responded with "in-store" bakeries and a much wider variety of breads inspired by French and Italian recipes.

Bread could be baked in the home in a gas or electric oven, but during the early 1990s manufacturers developed countertop breadmakers designed to give good results every time. West Bend produced the first American model in 1993. The company moved very quickly to enter this niche market, completing the project in only thirty-five weeks, from concept to shipping. A breadmaker is essentially a mixer, proofing oven, and mini-oven in one. They have plastic "cool wall" cases, usually with viewing windows. They automatically knead, proof, and bake, and they can take up to three sizes of loaf tin. The ingredients are placed in a nonstick baking tin, a cycle is selected, and the machine does the rest. A paddle in the bottom of the bread pan kneads the dough, stopping two or three times to let it rise. Settings are usually for overnight baking, but some models offer high-speed programs that deliver a loaf in less than two hours. Most models have midcycle indicators to allow extra ingredients such as fruits and nuts to be added. They come in a variety of sizes, which usually relate to the size of the loaf, from one to two and a half pounds.

Popular manufacturers include Black & Decker, Breville, Oster, Panasonic, Prima, Sunbeam, Toastmaster, and West Bend.

See also Black & Decker; Convenience Foods; Sunbeam; West Bend

Brillo Pads

Brillo pads are steel wool pads impregnated with a special soap containing jeweler's rouge. They were introduced by the Brillo Company of Brooklyn, New York, in 1930.

The company was the result of what would now be called a "market-led" approach. A Mr. Brady, a New York door-to-door salesman, was selling aluminum pots and pans and noted that his customers complained about how difficult they could be to keep clean. Brady consulted his brother-in-law, Mr. Ludwig, a costume jeweler. It was Ludwig who struck upon the idea of combining soap with jeweler's rouge to produce the required shine. Brady then found that his soap was beginning to out-sell the pans. Brady and Ludwig approached a lawyer, Milton B. Loeb, for advice on establishing a company to begin commercial production. Loeb must have seen the potential as he joined them, as well as providing the brand name, Brillo, after the Latin beryllus (shine). Loeb went on to become treasurer and president of the company. The Brillo soap was patented and registered as a trademark in 1913.

Brillo's main product was the soap that was sold with pads of steel wool. Initially sold by door-to-door salesmen, they were soon taken up by grocery and hardware stores and chains such as Woolworth's. The steel-wool pads, impregnated with the soap, were introduced in 1930. Brillo remains one of the world's best selling pan cleaners, along with its main rival SOS. They have survived the arrival of motorized scouring pads in the 1960s. The Kent Kordless of 1962 was one such product, but was deemed not worth its cost by *Consumer Reports.*

Thanks to Andy Warhol's oversize replicas, the Brillo pad's bright and simple packaging, along with the Campbell's Soup can, became an icon of 1960s pop art. The company is now a part of Church & Dwight Co. Inc.

See also Cookware

References:

Hillman, David, and David Gibbs. *Century Makers.* London: Seven Dials, 1999.

British Electrical Development Association (BEDA)

The BEDA was set up in 1919 and funded by a coalition of four associations of engineers from the electrical supply and manufacturing industries. Its mission was to promote the use of electricity and its primary target was the householder. The domestic market offered the greatest potential and, critically, the anticipated daily pattern of domestic demand would complement the different industrial demand pattern, reducing unit generating costs and supply prices. However, electricity had to overcome its competitive disadvantage to gas in terms of the higher costs of installation, supply, and appliances. While electric utilities pursued a loss-leader strategy, by renting out appliances at below-cost rates, the BEDA's role was as a propaganda machine.

In the 1920s and 1930s, the BEDA ran advertising campaigns extolling the labor-saving and life-changing potential of electricity. It constructed a vision of a Utopian future in which the all-electric home offered comfort ("healthy radiant heat"), efficiency ("freedom from domestic worry"), and hygiene ("protect your family from food danger"). Electricity was described as "a universal servant with an eternal willingness to work." This astute strategy simultaneously played on women's fears about their proficiency in the home and appealed to their aspirations to spend less time on housework. The BEDA was at least partly successful: there was an increase in households with electricity in the United Kingdom from 6 percent in 1919 to 65 percent in 1939. However, many houses only used electricity for lighting, and ownership of electrical appliances was very restricted.

The BEDA's role began to change during World War II when it had to reverse its message and encourage people to save electricity. It resumed its championship of the domestic appliance in the 1950s with campaigns such as the "Four Foundations of Modern Living" (cookers, water heaters, refrigerators, and washing machines). Success was more achievable against a background of rent–purchase schemes and growth in disposable income. In 1957, the Electricity Council was set up as the central coordinating body for the supply industry; in 1968, it absorbed the BEDA as its marketing department.

References:

Hannah, Leslie. *Electricity before Nationalisation: A Study of the Development of the Electricity Supply*

Four examples of BEDA promotional leaflets from the 1920s and 1930s (The Museum of Science and Industry in Manchester)

Industry in Britain to 1948. London: Macmillan, 1979.

Hannah, Leslie. *Engineers, Managers, and Politicians*. London: Macmillan, 1982.

Roberts, Gerrylynn K. "Electrification." In *Science, Technology, and Everyday Life, 1870–1950,* edited by Colin Chant. London: Open University/ Routledge, 1990.

British Telecom

British Telecommunications plc, more familiarly known as British Telecom or BT, was formed by the privatization of Britain's national telephone system in 1984. As a private-sector company, it not only retained its core business of supplying local, long-distance, and international telecommunications services and equipment in Britain, albeit in a new competitive environment, but also gained new international business opportunities. British Telecom now operates joint ventures in thirty other countries worldwide, including Spain, India, and South Africa.

The industry had been run as a state monopoly since 1912. Telephone services were introduced in Britain in the late 1870s and 1880s by privately owned companies, including the United Telephone Company, which was jointly owned by America's National Bell and Edison. In 1880, the government awarded licensing control over telephone services to the state-owned Post Office, which already had a monopoly of telegraph services. In 1889, when a number of competing private companies merged to form the National Telephone Company, the Post Office took over the operation of long-distance lines. The Post Office assumed full control of telephone services following the nationalization of the industry in 1912, although a few local telephone services continued to be owned and operated by local authorities.

In the late 1960s, a combination of factors prompted recognition of the need for institutional change within the British telephone industry. The introduction of automatic distance dialing, known as subscriber trunk dialing (STD), in 1959 and international subscriber dialing in 1963 put pressure on telephone exchanges, many of which needed upgrading. As a government department, the Post Office was subject to Treasury constraints on investment. In 1969, the Post Office gained a greater degree of financial autonomy when it became a national industry rather than a government department operating on a budget allocated by the Treasury. Consequently, it was able to commission a consortium of three British companies, GEC, Plessey, and STC, to develop a computer-controlled digital telephone exchange system, named System X. After prototype testing in 1978, it was introduced in London in 1980 and gradually extended.

Privatization came as a result of the 1979 election of a Conservative government on a free-enterprise platform. Under the ensuing British Telecommunications Act of 1980, the Post Office lost its monopoly of telephone services. In preparation for privatization, it was restructured in 1981 into two independent divisions, mail and telecommunications. A second telephone service supplier, Mercury Communications, was granted a license in 1982. In 1984, the newly privatized British Telecom opened its first digital international exchange, installed by Thorn-Ericsson Telecommunications Ltd.

Privatization and deregulation coincided with the introduction of cellular phone services. In 1982, the British government decided to grant two nationwide licenses for cellular phone services. One license was awarded to the Cellnet consortium, led by British Telecom in partnership with the security company Securicor, and the other went to the Vodafone consortium led by Racal Electronics. Both cellular phone services became operational in 1985. A second phase of telecommunications deregulation followed the release of the government's 1990 discussion paper "Competition and Choice: Telecommunications Policy for the 1990s." The duopolies in both the fixed telephone and cellular phone systems were discontinued, opening the market to new operators.

British Telecom's response to increased competition was to strengthen its commitment to customer care by launching a new BT mission in 1991 that promised to put customers first. While British Telecom has retained its overall leadership in British telephone services according to market share, in terms of financial success, it has been overtaken by the mobile phone company, Vodafone.

See also Ericsson; General Electric Company; Mobile Phones; Telephones; Vodafone

References:

Forester, Tom. *High-Tech Society: The Story of the Information Technology Revolution.* Cambridge, MA: MIT Press, 1987.

Robertson, J. *The Story of the Telephone: A History of the Telecommunications Industry of Britain.* London: Pitman, 1947.

www.bt.com

Broilers

Broiling, the cooking of meat on a fire or on a grid over it, is one of the most ancient forms of cooking. The first electric broilers appeared in 1916. The first table model was the Broil King, manufactured by the International Appliance Company in 1937. Table broilers were usually cylinders with hinged or removable lids. The meat sat on a perforated metal tray while the cooking element was housed in the lid. Farberware introduced the Open Hearth broiler in 1962. This featured a heating element placed below the food.

A variant was the Rotissimat of 1946, produced by the Rotissimat Corporation, which, as its name implied, featured a rotisserie for poultry. The product was promoted in supermarkets by using them to roast chickens, and it has been suggested that this stimulated the introduction of shop-roasted chicken. Rotissimat went into liquidation in 1954, but the name remains as a generic term.

Manufacturers in the United States also produced open gas broilers or indoor barbecues in the 1940s and 1950s that were companions to gas or electric stoves. Broilers were not so popular within the United Kingdom and Europe, where the home rotisserie became combined with a grill. Moulinex produced such table models in the 1970s. The 1960s saw "top-of-the-range" electric ovens having rotisserie attachments within their "eye-level" grills. Table broilers are no longer such popular items due to the increasing speed and sophistication of ovens, grills, and microwave ovens.

Paradoxically, the classic method of broiling food on a gridiron has become more popular with the outdoor barbecue. Here traditional charcoal, whether ignited by firelighters or gas jets continues to be the popular fuel.

See also Barbecues

Reference:

Goldberg, Michael J. *Groovy Kitchen Designs for Collectors, 1935–1965.* Atglen, PA: Schiffer Publishing, 1996.

Brother Industries

The Japanese company Brother Industries manufactures a range of consumer products, including sewing and knitting machines, business machines, and home electrical appliances. Brother has thirty-three foreign subsidiaries, and 90 percent of its sales are outside Japan.

Brother Industries began as the Yasui Sewing Machine Company, which was set up in Japan in 1908 to repair sewing machines and produce parts. In 1928, the company produced its own sewing machine, an industrial model, and adopted Brother as a brand name. It began producing domestic sewing machines in 1932. Two years later, it was incorporated as the Nippon Sewing Machine Manufacturing Company. In the 1950s, the company began to expand both its product line and its markets. In 1954, it produced its first knitting machine and entered the domestic electric appliance field. To stimulate overseas sales, the Brother International Corporation was set up in the United States in 1954; a European sales subsidiary followed in 1958.

In 1961, the company diversified into the machine tool and business machine fields by producing a small lathe, intended for the school market, and its first portable typewriter. The company changed its name to Brother Industries Ltd. in 1962. In 1968, Brother acquired the leading British sewing machine company, Jones, which had been set up near Manchester in 1859. Jones and Brother had developed a mutually beneficial partnership shortly after the end of World War II. Brother reached the production landmark of 10 million sewing machines in 1971. This was also the year that the company introduced its first high-speed printer. Brother began to manufacture sewing machines in Taiwan in 1979. Fourteen years later, it set up a domestic sewing machine factory in China.

In the 1980s and 1990s, while Brother continued to be a highly successful manufacturer of sewing and knitting machines, business machines became the company's major growth area. In 1980, Brother produced its 10-millionth typewriter and launched its first electronic typewriter. It began manufacturing electronic typewriters in Britain at a new factory in Wrexham in 1985. In the following year, Brother Industries (USA), Inc., was set up in Bartlett, Tennessee, to manufacture electronic typewriters. Production at the Wrexham factory was diversified in 1987 with the start of microwave oven and printer manufacture.

Brother launched its first fax machine in 1987. Another new product line in the business machine sector, electronic labeling machines, followed in 1988. Growing demand for Brother's products led, in 1989, to the construction of new factories to manufacture parts in Ireland and Malaysia. In the 1990s, the fax machine became the company's fastest-selling product ever. In order to meet demand, Brother began to manufacture fax machines in Malaysia in 1994. It took six years for the production of fax machines to reach the 1 million mark in 1993; accelerating sales meant that total fax machine production reached 2 million in 1994, 5 million in 1996, and 10 million in 1999. Today, two-thirds of company revenue comes from business machines, such as fax machines and computer printers, while about a fifth of revenue comes from sewing machines, knitting machines, and home electric appliances.

See also Computer Printers; Fax Machines; Sewing Machines; Singer Corporation

References:

Godfrey, Frank P. *An International History of the Sewing Machine*. London: Robert Hale, 1982.

www.brother.com

C

Calculators

The term "calculator" may be applied to any device that assists the process of calculation. However, in practice it has become shorthand for one such device, the electronic calculator, which was the first to achieve widespread ownership beyond the workplace.

The ancestors of the electronic calculator were the desktop mechanical calculating machines developed in the late nineteenth century. These were based on principles established in the seventeenth century by the French mathematical philosopher Blaise Pascal and the German Gottfried Wilhelm Leibniz and were commonly known as adding machines. By 1900, two main types had emerged: machines operated by setting levers; machines with numerical keyboards, of which the first was the Comptometer of 1886, invented by the American Dorr E. Felt. The American inventor William S. Burroughs developed an adding and listing machine in 1892, with a built-in printing facility for producing a paper record. Adding machines soon became a standard piece of office equipment.

The purely mechanical adding machine evolved into the more compact, electrically powered version, which became typical in the 1950s. The first commercial electronic calculator was a transistorized desktop model introduced by the American Bell Punch Company in 1963. Texas Instruments produced a hand-held electronic calculator in 1967. Early American and Japanese handheld calculators were still large by today's standards. Home ownership remained low because people's needs outside the workplace could be met more cheaply and conveniently by the use of "ready reckoner" tables and slide rules, or simply by mental effort.

The image and role of the calculator changed only when the advent of microelectronics enabled the production of small, cheap calculators. The world's first true pocket calculator was the Sinclair Executive calculator, designed by the British inventor Clive Sinclair and launched in 1972. It featured an LED (light-emitting diode) display. In the same year, Hewlett-Packard pocket calculators became available in the United States. In 1973, the Japanese company Sharp introduced the first electronic calculator with a liquid crystal display. Within five years, the price of pocket calculators had fallen dramatically. In 1979, the pocket calculator became the card-size calculator when Sharp developed a super-thin model.

The pocket calculator is an example of a product that created demand where it did not previously exist. Today, the sophistication of the pocket calculator has reached such a level that even cheap models incorporate a range of scientific functions well beyond the needs of the average user. More expensive models have larger displays so that results can be presented graphically. The leading manufacturers are Japanese companies such as Casio and Sharp. In environmental terms, the pocket calculator has another distinction: it is the only commonplace device available in a solar-powered form. The solar unit in a calculator is a semiconducting photoelectric cell, which converts light energy into electric energy, thus removing the need for batteries. Pocket calculators may be wholly solar-powered or dual-powered, with back-up battery power to compensate for low light levels.

See also Computers; Hewlett Packard; Sinclair Radionics and Sinclair Research; Texas Instruments

References:

Haddock, Thomas F. *A Collector's Guide to Personal Computers and Pocket Calculators.* Florence, AL: Books Americana, 1993.

Tweedale, Geoffrey. *Calculating Machines and Computers.* Princes Risborough, UK: Shire Publications, 1990.

Camcorders

The camcorder, or video camera, captures moving images and sound on videotape. Camcorders targeted at the amateur user came on the market in the 1980s within a few years of the first professional models. Although camcorders were initially a luxury item, reductions in their price and size boosted ownership, particularly in their birthplace, Japan.

The camcorder's predecessor, the motion picture (cine) camera, was never found in more than a small minority of homes. In the late nineteenth century, Thomas Alva Edison in the United States and the Lumière brothers in France pioneered the development of equipment for recording and playing moving pictures for public entertainment. Early motion picture cameras were hand-cranked, which required skill and was therefore a deterrent for amateur users. To create the illusion of continuous motion, the cameras had to capture 24 images per second, each of which was shown twice (i.e., at 48 frames per second) when the film was projected for viewing. Any change in the rate of hand-cranking would ruin the illusion.

The motorization of the motion picture camera made it more user-friendly, so cheaper models designed for amateur use were marketed in the 1920s by makers such as Kodak and Pathé. Although 8 mm motion picture film was available from 1932, 16 mm remained the standard for amateur cine cameras until the 1950s when more compact 8 mm models appeared, mainly produced by Japanese companies such as Canon. Motion picture cameras still had distinct disadvantages for leisure use. One disadvantage was the need for a projector and screen to show the films, and another was the absence of sound. Although professional motion picture film incorporating a sound track was developed in the 1920s, the equipment was not economically feasible for the amateur market.

The camcorder followed in the wake of the videocassette recorder, which gave the television set a new role as a playback device rather than just a broadcast receiver. As a sophisticated piece of technology, the camcorder was initially expensive and designed as a portable tool to meet professional broadcast standards. The conventional television camera owed much of its bulk to the size and shape of the orthicon electron tube. The first generation of camcorders contained a vidicon tube, which was much shorter and slimmer than the orthicon tube. Inside the camcorder, light entering the lens strikes the faceplate of the vidicon tube. The faceplate's photoconductive lead-oxide coating converts light to an electric charge, which is picked up by the scanning electron beam and delivered as an output signal to the video recording head. The video track is recorded diagonally across magnetic tape, whereas the sound track, recorded simultaneously through a mi-

A leather-vested biker using a camcorder at the Sturgis Rally and Races, SD, 1993 (Kevin Fleming/Corbis)

crophone, is placed along one edge. A small screen allows the user to preview shots and to play back the recording.

As Japanese companies had become dominant in the motion-picture-camera market in the 1950s, predictably they have also dominated camcorder production. In 1982, Sony released the Betacam professional camcorder, which used half-inch tape. Sony recognized that the Betacam format, which had already been overtaken by VHS in the videocassette recorder market, was too large to be successful for consumer camcorders. In 1982, a group of electronics manufacturers, including Sony and the Dutch company Philips, agreed to work on developing a standard miniature format, Video8, based on an 8-mm tape cassette. The Japanese company JVC (Japanese Victor Company), developer of the VHS format, soon pulled out of the Video8 consortium to concentrate on a compact version of VHS. In 1984, JVC launched the world's first compact VHS camcorder, the GR-C1. The Compact VHS cassette (VHS-C) had a running time of one hour and was only a third of the size of a standard VHS cassette, but could be placed in a special adaptor shell for playback on VHS videocassette recorders. The Video8 specification was agreed in 1983, and the first Video8 camcorders appeared in 1985. In the United States, Kodak launched the KodaVision 8 mm camcorder, which was manufactured for Kodak by Panasonic, a subsidiary of the Japanese company Matsushita. Sony's Handycam solid-state camcorder was more compact, weighing only 1 kg (2.2 lb), and the running time of the 8 mm cassettes was ninety minutes. In the case of camcorders, absolute standardization of tape format proved to be less critical than it had been in the case of videocassette recorders, largely because there was no prerecording issue and no need to use a videocassette recorder for playback.

Camcorders had far greater inherent consumer appeal than motion picture cameras, partly because of the convenience of playback via the television set. Other advantages were that videotape entails no external processing costs and is reusable. Recordings can be viewed immediately and then shot again if the results are not satisfactory. Since the mid-1980s, camcorders have evolved rapidly. In 1989, Sony brought out Hi8, a higher resolution 8 mm tape. The replacement of the vidicon tube with solid-state imaging devices not only reduced the size and weight of camcorders, but also improved the video quality and reduced power consumption. There are two types of solid-state video pickups—the metal-oxide semiconductor (MOS) and the more popular charge-coupled device (CCD). Both consist of an array of tiny photodiodes that convert light to electrical energy, but CCDs employ a scanning method that produces a higher output signal. The CCD was invented at American Telephone and Telegraph's Bell Laboratories in 1969 by George Smith and Willard Boyce. It was first used in Sony's Handycam camcorder.

Digital camcorders arrived in 1993 when Sony digitized its Betacam format to produce broadcast quality DigiBeta. Two years later, the JVC GR-DV1 CyberCam was announced as the world's smallest digital video camera, smaller than any nondigital camcorder. Again, different formats have emerged, with Sony using the higher quality DVCAM format and JVC using MiniDV. With the rapid increase in home computer ownership, digital camcorders will undoubtedly grow in popularity because of the comparative ease of on-line editing, but currently suffer from their relatively high purchase price.

See also AT&T; Cameras; Matsushita; Sony Corporation; Videotape Recorders

References:

Goodman, Robert. *How Electronic Things Work . . . and What to Do When They Don't*. New York: McGraw-Hill, 1999.

www.popadom.demon.co.uk

Cameras

The camera is a device for capturing an image, created by light entering a small aperture in a box forming an inverted picture of the subject in front of the aperture, on a light-sensitive (photosensitive) material. The basic principles of photography were known in mediaeval times, but it was not until the early nineteenth century that photographic pioneers discovered photosensitive chemicals that allowed the image to be retained on a metal plate. Photography became an affordable and popular hobby for the masses only at the end of the nineteenth century. The immense popularity of photography today is indicated by the fact that in 1995 an estimated 60 billion photographs were taken worldwide.

Early Photography

"Camera" is the Latin word for vault or chamber. It was adopted as the name for the photographic device because early demonstrations of the basic photographic principle involved a dark chamber, hence "camera obscura," with a pinhole in one wall, letting in light and forming an image of an external scene on the opposite wall. The Italian artist and inventor Leonardo da Vinci was the first person to describe the camera obscura in 1515. In the form of small wooden box with a simple lens instead of a mere pinhole, the camera obscura became an optical toy. The addition of a prism to reflect the image downward onto paper created the camera lucida, which could be used for tracing a scene or subject. In the eighteenth century, chemical experimenters discovered that certain salts, such as silver nitrate, reacted to light. The next discovery was that an image could be retained by placing an item such as a leaf on a surface coated with an emulsion of silver salts and exposing it to light. The exposed surface darkened, and the covered surface remained unchanged. Images produced by this contact process were called photograms.

In the early nineteenth century, pioneers in England and France invented true photography. In the early 1820s, the French doctor Joseph-Nicéphore Niepce developed his "heliographic" process, whereby a pewter plate coated with bitumen was placed in a camera obscura and exposed to light for eight hours. Another Frenchman, Louis-Jacques Mandé Daguerre, developed the first commercially successful photographic process in 1838. He captured the image on a metal plate coated with silver iodine and used mercury vapor to "fix" the image. However, daguerreotypes were delicate and needed careful handling. They were typically mounted in hinged cases to prevent fading and protect the surface. In 1839, Alphonse Giroux of Paris developed an improved camera for Daguerre, and exposure times began to shorten.

The major disadvantage of the daguerreotype was that the image could not be reproduced. The one-step process was soon superseded by the calotype process, patented by the Englishman William Henry Fox Talbot in 1841. Fox Talbot's process produced a negative image, with the black and white portions reversed, which was developed and fixed.

The image was then reversed to produce one or more positive images by placing the negative in contact with photosensitive paper and re-exposing it to light. In 1843, Fox Talbot invented an enlarger, which created a positive photographic print larger than the negative. However, his calotype paper was soon supplanted by more durable and fast-exposure glass-plate negatives. The wet-plate or wet collodion process, developed by the Englishman Frederick Scott Archer in 1851, required the photographer to coat the glass plates immediately before use and develop them straight afterward, a rather messy procedure that tended to discourage amateur interest. The introduction of dry gelatin plates, consisting of glass with a coating of silver bromide, in the 1870s made outdoor photography easier and stimulated amateur photography. It was the invention of roll film in the 1880s, however, that made photography attractive to a mass audience.

The Popularization of Photography

A number of American and European inventors developed coated paper films, but the big breakthrough was the invention in 1889 of celluloid roll film by the American chemist Henry Reichenbock. George Eastman, former bank clerk and founder of the Eastman Kodak Company, was Reichenbock's employer. Based in Rochester, New York, Eastman Kodak revolutionized the camera industry by concentrating on the mass market. Before 1888, when Eastman launched his first camera, camera developments were geared to the needs of the professional user. Studio cameras had large, heavy wooden bodies, while field cameras, with folding bellows, were more portable, but expensive. The first Kodak camera, loaded with a 100-frame paper roll film, was advertised with the slogan, "You press the button, we do the rest." Once the film was completed, the owner returned the camera to Eastman Kodak for processing. The camera plus film cost $25, and the cost of the prints and a new film was $10.

Eastman's next step was the introduction of the Pocket Kodak box camera, the first truly portable camera, in 1895. When the Kodak pocket bellows camera followed in 1898, Eastman Kodak had sold one and a half million cameras. By 1900, about 10 percent of the population in both the United States and Britain owned a camera. A small, cheap box camera, the Kodak No. 1 Brownie was introduced in that year and sold for just $1, plus 15 cents for a six-frame film. Designed by Frank Brownell and made from wood and cardboard, the No. 1 Brownie brought photography within the means of the average person.

By the 1920s, the amateur camera market was more competitive, so makers began to use design to increase the desirability of their products. In 1927, Eastman Kodak employed the American designer Walter Dorwin Teague to redesign the Box Brownie. Teague's Beau Brownie design was launched in 1930 and featured a two-color Art Deco geometric front panel. Teague also designed

An advertisement for the Kodak Brownie camera (Courtesy, George Eastman House)

the 1928 Vanity Kodak bellows camera, which came in a range of colors and with a matching case. The availability of Bakelite and other new plastics made it possible to produce cheaper cameras in a variety of shapes and colors. The Kodak Baby Brownie of 1933, again designed by Teague, had a squat Bakelite shell with rounded edges and a distinctive ribbed lens panel. A particularly innovative use of plastics was made by the American designer Raymond Loewy, in his 1937 Purma Special camera design for the British company R. F. Hunter. The streamlined black Bakelite shell incorporated an integral viewfinder and wind-on mechanism, while the lens was not glass, but Perspex, thus reducing the costs. Fun cameras are exemplified by the Coronet and Corvette Midgets, made by Britain's Coronet Camera Company. These miniature cameras had rounded Bakelite bodies with domed tops housing the viewfinder and came in a range of striking, mottled colors.

Miniaturization of the Precision Camera

The Kodak Brownie was technically very basic. The cheap fixed focus lens was adequate for snapshots of places and people, but was incapable of close-up photography. Meanwhile, a number of technical advances benefited professional photographers. German camera manufacturers led the way in this sector of the market. The first anastigmatic lens, the Protar f7.5, was developed in 1889 by the German physicist Paul Rudolph for Carl Zeiss, manufacturers of optical equipment based in Jena, eastern Germany. An anastigmatic lens guarantees that all points of the image are accurately aligned in both the vertical and horizontal planes. The Zeiss Tessar lens of 1902 reduced the maximum aperture to f4.5, which improved the depth of field. In 1898, the American inventor William F. Folmer developed the Graflex camera, the first camera capable of high-speed photography in split focal planes. The first compact precision camera, the 35 mm Leica, was manufactured by the German company Leitz, based in Wetzlar. Although Oskar Barnack developed the prototype in 1914, the production model was introduced at the Leipzig trade fair only in 1925. With its rangefinder optical viewing system, interchangeable lenses, and range of accessories, the Leica became an industry standard, and the 35 mm format is the dominant format today. The Zeiss Ikon, marketed from 1932, was another popular professional 35 mm camera.

The portability of the Leica also encouraged the growth of photojournalism, although many publishers insisted on contact printing from large-format negatives. A number of magazines with high photographic content, including *Life* in the United States and *Picture Post* in Britain, were launched in the 1930s. Reflex cameras were invented in Britain in the nineteenth century but became a standard camera type only in the 1930s. The key advantage of the reflex camera is that the photographer sees the same image that the lens "sees," enabling accurate focusing. This is achieved through an arrangement of angled mirrors and prisms that reflects the image entering the lens to the viewfinder. The single-lens reflex (SLR) camera, invented in 1861, was followed by the twin-lens reflex (TLR) camera in 1880, which had a separate lens supplying the viewable image above the main lens. In 1929, the German company Rolleiflex produced a TLR camera with a large viewing screen in the top panel, which became a popular professional model.

SLR cameras, which were compact and more suitable for amateurs, were available from the mid-1930s but did not became common until the 1960s when Japanese camera makers brought out more affordable models. They were easier for hand-held, rather than tripod, use because they had a conventional front-facing viewfinder, which received the image from a hinged mirror that swung back out of the path of the lens when the shutter release was activated. A subminiature precision camera, the Minox, was produced from 1937 by the Latvian company V.E.F.

Two men photographing each other with hand-held cameras while perched on a roof, about 1920 (Library of Congress)

Film and Flash Technology

Photograph quality relied as much on the performance of film and the lighting of the subject as on lens technology. In the days of long exposures, professional photographers became proficient at calculating exposure times through trial and error and controlled the light entering the camera merely by removing and replacing a lens cap. Exposure tables, calculators, and strips of photochromatic paper assisted the amateur photographer. Cameras began to incorporate simple mechanical shutters to control light input, and the first leaf shutter, the Deckel, was developed in Germany in 1902. Accurate short exposures only became possible when the photoelectric cell (or photocell) was invented. The first practical photoelectric cell was invented by the German physicists Julius Elster and Hans Friedrich Geitel in 1904. The photoelectric cell was first incorporated in a separate exposure meter in 1931 by the American William Nelson Goodwin Jr. and was developed commercially by the Weston Electrical Instrument Company of Newark, New Jersey, in 1932 as the Photronic Photoelectric Cell. The U.S. Time Corporation (Timex) produced an "electric eye" camera in 1950, but the first camera to feature a built-in photoelectric cell positioned behind the lens was the Pentax Spotmatic, introduced in 1964 by the Japanese Asahi Optical Corporation. Built-in exposure meters meant that cameras could be operated by the selection of foolproof automatic settings. The first automatic-exposure SLR camera was the Pentax ES model of 1971.

From the 1880s, electric lights could be used to enhance indoor lighting, and flash photography was possible through the use of flash powders and magnesium ribbon. Flash photography became easier with the invention of the electric flashbulb, patented by the

German inventor Johannes Ostermeier in 1930. The electronic flashbulb was invented in the United States in 1935. Flashbulbs were initially mounted in separate flashguns, but as they became smaller, they were embodied in attachments that fitted directly onto the camera. The hot shoe flashgun attachment, a metal "shoe" on top of the camera into which a metal "foot" on the flash was slid, was introduced in 1939. As cameras became fully electronic, the flash became an integral part of the camera.

In the 1930s, highly flammable celluloid film was replaced by nonflammable cellulose acetate. While the Scottish physicist James Clerk Maxwell expounded the principles of the three-color photographic process in 1861, it was another thirty years before the French doctor Gabriel-Jonas Lippman developed a process for color photography. In the early twentieth century, a number of color processes were developed, including the Lumière brothers' Autochrome process, but these were for plates rather than film and were time-consuming and expensive. In the 1930s, Eastman Kodak and the German company Agfa (A.G. für Anilin, the Aniline Company) began to develop color film in transparency form. Kodachrome film became available in 35 mm cartridges and roll film in 1936, and Agfacolor film was available in 35 mm cartridges the next year. The first color negative film, Kodacolor, did not come on the market until 1944 and was quickly followed by faster Kodak Ektachrome transparency film until 1946 and negative film in 1947. While Eastman Kodak's domination of the camera market began to wane when Japanese companies moved into the camera industry in the 1950s, the company continued to be a leading force in film technology.

After the introduction of color film, the main improvements in photographic film lay in the development of faster films. Film speed is a measure of light-sensitivity. The first system for measuring film speed was developed in Britain in 1890, but from 1947, the American Standards Association (ASA) ratings became the industry norm. Eastman Kodak has continued to be a leader in film technology: for example, it launched a series of high-speed X films, starting with Tri-X black-and-white roll film in 1954. A significant advance in 1963 was the development by the Swiss company Ciba-Geigy of the Ultrachrome process, which allowed prints to be made from transparencies.

Instant Photography

The American Edwin Herbert Land became involved in the photographic industry when he developed a light-polarizing film in 1932 and began to produce polarizing camera filters from 1935. He founded the Polaroid Corporation in 1937. In 1947, he patented the innovative Polaroid Land camera, which produced instant photographs. The special Polaroid film consists of a sandwich of negative film and positive paper with a thin filling of developer gel encased in a fragile membrane. After exposure, the film pack passes between rollers, rupturing the membrane and dispersing the developer. The Polaroid Land camera marketed in 1948 produced sepia photographs, but a true black-and-white model became available in 1950. By 1952, half a million Polaroid cameras had been sold. In 1959, the developing time was reduced by a ten-fold increase in film speed to 3000 ASA. The first color film Polaroid camera was introduced in 1962, with a developing time of a minute, compared to 10 seconds for black-and-white. The SX-70 color model of 1972 was the first Polaroid camera of the SLR type and the first where the film pack was ejected immediately, before the positive had developed. Early Polaroid cameras had a novelty appeal, but the high price of Polaroid films and the one-off nature of the process (as with daguerreotypes) limited the market potential.

Japanese Domination of the Camera Market

After World War II, Japanese companies began to compete very effectively in the international camera market. At the top end of the professional camera market, Rolleiflex and other European companies, including Sweden's Hasselblad and Austria's Voigtlander, maintained their supremacy. For example, Voigtlander introduced the zoom lens, widely used in motion-picture photography since the 1930s, for still photography in 1958. However, Japanese companies made serious in-roads into the lower end of the professional camera market. Nippon Kogaku launched its first camera, the Nikon I rangefinder 35 mm camera, in 1948. Exceeding a million units in sales, its Nikon F SLR camera of 1959 was the first commercially successful SLR model. In 1953, Nikon was also the first maker to produce a camera with motorized drive, but motorized drives were uncommon until the 1960s. Another Japanese company, Olympus, produced the first compact SLR camera, the Olympus Trip, in 1968. Twenty years later, sales of the Olympus Trip reached 10 million.

Even the smallest SLR cameras could not be described as pocket-size, so there was a gap in the market between low-performance cartridge cameras that were small and lightweight and the bulkier SLR cameras. In the late 1970s, this gap was filled by the introduction of fully automatic, compact 35 mm cameras. These cameras improved in the early 1980s as a result of the development of auto-focus lenses. As cameras gained more electronic functions, their styling reflected this transition by becoming increasingly high tech. The harder lines of the older metal-bodied mechanical SLR cameras gave way to the sleek lines of plastic-bodied electronic compact and SLR cameras.

The serious image of the camera was only challenged by the modern equivalents of the Brownie. Eastman Kodak had continued to periodically reinvent the simple "point and press" camera. The cheap Kodak Instamatic camera of 1963 used easy-to-load film cartridges and achieved sales of 50 million units by 1970. Its successor, a pocket-size model introduced in 1972, was equally successful. In 1982, a new Kodak format, the film disc, was launched. Single-use disposable cameras followed in the late 1980s. Kodak's colorful Fun Saver disposable cameras achieved dramatic market penetration, reaching sales of 50 million units by 1995. In the United States, single-use cameras accounted for 75 percent of annual camera sales. In keeping with environmental concerns, Eastman Kodak recovers more than 80 percent by weight of the materials in disposable cameras by reuse or recycling.

APS and Digital Photography

Both of the latest developments in photography draw on electronic technology. It was almost inevitable that digital photography would emerge in the wake of digital computers and digital sound recording. The key component of a digital camera is a powerful sensor capable of translating the light components of an image into a sufficient number of pixels for it to be reproducible as a continuous image. In 1986, Eastman Kodak developed the first megapixel sensor, with a capacity broadly equivalent to the content of a 13 by 18 cm (5 by 7 inch) photograph. Digital cameras soon followed, with Japanese companies predictably leading the way in making the technology affordable. All digital cameras have the common feature of being filmless, but there are several methods of storing images, including computer floppy disks, removable memory cards, and built-in memory. Eastman Kodak's PhotoCD pioneered digital technology whereby negatives on film can be scanned and stored digitally on compact disc. The PhotoCD system was introduced in 1990. For the private consumer, the main disadvantage of digital photography is that the quality of output from the typical home printer is far below that of conventional photographic printing.

APS, the Advanced Photo System, was developed by a consortium of five companies—Eastman Kodak and four Japanese companies, Fuji, Nikon, Canon, and Minolta. The APS project began in 1992, and APS cameras and film went on the market in 1996. APS cameras take cartridges of 24 mm film, which incorporates magnetic strips for storing information from sensors about date and time, exposure length, and size settings. Date or message imprinting and normal/panoramic size settings were features already incorporated in some 35 mm cameras; the additional advantages of APS are that the film is thinner and hence the cartridges are slimmer, permitting corresponding slimmer and lighter cameras, and the developed negatives are rewound into the cartridge after printing, so there is less danger of damage through mishandling. A reference set of contact prints is provided to facilitate reordering of prints.

See also Asahi Optical Corporation; Camcorders; Eastman Kodak

References:

Coe, Brian. *Cameras from Daguerreotypes to Instant Pictures.* London: Marshall Cavendish, 1978.

Hick, Roger. *A History of the 35 mm Still Camera.* London and Boston, MA: Focal Press, 1984.

Can Openers

Tinplate canisters or "cans" for food were developed by the Frenchman Nicholas-François Appert and the Englishman Peter Durrand in the late eighteenth and early nineteenth centuries. It was thanks to Louis Pasteur's work on bacteria and improvements in sterilization that by 1860 canning had become commercially viable. It established the emerging food-processing industry, delivered more reliable food supplies, and led to new forms of food retailing. The earliest can openers were of the "spike and blade" variety. A spike was driven into the top of the can and the lid cut off with the blade. Although still on sale until the late 1930s they were superseded by the "blade and cogwheel" type with "butterfly" handles. These cut off the lid far more cleanly. They were usually made of either all steel or with wooden handles and are still on sale today, usually with plastic bodies and steel blades and cogs.

The 1950s and 1960s saw the variety of canned goods increase. Wall-mounted can openers were introduced with gear-driven cutting wheels operated by a handle. Most featured a magnet to hold the lid once it had become separated from the body of the can. They were available in a range of colors to match the increasingly brighter kitchen units of the time. In 1968 Sunbeam produced an electric combination can opener/knife sharpener in avocado green.

Electric openers became available in the 1960s, either as wall-mounted or free-standing appliances. The can is placed against the cutting wheel and held in place by a lever. The motor drives the blade around the can, switching itself off automatically once the can is open. Black & Decker currently produce seven types, some with built-in knife sharpeners and bottle openers.

Carpet Sweepers

The carpet sweeper is manually operated. Its rotating brushes pick up dust, which is then deposited in the pan above. In the late 1850s, many patents for carpet sweepers were lodged in the United States. They were based on the same principles as the first street-sweeping machine patents granted to the British engineer Joseph Whitworth in 1840 and 1842. These early patents did not result in commercial production.

In 1876, Melville Reuben Bissell, of Grand Rapids, Michigan, patented an improved design of carpet sweeper. He began production and the Bissell carpet sweeper became the first commercially successful model. It consisted of a long pivoted handle, a wooden dust box on wheels and a set of rotating brushes. Bissell's innovation was the central bearing brush, which allowed the sweeping brushes to self-adjust to suit different surfaces. By 1906, annual production of Bissell carpet sweepers had exceeded the one million mark. In Britain, similar carpet sweepers appeared in the 1880s. Carpet sweepers have become even more portable since being made of plastics and lightweight metals but their design has changed little, except for minor details such as the addition of corner brushes. They have retained a market niche because of their convenience for small cleaning jobs.

See also Vacuum Cleaners

References:

Giedion, Siegfried. *Mechanization Takes Command.* New York: Oxford University Press, 1948.

Matranga, Victoria Kasuba, and Karen Kohn. *America at Home: A Celebration of Twentieth-Century Housewares.* Rosemont, IL: National Housewares Manufacturers Association, 1997.

Casco Products Corporation

Casco Products, an industrial-products company, began producing domestic appliances in 1949 with a successful electric iron. Soon after the company was aquired by Standard Kollsman Industries in 1960 it introduced the Lady Casco range of appliances. Pots and pans had long been sold as matching sets that could also be displayed, especially if enameled or colored; the Lady Casco range took this concept one step further. It attempted to "theme" appliances to match the American "dream kitchen," an inducement to replace existing appliances with a new set all by the same manufacturer. The line of ten matching appliances centered on the Chef Mate, a motor-driven base with a range of attachments, including a mixer and a blender. The rest of the Lady Casco set consisted of a toaster, coffeepot, iron, and frying pan.

If this approach was novel, so was the method of marketing the products. The line was to be offered to stores on a franchise arrangement with the added gimmick that each set carried an exclusive five-year guarantee backed by Lloyd's of London! By the close of 1961 over 2,000 stores had signed up to the franchise deal and sales were encouraging. Then in 1962 the parent company decided to abandon the project and the Lady Casco program was discontinued. In 1963 the appliance section of the business was acquired by Hamilton Beach.

Most domestic appliances follow a "house style," and smaller pieces such as toasters and kettles have been themed to complement each other, even if sold separately. A recent trend has been the marketing of "double" or "triple packs" of products by manufacturers like Hinari, Breville, and Morphy Richards. These usually consist of a toaster and a kettle, the third element being either a sandwich toaster or a coffeemaker.

Reference:

Goldberg, Michael J. *Groovy Kitchen Designs for Collectors, 1935–1965.* Atglen, PA: Schiffer Publishing, 1996.

Ceramics

The use of ceramic vessels in the home is, of course, age-old, and most of the major developments in production techniques had been achieved before the twentieth century. Nevertheless, there were a number of processes that helped to further democratize the range and style of products available.

Photolithographic images for ceramics were developed during the late 1930s. This process allowed exact copies of original artwork to be reproduced on a piece via a transfer. It was first successfully exploited in the United States in the 1950s. A further improvement was the Murray-Curvex offset litho process that became available during the mid-1950s. This process transferred the still wet print onto the ceramic article via a gelatin pad or "bomb," allowing the print to cover the sides of bowls and tureens, producing an "all-over" pattern.

These techniques were exploited by American and European designers and manufacturers and gave rise to a new wave of brightly patterned wares, often influenced by current artistic movements such as abstract expressionism. Some products were criticized as simply having new surface decoration applied to older shapes, but others were genuinely new combinations of exciting shapes and patterns, available at affordable prices.

The 1960s and 1970s saw less creativity in design, but technological development continued with tougher glazes able to withstand electric dishwashing.

See also Dishwashers

References:

Jackson, Lesley. *The New Look: Design in the Fifties.* London: Thames & Hudson, 1991.

McLaren, Graham. *Ceramics of the 1950s.* Princes Risborough, UK: Shire, 1997.

Clothes Dryers

There are two types of mechanical aid to drying washed clothes: appliances for extracting water by pressure and appliances for producing evaporation through heat. In 1900, with the exception of very large households, the only equipment available for drying clothes was the mangle or wringer, where wet clothing was passed between heavy rollers to squeeze out water. The production of new electrical appliances for drying clothes began in the 1920s but only reached the mass market from the 1950s onward.

The first widespread improvement of the twentieth century was the fitting of powered wringers to electric washing machines. The wringer was connected to the electric motor at the base of the washing machine by a vertical driveshaft. The user still had to lift out the wet clothing and guide it through the rollers. Washing machines with wringers were the most common type of washer from just before World War I until the late 1950s.

Electric spin dryers were introduced in the 1920s but did not find an immediate market as domestic appliances, although larger models for public laundries were more commercially successful. The spin dryer is based on the principle of centrifugal force: wet clothing is placed in a perforated drum that rotates about a vertical axis, forcing the clothing against the walls and pressing out the water, which drains downward naturally or can be pumped out upward to a sink. In the late 1940s, the relaunch of the automatic washing machine with its integrated spin-drying function provided new impetus. By the mid-1950s, manufacturers had begun to

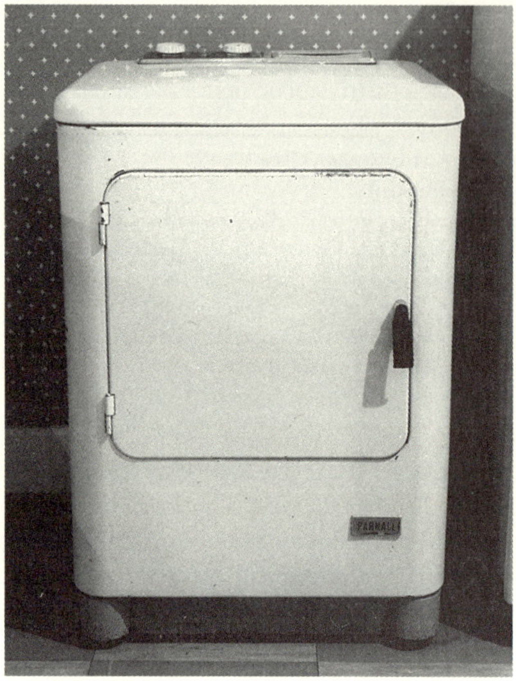

1958 Parnall tumble dryer (The Museum of Science and Industry in Manchester)

exploit the domestic potential of the spin dryer both as a separate appliance and in combination with the washer. In Britain, where ownership of automatic washing machines with integral spin drying grew slowly, the first spin dryer aimed at the mass market was introduced by Creda in 1956. The spin dryer has changed little in essence except in being made lighter (by replacing steel casing with plastic) and more compact.

The drying of clothes by applying heat has evolved from the practice of placing damp clothes on racks in front of a fire or in airing cupboards near a hot water tank. Some versions of the lamp radiator type of electric room heater (circa 1905–1915) had rails at the top for hanging towels or clothes on. This was the only type of room heater that was safe for placing in direct proximity to damp clothes. Electric fans could also assist in the drying of clothes, but these were scarcer than room heaters. The next step was separate heated towel rails and drying cabinets, which appeared in the 1920s. The standard design was for the heater to be placed at the bottom of the cabinet with racks above. Drying cabinets had vents to disperse steam. Some models also had fans to improve the circulation of hot air, but clothes still needed to be turned over regularly. Electric drying cabinets reached their peak of popularity in the 1950s.

The tumble dryer was invented by the American J. Ross Moore in 1938. Like the automatic washing machine of the same year, it was ahead of market demand and received little attention until the 1950s, when it was marketed more vigorously both in the form of the composite washer-dryer and as a separate companion to the automatic washing machine, to sit alongside or to stack above. In the United States, the Whirlpool Corporation launched its first automatic dryers in 1950, and Maytag followed suit in 1953. In Britain, Parnall introduced a tumble dryer in 1958. However, tumble dryers have only achieved significant sales over the last twenty years. In the tumble dryer, damp clothes are placed in a heated drum that is gently rotated to circulate the heat. The main problem that had to be overcome with tumble dryers was the extraction of water vapor from the drum as the air becomes moist. Removing water vapor by venting through tubes was cumbersome and restricted the placement of tumble dryers. Most modern tumble dryers have condensers that cool the vapor back to its liquid form so that it can drain away.

See also Washing Machines

References:

Byers, Anthony. *Centenary of Service: A History of Electricity in the Home.* London: Electricity Council, 1981.

Coffee Grinders

Coffee grinders developed in Europe with the arrival of the drink in the seventeenth century. These were manually operated mills with steel or cast-iron grinding mechanisms within hard wood or cast-iron bodies. Some had adjustable mechanisms. Commercial electric coffee grinders appeared in the 1920s and 1930s, with models such as the Hobart. Domestic grinders arrived in the 1950s. The basic process remained the same, but electric models offered extras such as settings to regulate the fineness of the grains for either percolator or filter coffee and airtight hoppers to store the beans. Smaller models used fast rotating chopping blades rather than crushing mills.

To make a really good cup of coffee requires freshly roasted beans that should be freshly ground just before it is made. Despite the improvements in vacuum-packed preground coffee, grinders remain popular with coffee aficionados.

See also Coffeemakers

Coffeemakers

The application of technology to assist the art of making a good cup of coffee began in the nineteenth century with the invention of the percolator by the American-born Count Rumford in Germany in 1806, with at least one aim being to discourage the heavy drinking of Munich workmen. His invention improved upon the traditional Turkish method of heating both ground beans and water in the same container. Water trickled down through a central cylinder that contained the coffee and filter and then up into the outer body of the vessel. The other long-favored method was the tinplate or enamel "drip-pot" that simply filtered the hot water from an upper vessel, through the ground coffee and into a lower one.

By the mid-1830s the first true coffee machines, large alcohol-heated drip machines, had been developed, primarily for cafés. Steam pressure was popular for smaller domestic models, especially in Italy. Simple steam-pressure machines featured a water container with a filter for the ground coffee. A metal tube dipped into the water, and as it heated, the pressure of the resulting steam forced the water through the coffee and out of the tube. Italian companies like Pavoni and Snider produced a variety of these models in

Grinding coffee in a general store, Lamoille, Iowa, 1939 (Library of Congress)

the early twentieth century. Later models were electrically heated.

Coffee was a popular drink in Europe and America, and it was here that the major developments took place. The earliest electric appliances were percolators (the first introduced in 1908 by Landers, Frary & Clark under their Universal trade name) with a heating element attached to the base. Two types of vacuum coffeemakers were developed in Britain. The Siphon percolator of the 1850s used the principle of the vacuum siphon patented by Robert Napier in 1830. The apparatus consisted of two flasks linked by a pipe. Boiling water was poured onto ground coffee in a glass flask. The steam generated by hot water in another flask, usually of china, created a vacuum that drew the liquid coffee through. It was then served from a tap on the side. Another nonelectric solution was the Cona vacuum system developed by Alfred Cohn in London in 1910. It consisted of two glass vessels. The bottom one held the water and was connected to the top one, which held the ground coffee. Once heated the water rose into the top to infuse the coffee, while the cooling lower half created a partial vacuum, which drew the liquid coffee back down. The Cona remains popular today. The Danish Bodum company introduced their version, the Santos, designed by the architect Kaas Klaeson, in 1958 and it is also still in production.

The United States and Germany, both coffee-drinking nations, continued to develop electric models that were effectively percolators with electric heating elements in the base. AEG produced an electric siphon model during the 1920s. West Bend developed the Flavo-Drip coffeemaker that did not require a filter in 1922. Its popularity led to a stovetop percolator called the Flavo-Perk. The popular American Silex of the 1930s was a glass, two-bowl drip model that sat on a separate electric burner. Like kettles of the period, few were automatic. In 1937, S. W. Farber introduced the Coffee Robot that proclaimed to "do about everything but buy the coffee." It was a vacuum type with an automatic shut-off and a thermostat to keep the coffee warm. Its success tempted other American appliance manufacturers into the market.

The postwar trend in the United States was for sleeker all in one, automatic electric coffeemakers. Glass was replaced by chrome bodies with Bakelite handles. Many had simple engraved patterns on their sides. Popular models included the Sunbeam vacuum Coffeemaster and the Universal Coffeematic percolator.

Meanwhile, Italy produced more important developments, both in 1933. Alfonso Bialetti designed and produced the Moka Express, a two-part machine that forced the heated water up through the coffee into the upper vessel. Made of cast aluminum, it is still popular today, and it still carries the distinctive trademark of the cartoon caricature of its inventor. If the Moka was uncomplicated, the cafetiere designed by fellow Italian Calimani was simplicity itself. Its now familiar form is that of a glass vessel with a plunge filter that is pushed down through the infusing coffee. It began to be used in French cafés after 1945 and became popular in the 1950s. The cafetiere is now ubiquitous on both sides of the Atlantic.

Italy was also the birthplace of espresso, a coffee produced through pressurized machines based on the 1901 patent of the Milanese engineer Luigi Bezzera. The main drawback of these machines was that the steam was forced through the coffee at a relatively slow rate, resulting in a bitter flavor. A Milanese man, Cremonesi, experimented with a piston mechanism to increase the pressure. He fitted it to the machine in Achille Gaggia's bar in Milan. The piston method of forcing water through a bed of coffee at high pressure resulted in a fresher cup of coffee with a creamy head or crema. Cremonesi died during World War II, and Gaggia went on to develop the idea, with the Gaggia machine going into production in 1948. This machine was synonymous with the rise of coffee bars in Europe and America during the postwar period and stimulated the desire for authentic espresso at home. Gaggia produced the first domestic electric espresso machine in 1952. It was named Gilda, after the film that starred Rita Heyworth. A further improvement was the pump system developed by the Faema Company of Milan in the 1950s. A pump forced the water directly through the coffee at a constant temperature of 200°F. This method produced espresso very quickly and was adopted as the preferred method for domestic machines.

During the 1960s and 1970s these European methods began to make headway in the United States and Britain. Fresh filtered coffee was simple to make, and there was less chance of overheating it, which could happen with percolators. Manufacturers like Braun, Philips, and Rowenta produced well-designed automatic filter coffeemakers with plastic cases. The cafetiere was also successfully marketed by Bodum, which introduced their Bistro cafetiere in 1974, beginning their successful Presso line. Coffee was now one of the world's favorite beverages, although it must be remembered that the majority of sales were for the instant granulated variety. Instant coffee was the result of eight years research by the Swiss Nestlé Company and was introduced in 1938. The coffee was freeze-dried to eliminate the water but leave the oils that gave the taste. By the mid-1990s, it accounted for 90 percent of all coffee drunk in the United Kingdom, over 70 million cups a day.

Nevertheless, the 1980s saw the manufacturers respond to an increasingly sophisticated market. The domestic espresso machine came of age with sleek matt black miniatures from the likes of Braun, Bosch, Gaggia, Krups, and Siemens, fully equipped with steam pipes to froth up milk for cappuccinos. Initially expensive, these models forced the water through the coffee with either an electric pump or a centrifugal system that spins the water at high speed. In the early 1990s Russell Hobbs, Tefal, and Krups produced combination machines featuring an espresso maker, milk frother, and filter coffeemaker.

As the kitchen had become both a stylish room and a workspace, the coffeemaker, like the kettle has not escaped the attentions of contemporary designers, especially those working for Alessi. Aldo Rossi produced an espresso maker and a cafetiere, Richard Sapper an espresso maker, and Michael Graves a cafetiere.

Coffee remains popular throughout the world and the public taste for distinctive coffee has been stimulated by the growth of specialist coffee shops and cafés. Such is the market that brands like Starbucks are becoming global. In this environment appliances that replicate the coffee shop taste remain in demand.

See also Alessi; Braun; S. W. Farber, Inc.; Sunbeam; Tefal; West Bend

References:

Fumagalli, Ambrogio. *Coffeemakers.* San Francisco: Chronicle Books, 1995.

Goldberg, Michael J. *Groovy Kitchen Designs for Collectors, 1935–1965.* Atglen, PA: Schiffer Publishing, 1996.

Katz, Sylvia, and Jeremy Myerson. *Kitchenware.* London: Conran Octopus, 1990.

Compact Disc Players

Launched in 1983, the compact disc player (or CD player) is the digital recording equivalent of the gramophone (or record player). It is a device for playing back sound recorded on a small optical disc. The term "compact disc" was used because the first commercially produced optical discs were 30 cm (12 in) videodiscs, whereas audio optical discs, which had less information to record, were only 12 cm (4.75 in) in diameter.

Optical disc technology uses a laser both to embed the recording and to decode it for playback. During recording, a laser beam removes tiny dots from the etch-resistant chemical coating of the glass master disc. The dots vary in length according to the digital sound input and form a spiral track up 5 km (3.3 miles) long as the disc rotates. The master disc is then placed in a bath of hydrofluoric acid, which etches pits in the glass where no coating remains. In the mass production process, the master disc is replicated as plastic discs with a thin aluminum coating. The disc provides up to 100 minutes of sound. Inside the CD player, the disc rotates on a turntable and is scanned by a laser beam that detects reflection from the nonpitted surface and its absence from the pitted track. The laser transmits pulses of light to a photodiode, which converts the light to electrical pulses for transmission to an amplifier and loudspeakers.

The compact disc was developed through a joint venture between the Dutch company Philips, the pioneer of videodisc technology, and the Japanese company Sony. When the joint venture was agreed upon in 1979, both companies had reason to pool their resources

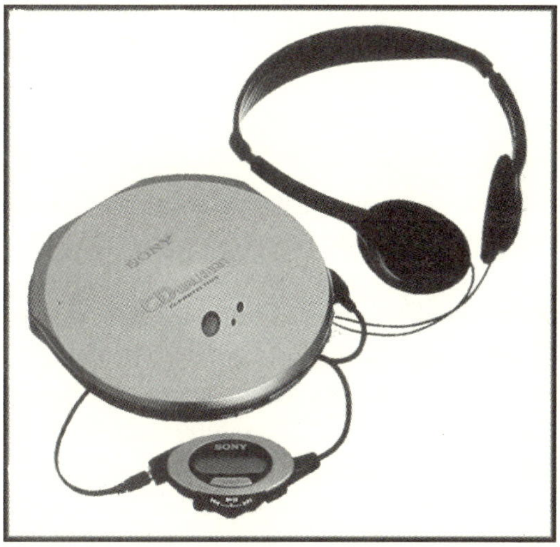

A SONY CD player, 2000 (Courtesy of SONY Electronics, Inc.)

rather than go ahead independently. Both had recently lost out to Matsushita, the world's leading electronics producer, when the VHS format outsold their separate videocassette formats. Moreover, Philips was then involved in a costly videodisc rivalry with JVC (Japanese Victor Company) and the U.S. company RCA. By 1980, the two companies had agreed on the standard for audio CD and began to develop their products independently. In 1982, Sony launched the CDP-101 CD player in Japan. It was designed to fit in with existing hi-fi stacking systems.

The CD player became the fastest-selling machine, as of then, in the history of consumer electronics, although it has recently been surpassed by the digital versatile disk (DVD) player. In the United States, the sales of CD players grew from 35,000 in 1983 to 700,000 in 1985, while CD sales grew from 800,000 to 15 million. The introduction of portable CD players increased the popularity of the CD format. Sony launched its first portable CD player with headphones, the D-5, in 1984. The Sony D-88 Pocket Discman, a slimmer model based on the successful Walkman personal cassette player, arrived in 1988. CD players were also incorporated in ghetto blasters, or boom boxes. Two new variants of the audio compact disc format were introduced in 1999: the DVD-Audio format, developed by the Japanese company Matsushita, and the Super Audio Disc format, developed by Sony and Philips. Both offer enhanced sound quality through increasing the rate of digital sampling.

The CD player was marketed as a major advance in the quality of sound reproduction on several grounds: greater dynamic range (essentially loudness), the inherent superiority of the digital copying that permits the master recording to be exactly reproduced, and the absence of wear and surface noise compared to the gramophone (phonograph) or tape recorder. Other factors favoring the CD player include the convenience of operation by remote control and programmed track selection. While studies have shown that the majority of people, including trained musicians, cannot reliably distinguish between analogue recordings (LPs) and digital recordings (CDs), by 1988 CDs were outselling LPs. Today, the CD player has supplanted the record player in the majority of homes.

See also Computers; MiniDisc Recorders; Record Players; Videodisc Players

References:

Schoenherr, Steve. "Recording Technology History." On http://ac.acusd.edu/history/recording/notes.html.

Compasso d'Oro

The Golden Compass Awards (Il Compasso d'Oro) are a series of industrial design awards that originated in Italy in 1954 when Aldo Borletti of the Milan department store La Rinascente founded them as a one-off. It was an immediate success, attracting 5,700 entries. The Compasso d'Oro was important in promoting good design in everyday things at a time when Italian industry was reestablishing itself after World War II. It also encouraged Italian designers to consider such mundane objects as being worthy of their attention as well as encouraging manufacturers to invest in designers. The 1954 winners set the pattern for future competitions; they included a sewing machine, an electric fan, a typewriter, and kitchen components.

The awards continued in 1955, 1956, 1957, and 1959. Due to organizational or economic reasons, the awards are not given every year. The panel of judges is small, usually consisting of around six people drawn from the relevant industries. Judges have included designers such as Marco Zanuso, Vico Magistretti, and Philippe Starck. Awards were given four times in the 1960s, only twice in the 1970s, and four times in the 1980s. Winners have included plastic buckets, sewing machines, lemon squeezers, collapsible dish-racks, washing machines, lamps, gas cookers (stoves), and telephones as well as cars and furniture.

Il Compasso d'Oro is now run by the Associazione Design Industriale (ADI), an association of 750 manufacturers, architects, and designers working in Italy.

Reference:

www.dolcevita.com

Computer Printers

Information stored on computer can be read off the screen, but many people find that reading from a computer screen is more physically wearing in terms of eye strain and postural fatigue than reading from the printed page. In the days of mainframe computers, printing was a batch job and required large, durable machines. The development of the first personal computers in the mid-1970s led to the corresponding development of desktop printers.

In the period 1976 to 1979, several types of printer became available. The best print quality was delivered by printers that used the same printing technology as contemporary typewriters. Indeed, some models were converted typewriters that retained the keyboard for dual-purpose use. The print head was a daisy wheel, a disk with spokes and raised characters around the circumference, and the printing medium was a carbon tape. Daisy-wheel printers were comparatively slow and noisy. An acoustic hood could be placed over the printer to deaden the noise, but this made the printer more bulky.

The alternatives to the daisy-wheel printer were cheaper, faster, and quieter, but delivered much lower print quality. In the late 1950s, dot-matrix printers were developed for use with mainframe computers. Dot-matrix printers use carbon tape, but the printer head consists of tiny pins that are selectively used, as instructed by the built-in microprocessor, to form characters. The print quality of desktop models improved somewhat in the early 1980s, when 24-pin heads superseded the original 9-pin heads. The cheapness of dot-matrix printers made them a popular choice where print quality was not the main consideration. Thermal and electro-sensitive printers were quieter still because they were nonimpact printers. Instead of a printer head, they used a stylus and the printing medium, carbon, was impregnated in the paper. The carbon was released in response to electric current flowing through the stylus. The print was fainter and less crisp than that produced by a daisy wheel. Low print quality together with the high cost of the special paper limited the sales of these printers. However, thermal printing is still used in many fax machines.

Two methods of nonimpact printing have proved very successful: the ink-jet printer and the laser printer. The ink-jet printer first appeared on the market in the early 1980s. The Japanese optical and electronics company Canon pioneered "bubble jet" ink printers in 1981. In 1984, the American electronics company Hewlett-Packard introduced the first of its ThinkJet series of ink-jet printers. Ink, supplied in cartridges, is sprayed through a matrix of perforations in the printer head. As with the dot-matrix printer, each character is a composite of dots. In the early days of ink-jet printers, there was a tendency for the ink to "bleed," creating a fuzzy effect. Bleed-resistant papers were created but these, predictably, were more expensive than ordinary computer paper. By the late 1980s, improvements made to reduce the bleed problem and a steep drop in prices established the ink-jet printer as the favored budget purchase, in place of the dot-matrix printer. Ink-jet printers also have the advantage of compactness. When the notebook generation of portable computers emerged in the late 1980s, complementary portable models of ink-jet printers followed.

The only printer to match the daisy-wheel printer in terms of quality is the laser printer, which has more in common with the photocopier than the typewriter. The world's first laser printer, the IBM 3800, was introduced by the U.S. office machine giant in 1976, but it took another ten years for the price to fall sufficiently for laser printers to become commercially competitive. Laser printers use powdered ink known as toner and contain a light-sensitive drum, a laser, and a rotating mirror. Where light from the laser beam falls on the electrostatically charged drum, the charge is dissipated; where no light falls, the charge remains and toner is attracted. The toner is transferred to paper and fused in place by heating. While characters and images are formed as patterns of dots, as with dot-matrix and ink-jet printers, the laser printer dots are so small and closely spaced that lines appear to be continuous.

Both ink-jet and laser printing technologies brought another advance—color printing. Canon introduced its first color ink-jet printer in 1982, only a year after its first monochrome model. It is now standard for ink-jet printers to operate as dual monochrome or color printers. Color printing is slightly more expensive because a tri-color ink cartridge has to be replaced more often than a black ink cartridge. Ink-jet models have a huge price advantage over color laser models. Hewlett-Packard, the company that has set the standards in laser printer technology since the launch of its first LaserJet printer in 1984, did not introduce a color model until 1994.

As printer technology changed, different manufacturers became involved. Makers of daisy-wheel printers included major typewriter manufacturers such as IBM and Olivetti as well as Xerox and Tandy/Radio Shack. Since then, Japanese companies have taken over the printer market. Epson became a leading maker of dot-matrix printers, and Canon, with a pedigree in the unrelated field of camera manufacture, is a leading maker of ink-jet printers. In the laser printer field, Japanese companies such as Canon and Panasonic dominate the lower end of the market, but Hewlett Packard of the United States is still a major supplier of top quality models.

See also Computers; Fax Machines; Hewlett Packard

References:

Haddock, Thomas F. *A Collector's Guide to Computers and Pocket Calculators.* Florence, AL: Books Americana, 1993.

Computers

Since the creation of the first electronic computer in 1946, computer technology has evolved with unparalleled speed. Conceived as a machine to automate and accelerate the calculation of complex sums, the computer became the universal machine for business and personal use because of its ability to process verbal as well as numerical data. Ownership of computers in the home became feasible in the late 1970s when computers of desktop size were developed. As the price of personal computers plummeted and the functionality of the computer became more diverse, home ownership rose. In 1995, the United States led the home ownership rankings with a 37 percent home ownership rate, while Britain, ranked sixth, had a 25 percent home ownership rate. Today, with appropriate software and peripheral devices, the home computer can provide many services, including processing of household financial accounts, word-processing, electronic mail, entertainment, and information.

Early Computing

The origins of the computer lie in the development of large mechanical calculating ma-

The ENIAC machine at the University of Pennsylvania, Philadelphia, April 1946 (Bettmann / Corbis)

chines from the early nineteenth century. The English mathematician Charles Babbage is considered to have invented the concept of the programmable computer when he devised his Analytical Engine, which was never completed. Babbage's idea of storing instructions on punched cards was adopted with commercial success by the American inventor Herman Hollerith. Hollerith's first punch-card data-processing machine was developed specifically for tabulating U.S. census returns in 1890. From 1896, Hollerith's company, the Tabulating Machine Company (which later became part of the International Business Machines Corporation), built similar machines for a range of uses.

While punched-card calculating machines proved an effective means of speeding up lengthy tabulations, they were not suitable for carrying out more complex mathematical tasks, such as differential equations. In 1876, the Irish physicist Sir William Thomson (later Lord Kelvin) put forward the concept of a mechanical differential analyzer for solving differential equations. However, as with Babbage, Thomson's ideas were too advanced for contemporary engineering capabilities. The idea of the differential analyzer resurfaced in the 1930s. In the mid-1920s, the American scientist and electrical engineer Vannevar Bush began work on a mechanical-electrical differential analyzer, which he called the product integraph. As the product integraph could only solve the simplest differential equations, in 1930 Bush began to develop a more complex differential analyzer that could handle eighteen independent variables.

The leading European computing pioneers of the 1930s included Douglas

Hartree, who constructed the first British differential analyzer, and Konrad Zuse, a German engineer who built the first binary calculator, fed by a punched-tape reader, in 1938. The significance of Zuse's work is that it laid the foundations for digital computing. Earlier mechanical and electromechanical calculating machines were analogue computers, meaning that each of their components yielded a range of values, which combined to produce a result. Zuse's binary calculator was based on the binary algebraic method of the nineteenth-century English mathematician George Boole, who demonstrated that equations could be reduced to a series of true or false propositions. This is known as Boolean logic, and in binary code the values of 0 and 1 are used to represent false and true. The advantages of the binary system became more apparent when electronic computers were developed in the late 1940s. The binary system lends itself perfectly to circuits where the state at any point depends on the presence or absence of a pulse of current or the low or high voltage of a component. A long series of bivalue electronic transactions is much simpler to engineer reliably and much more flexible in terms of program routines than fewer transactions with many possible values. In 1939, John V. Atanasoff and Clifford Berry of Iowa State University built the world's first electronic calculator, which had an external magnetic drum to store a binary code program.

The First True Computers

World War II stimulated computer development as military advantages could be gained through designing weapons according to more sophisticated ballistic calculations and deciphering the encoded communications of the opposing side. In the early 1930s, the U.S. Navy Board of Ordnance sponsored the American mathematician Howard Aiken, and in 1939, in collaboration with engineers at the International Business Machines Corporation (IBM), he was contracted by the Navy to develop a machine for ballistic calculations. Aiken's electromechanical Automatic Sequence Controlled Calculator, also known as the Harvard Mark I, was completed in 1944 at a cost of $500,000. It was operated by a punched-tape program and weighed 5 tons.

In Britain, computer research efforts were concentrated on code breaking. Alan Turing, the British mathematician who in 1936 had formulated his vision of a "universal computing machine," was one of the team that created the Colossus code-breaking machine. Colossus succeeded in breaking the supposedly impregnable German Enigma code, but, for obvious reasons, the project was kept top secret. The most influential of the computers developed in the course of military research was not completed until 1946. This was the Electronic Numerical Integrator and Calculator (ENIAC), commissioned by the U.S. Army Ordnance Department. ENIAC was built by a team at the University of Pennsylvania, led by John Presper Eckert and John William Mauchly. Drawing on Atanasoff and Berry's design, ENIAC was the world's first electronic computer. Weighing 30 tons and occupying 160 square meters (1,600 square feet) of floor space, it contained 19,000 thermionic valves, which acted as gates controlling the flow of electric current. Each calculation was programmed by operators feeding in punched cards, and the results were also presented on punched cards.

Feeding in punched cards was a slow and laborious process, so university scientists elsewhere began working on methods of internal program storage. In 1945, the eminent Hungarian-born American mathematician John Von Neumann outlined his theory of a stored-program computer with a central unit to control and process operations in sequence and with read-write random access memory. In Britain, teams at the Universities of Manchester and Cambridge were also addressing this issue. The Manchester team was led by Freddie Williams and Tom Kilburn and assisted by Alan Turing. In 1948, the Manchester electronic computer, known as the Small

Scale Experimental Machine (SSEM) and nicknamed the Baby, ran the world's first stored program, which was stored on cathode ray tubes. Von Neumann's ideas first came to fruition in the Electronic Delay Storage Automatic Calculator (EDSAC), built at Cambridge University and operational from 1949. EDSAC used mercury delay line storage, a technology developed at the Massachusetts Institute of Technology. EDSAC was completed in advance of the Von Neumann computers developed in the United States, namely the Electronic Discrete Variable Computer (EDVAC) at the University of Pennsylvania and the MANIAC-1 computer at the Institute for Advanced Study at Princeton.

The Mainframe and Mini Computers

In 1948, IBM decided not to manufacture computers commercially, believing, based on market research, that expense and size were prohibitive factors. Howard Aiken, who had joined IBM, remarked in 1950 that he could not ever see the need for more than six computers in the world. However, other scientists who had built prototype computers thought otherwise and assisted in the development of commercial models. The Manchester University team collaborated with the Manchester-based electrical engineering and electronics company, Ferranti, to create the first commercial computer, the Ferranti Mark I, launched in 1951. Eckert and Mauchly set up in commercial partnership in 1947, but sold their business to Remington Rand three years later. They developed the first commercial American computer, UNIVAC, for Remington Rand in 1951. The original UNIVAC model, supplied to the U.S. Bureau of Census, was the first computer to use magnetic tape for storage. More unusually, the Cambridge University team entered into collaboration with the catering company J. Lyons, which operated a chain of tea shops, to develop the LEO (Lyons Electronic Office) computer for processing business data.

IBM soon reassessed its position and in 1952 Aiken designed its first commercial computer, also its first electronic computer, the model 701. IBM soon acquired a reputation for innovation in computing and overtook Remington Rand's early lead in the U.S. computer market. It recognized that the high power consumption (up to 100 kilowatts) and heat output of valve computers were disadvantageous, causing valves to burn out too frequently and creating uncomfortable working conditions. The alternative to the valve was the smaller and more resilient solid-state transistor, invented in December 1947 by John Bardeen, Walter Brattain, and William Shockley at Bell Laboratories of AT&T. At first, commercial transistor production had a high failure rate, so transistors were expensive. As an intermediate measure, IBM began to manufacture hybrid computers incorporating valves and transistors, which brought some gains in size and power reduction. The experimental hybrid model 604 computer, built in 1954, led to the commercial model 608 computer of 1957. IBM contracted Texas Instruments, a company that began as a manufacturer of geophysical instruments in the 1930s and moved into the semiconductor industry in the 1950s, as its transistor supplier. Two years later, the IBM model 7090 computer was fully transistorized. Reductions in size were not only beneficial to customers in terms of space savings, but also increased the speed of data processing because the electric impulses had less distance to travel.

Computer storage capacity also improved during the 1950s. In 1953, Jay Forrester of the Massachusetts Institute of Technology installed the first magnetic core memory in the Whirlwind computer, which had been developed specifically for the U.S. Navy in the 1940s. IBM's contract to develop a successor to the Whirlwind, the SAGE computer of 1956, provided the opportunity to work on magnetic core memory and magnetic drum storage. The magnetic drum evolved into the magnetic disk. In 1957, IBM's 305 Random Access Method of Accounting and Control (RAMAC) was the world's first commercial computer disk storage system. In the 1950s, there was no concept of generic software, as each computer was programmed to perform the specific tasks required by the individual client. The programming process was simplified by the development of high-level computer languages that were designed for particular programming purposes. The high-level languages were supported by interpreter or compiler programs, which translated the language into binary machine code. The first of these languages, introduced by IBM, in 1956 was FORTRAN (FORmula TRANslation), which was intended for scientific and mathematical programs. For business applications, COBOL (COmmon Business Oriented Language) was introduced in 1959.

These large computers running specialized programs became known as mainframe computers. IBM had sold 1,800 mainframe computers by 1960 and 12,000 by 1964. IBM's sales philosophy placed great emphasis on a continuing close relationship with customers. However, by the early 1960s, it became clear that smaller customers might favor a more generic approach. In 1963, the American company Digital Equipment Corporation (DEC) introduced the PDP-8, the world's first minicomputer. The launch of the more generalist minicomputer was closely followed by the development of the first general-purpose computer language, BASIC (Beginner's All-purpose Symbolic Instruction Code), in 1964. BASIC was written by John Kemeny and Thomas Kurtz at Dartmouth College. IBM did not immediately embrace the change in business strategy that the minicomputer represented, as it had too much invested in its mainframe strategy. However, it did respond by developing a more flexible type of mainframe architecture. In 1964, IBM launched the System/360 computer, which was conceived of as a "family" of mainframe equipment. System/360 was modular rather than highly tailored and offered a choice of processors, peripherals, and complementary software packages, allowing upgrading or expansion over time. It was a commercial success and total sales of IBM computers rose to 35,000 by 1970.

The long-term future of the mainframe was threatened by developments that made it possible to link up, or network, separate computers. AT&T's core business gave it a vested interest in computer systems that were interoperable and accommodated multiple users, such as individual telephone exchanges. In 1969, Bell Laboratories developed the UNIX operating system, which became widely used for networking computers. Bell researchers developed a high-level, general-purpose computer language, C, which made UNIX compatible with virtually any of the existing minicomputers. When C became too restrictive for more demanding computer applications, it was modified by a Bell Laboratories researcher, Bjarne Stroustrup, to become C^{++}, introduced in 1983. C^{++} incorporates object-oriented programming, a more flexible way of modeling data relationships and has become one of the most widely used programming languages.

Microelectronics

While the minicomputer widened the market for computers, they were still too expensive and complex for small businesses, let alone individuals. For computers to be brought within the reach of a mass market, they needed to become still smaller, cheaper, and easier to use. The next advance in fundamental electronic technology after the transistor was the integrated circuit. While it had taken the research resources of the world's largest company, AT&T, to invent the transistor, within ten years transistor manufacture was dominated by new specialist semiconductor companies. The first integrated circuit was created in 1958 by Jack Kilby of Texas Instruments. It consisted of five components on a single germanium chip. A year later, Robert Noyce of Fairchild Semiconductor, founded in 1957, produced the first planar transistor. The planar process involved oxidizing a silicon wafer, coating it with a photosensitive material, photographing a pattern onto it and etching the pattern into the oxide, washing off the coating, and selectively introducing impurities. It was a repeatable process that enabled complex circuits to be built on a silicon wafer. By 1970, the price of an integrated circuit, also known as a silicon chip, had fallen from about $30 to $1, and an integrated circuit might contain up to 100 components.

The use of integrated circuits meant that the printed circuit boards of devices such as calculators became much more compact. Integrated circuits began to be used in computers in the late 1960s, but the central processing unit of a computer required thousands of integrated circuits. In 1968, Robert Noyce cofounded Intel, which began to develop large-scale integrated circuits. While Noyce had predicted that the way forward would be to fit the whole central processing unit onto a single chip, it was one of his employees, Ted Hoff, who actually achieved that. Hoff developed the Intel 4004 chip, the world's first microprocessor, which made the pocket calculator possible. In terms of mathematical processing power, the Intel 4004 chip was virtually the equivalent of ENIAC. However, its limitation was that as a 4-bit chip (meaning that it could handle four binary digits simultaneously) it could not process alphabetical characters, because it could only define 16 4-bit characters, or bytes. The IBM 7030 computer of 1961 had established the 8-bit character, or byte, as the standard for general computing. Intel launched its first 8-bit microprocessor, the 8008 chip, in 1972, followed by the improved 8080 chip in 1973, paving the way for the first generation of home computers. The 8-bit chip could define 256 different 8-bit characters.

The Home Computer Arrives

In 1975, Micro Instrumentation and Telemetry Systems (MITS), a small firm based in Albuquerque, New Mexico, introduced the world's first microcomputer, the Altair 8800. Lacking its own monitor and keyboard, the Altair 8800 was intended for the serious home enthusiast. Bill Gates (William Henry Gates III) and Paul Allen developed a modified version of the BASIC programming language for the Altair. They registered the Microsoft trade name in November 1976 to market the new language as MS-BASIC. Steven Jobs and Stephen Wozniak, two computer enthusiasts based in Silicon Valley, the heart of the semiconductor industry, were inspired by the example of the Altair. Using a cheaper 8-bit microprocessor, the MOS Technology 6502, they built their own microcomputer. Encouraged by the response of fellow enthusiasts, they began small-scale production of the Apple I computer in 1976. Snubbed by the companies offered the commercial rights but convinced of the commercial potential of the microcomputer, Jobs and Wozniak raised venture finance and set up Apple Computer in 1977. The Apple II computer, the world's first commercial microcomputer, had generated $2.5 million in sales revenue by the end of the year.

The immediate success of the Apple II energized the computer industry. Other companies, particularly calculator manufacturers, were quick to see the potential of the standalone, desktop computer and began to develop rival products. Like Apple, they hoped to appeal simultaneously to the potential home user and the small business. The U.S. company Commodore Business Machines, founded by Jack Tramiel in 1958, introduced the PET 2001 only two months after the launch of the Apple II. By 1980, a number of U.S. companies were producing microcomputers (all of which were mutually incompatible) and companies such as Epson were selling compact, cheap printers to complement microcomputers. In Britain, Clive Sinclair, developer of the first pocket calculator, introduced the Sinclair ZX80 home computer in 1980. The ZX80 became the cheapest microcomputer on the market. It was designed to use a television set as a display screen rather than a dedicated monitor.

The fall in the price of microcomputers was largely due to the astonishing decrease in the costs of microchip manufacture. No other industry has matched the semiconductor industry for sustained reduction in costs coupled with faster performance. While U.S. companies such as Intel and Motorola dominated the microprocessor market, Japanese companies such as Fujitsu and NEC (Nippon Electric Company) began to make major inroads into the memory-chip market. In 1970, Intel's first RAM (random access memory) chip had a mere 1K (kilobyte) capacity. Over the next decade, the capacity of RAM chips rose to 4K in 1973, 16K in 1976, and 64K in 1979. Japanese manufacturers were able to rapidly penetrate the memory-chip market by taking an approach different from that of the U.S. memory-chip companies. Instead of investing time trying to get more memory on the same size of chip, they opted for the simpler approach of making bigger chips. They also championed the CMOS chip design, which consumed less power than the NMOS chip and was more resilient.

Personal Computers

In terms of mass-market potential, the problem with the microcomputer industry in the late 1970s was the proliferation of incompatible machines. No company was able to establish a sufficiently large market share to shape the direction of microcomputer production. IBM initially adopted a disdainful approach to the nascent microcomputer industry. However, once the demand for single-user computers became evident, IBM entered the market in 1981 with the launch of the 5150 PC. The key features of this IBM PC were an Intel 16-bit microprocessor, 64K RAM, and the Microsoft Disk Operating System (MS-DOS). IBM appropriated the term "personal computer," which—shortened to PC—became used to describe the system architecture. Reputation, marketing channels, and immense research and development resources soon gave IBM a decisive competitive edge in the business market, in spite of its relatively high prices. In 1983, IBM introduced an upgraded PC, the 5160 PC XT, which had a hard-disk drive as well as a floppy-disk drive, and the cheaper IBM PC jr, aimed at the home consumer. (The floppy disk had been introduced as a convenient portable storage medium in 1971.) By the end of 1983, IBM had sold 800,000 PCs. In 1984 came the IBM 5170 PC AT, which introduced the 16-bit ISA (industry standard architecture) data bus, which accelerated the flow of data.

PC architecture was soon cloned by other companies to create a range of IBM-compatible models. At first, would-be imitators had to use the practice of "reverse engineering," whereby they deconstructed an IBM PC to analyze its technical design. This became unnecessary when IBM decided to publish its system architecture in order to encourage software companies to develop PC applications and thus stimulate the growth of PC ownership. While IBM achieved its goal of making the PC the industry standard for microcomputers, it lost out in terms of computer sales to companies making cheaper clones. For example, the British Amstrad PC1512 personal computer, introduced in 1986, was both cheaper and faster than the IBM PC. In the United States, Compaq, a spin-off from Texas Instruments, was so successful with its IBM clones that in 1986 it superseded Apple as the fastest-growing American corporation ever.

Computer Software

The key to the mass-market success of the microcomputer lay, not in the hardware itself, however small or cheap it became, but in the development of a range of generic applications. Compatible computer hardware meant that there was a huge incentive for companies to develop software that would enable users to exchange data easily. From the beginning to the present day, the business-software industry has been dominated by American companies. The first mass-market applications provided the means of computerizing the tasks that were common to all businesses—accounting and word-processing. In 1979, Software Arts introduced VisiCalc, the first commercial spreadsheet, to run on the Apple II computer. Spreadsheets create files in the form of tables in which numerical data can be sorted and manipulated. Launched in 1982, the Lotus Development Corporation's 1–2–3 application for PCs, which added database and graphics display functions to the core spreadsheet functions, soon became the market leader. In the same year, Ashton-Tate released dBASE II, the first commercial relational database. While the pre-PC WordStar was the first commercial word-processing package, WordPerfect—launched in 1982—became the market leader of the 1980s.

The Graphical User Interface

By the mid-1980s, personal computers were becoming common in the workplace, but they were still rare in the home. Expense was not the only factor; other factors were operational skills and functionality. While the microcomputer was domestic in scale, it made few concessions to the casual user in terms of usability. Personal computers were marketed as "user-friendly," but many people were intimidated by disc operating systems that offered only an enigmatic prompt, signifying the active disk drive, on the opening display screen. Apple again demonstrated its inventiveness when it introduced the Lisa in 1983. The Lisa introduced the graphical user interface (GUI), a screen display that showed program options as graphic icons, pull-down menus from menu bars, and "windows," screens that could be overlaid and sized. It also offered a pointing device called a mouse as an alternative to the keyboard for navigation and activating menu commands. The computer mouse had been developed in the 1960s at the Stanford Research Institute by Douglas Engelbart, who obtained a patent in 1970. It was commercially developed by the Xerox Corporation in the 1970s, but only became a standard computer device when GUI displays arrived.

Although the Lisa was too expensive to have a major impact on the microcomputer market, the launch of its cheaper sibling, the Apple Macintosh, in 1984 established the GUI as the truly user-friendly face of computing. The Macintosh, familiarly known as the Mac, became particularly popular with graphic designers as it ran the first commercial desktop publishing (DTP) package, Adobe PageMaker. With its streamlined shell, the Mac was also the first microcomputer to be hailed as a design icon. While purist DOS users disparaged the Mac as a WIMP (windows, icons, menus, pointing device), Microsoft was quick to recognize the mass-market appeal of the GUI. As the developer of the Word and Excel applications for the Mac, Microsoft had privileged access to the Apple GUI program code, which became a bone of contention when Microsoft began to develop its own GUI operating system, Windows, for PCs. A legal judgment imposed restrictions on the design of the first version (1.0) of Windows, launched in 1985, but the restrictions ceased to apply thereafter. Nevertheless, it was only with the release of version 3.0 in 1990 that Windows achieved equivalent user-friendliness to the Mac interface. The later versions, Windows 95 and 98, improved the multitasking performance of the interface, which allows separate applications to be open at the same time.

Microsoft's monopoly of the PC operating system gave it clear advantage in the development of PC applications, as its applications programmers had first access to new code. Microsoft's first PC application was the PC version of the Excel spreadsheet, introduced in 1987. Since then, its suited Office and Office Pro packages of business applications have become the PC market leaders.

Size, Speed, and Price

The desktop personal computer has become the dominant computer type for business and home use. Hardware has grown in size in order to accommodate more devices, provide more storage capacity, and generally enhance performance. In the mid-1980s, the typical PC consisted of a 12-inch (30 cm) monitor, a keyboard, and a central processing unit (CPU) that accommodated a 20-Mb (megabyte) hard-disk drive and a 5.25-inch (13 cm) floppy-disk drive, and had a number of ports (connection sockets) for optional peripheral devices. The printer was the most common peripheral device. By the late 1990s, 14- and 17-inch (36 cm and 43 cm) monitors were standard, in order to provide improved display of pictorial content. The CPU, usually in a tower format, typically accommodated a several gigabyte hard-disk drive, a 3.5-inch (9 cm) floppy-disc drive, a CD-ROM (compact disc read-only memory) or DVD (digital versatile disk) drive, and a modem. It had sufficient ports to take a range of peripheral devices, including scanners, digital cameras, CD writers, loudspeakers, and extra storage drives, such as the Zip or Jaz drives. The CD-ROM, introduced in 1984, has become the standard format for applications software, games programs, and educational software such as multimedia encyclopedias. Recordable CDs (CD-Rs), requiring a CD writer, became available in 1990. Introduced in 1995, the DVD can hold a full-length motion picture.

However, the key determinants of performance, the microprocessor and the RAM chip, have grown in processing power, speed, and capacity without growing in size. Since its introduction in 1993, the 32-bit Intel Pentium chip, the dominant PC microprocessor chip, has evolved from running at a speed of 60 MHz (megahertz) to 600 MHz. A 64-Mb RAM chip is now regarded as no more than average. Therefore, while desktop models dominate, portable computers are now available in sizes ranging from the palmtop to the notebook. In the 1970s, portability was more of a relative concept. The first portable computer, the Baby suitcase computer of 1976, was a CPU without a monitor, like the Altair desktop. Even by 1981, when the Osborne I portable computer was introduced, portable computers were still suitcase-size and referred to as luggables rather than portables. Compaq, which introduced a portable PC in 1982, was the first company to really focus on the portable computer market. By 1986, the portable computer had shrunk from the luggable to the briefcase-size laptop, and by 1989, from the laptop to the thinner notebook. The notebook is the smallest type of PC that retains full functionality; it can accommodate hard disk, floppy disk, and CD-ROM drive as well as an internal modem. Subnotebooks and hand-held palmtops or personal organizers economize on size by having limited data storage facilities and small keyboards, but can transfer data to desktop or notebook computers by wired or infrared linkages. Some palmtop computers, such as the Apple Newton, introduced in 1993, omit the keyboard entirely and instead allow input to be written onto an LCD "notepad" using a stylus. They have built-in optical character recognition (OCR) software.

Prices have continued to fall, thanks to economies of scale, increases in production efficiency, and competitive market forces. While big American manufacturers, including IBM, Dell, and Compaq, still have sizeable market shares, the nature of computer retailing has allowed small companies to prosper. Purchasing direct from the manufacturer or from computer warehouses via mail order or electronic commerce has become a significant feature of the personal computer trade. Although personal computers may be ostensibly made by European, Canadian, or U.S. companies, in this context "making" means assembling, and the majority of the manufacturing process takes place in the Far East. Japanese companies, such as Toshiba, Sony, and Fujitsu, have been particularly successful in the portable computer market. Portable computers continue to be significantly more expensive than desktop models offering equivalent performance. This largely reflects the relatively high cost of flat liquid crystal display (LCD) screens in comparison with conventional cathode ray tube-based monitors.

Initially, the LCD screen was not only expensive, but also low-resolution. LCD technology originated at the research laboratories of the Radio Corporation of America (RCA) in the 1960s. The LCD consists of a layer of liquid crystals between polarizing filters. RCA developed a technique called the dynamic scattering mode, whereby an electric charge rearranged the crystal molecules to scatter light. This proved to be high on power consumption and low on image definition. In 1969, James Fergason of Kent State University discovered a more effective technique, whereby the electric charge twisted the crystals to reflect the polarized light. Unaware of Fergason's work and patent application, Wolfgang Helfrisch and Martin Schadt of the Swiss company Hoffmann La Roche announced the same discovery in 1971. The company subsequently purchased the rights to Fergason's patent. The early LCDs of this type used in portable computers became known as passive matrix displays, and the quality of graphics was far inferior to that of a conventional monitor.

Introduced in the early 1990s, active matrix displays are also known as thin film transistor (TFT) displays because the controlling transistors, one per pixel, lie within the crystal filling, increasing responsiveness.

Intelligence and Interoperability

In 1981, the Japanese government announced the launch of the Fifth-Generation Computer Project. While the preceding four generations were defined by their core electronic characteristics—valves, transistors, integrated circuits, and microprocessors—the fifth generation was a more holistic concept. The objective of the project was to develop computers with artificial intelligence over a ten-year period. The idea of artificial intelligence was not a new one. The term came into use in the mid-1950s, and long-term research had been undertaken in the United States at universities including Stanford and Carnegie-Mellon. However, the focus had been much narrower and progress had been limited. While the algorithmic logic advocated by Von Neumann was expressed in a sequential data processing architecture, artificial intelligence required parallel processing architecture, which was more akin to the heuristic reasoning patterns of the human brain. Heuristic is the process whereby we draw on our knowledge of many things to infer solutions to problems. The success of the solutions depends on our expertise, in terms of the quality and range of our knowledge. By 1980, only the most powerful computers, known as supercomputers, used parallel processing. In 1965, the U.S. company Control Data Corporation introduced the first supercomputer, the CD6600, designed by Seymour Cray, who went on to set up Cray Research, which became the leading producer of supercomputers. These computers were designed for complex tasks such as weather and aerodynamic modeling.

The Fifth-Generation Computer Project sought to bring parallel processing and artificial knowledge management into the domain of the microcomputer. It also placed great emphasis on image processing and voice recognition, which was seen as a faster, more efficient, and more natural way of communicating with computers. The ten-year target was not achieved because developing natural language and voice recognition systems proved to be as difficult as optical character recognition, another area of research that was slow to yield reliable technology. However, the Japanese have enjoyed a measure of success in smart chip technology. The smart chip, or fuzzy logic chip, was invented at Bell Laboratories by two Japanese-born researchers, Masaki Togai and Hiroyuki Watanabe. Japanese companies were quick to see the value of the fuzzy logic chip as a control device for motor cars and appliances such as washing machines and have since led the way in this area.

Aside from artificial intelligence, the other significant strategic trend of the 1980s and 1990s has concerned the development of software that allows greater interoperability of computers. While hardware compatibility is one way of achieving interoperability, it was evident that total hardware compatibility was extremely unlikely to occur. UNIX, the pioneering hardware-independent operating system, had provided a means of establishing a common platform across a network of nonmatched computers. In the early 1990s, another operating system on similar lines, Linux, named after its Finnish inventor Linus Torvalds, was released as "open source (code)" a term for nonproprietary systems that are made freely available. Further developments in open systems were stimulated by the introduction of the World Wide Web in 1993. As the whole concept of the Web is that it is accessible to all computers irrespective of platform (hardware and operating systems), it fostered new languages and applications. It has become accepted for Web applications, such as browsers and document readers, to be made available as freeware. One of the first of these applications was Adobe's Acrobat Reader, which allows documents to be downloaded onto any platform. The leading web browsers, Netscape Navigator and Microsoft's Internet Explorer, were introduced as freeware, in 1994 and 1995, respectively. Released in 1995, Sun Microsystems's Java programming language has become the main language for Web applications.

Empowering or Entrapping?

The sharp rise in home computer ownership during the 1990s and the associated growth in influence of the Internet provoked concerns about the social consequences of computerization. To some, the World Wide Web is a positive force for good worldwide as a virtual expression of the global village; to others, it is culturally imperialistic and a corrupting force. In many ways, the debate has paralleled earlier arguments about the role of television in society. On the one hand, computer technology was applauded as empowering and democratizing; on the other hand, it was denigrated as socially exclusive and escapist. As with television, concern centered on the potential negative effects on children of the home-computer generation. The image of the socially gauche computer nerd, more comfortable with the synthetic relationships of the Internet chat room than face-to-face interaction, has become familiar through movies such as *Weird Science*. However, it is the ready availability of pornographic material on the World Wide Web that has generated the greatest outrage. Like most technologies, computer technology is open to abuse, but only a minority of people would currently contend that the negatives of computerization outweigh the positives.

See also Apple Computer, Inc.; AT&T; Calculators; Compact Disc Players; Electronic Games; IBM; Microsoft Corporation; Texas Instruments; Typewriters; Videodisc Players

References:

Braun, Ernest, and Stuart Macdonald. *Revolution in Miniature*. New York: Cambridge University Press, 1982.

Brook, James, and Iain A. Beal, eds. *Resisting the Virtual Life: The Culture and Politics of Information*. San Francisco: City Lights Books, 1995.

Forester, Tom. *High-Tech Society: The Story of the Information Technology Revolution*. Cambridge, MA: MIT Press, 1987.

Gardner, Robert, and Dennis Shortelle. *From Talking Drums to the Internet: An Encyclopedia of Communications Technology*. Santa Barbara, CA: ABC-CLIO, 1997.

Haddock, Thomas F. *A Collector's Guide to Personal Computers and Pocket Calculators*. Florence, AL: Books Americana, 1993.

Rogers, Everett M. and Judith K. Larsen. *Silicon Valley Fever*. New York: Basic Books, 1984.

Tweedale, Geoffrey. *Calculating Machines and Computers*. Princes Risborough, UK: Shire Publications, 1990.

Williams, Trevor I. *A History of Technology, Vols. VI and VII: The Twentieth Century*. Oxford, UK: Clarendon Press, 1978.

Zerzan, John, and Alice Carnes. *Questioning Technology: A Critical Anthology*. Philadelphia, PA: New Society Publishers, 1991.

Consumers

The growth of industrialization in the nineteenth century was stimulated by, and linked to, a rising population that created bigger markets. The establishment of modern capitalism grew in association with many of these developments. The innovations within technology and science were not driven only by "pure" experimentation but also by the desire to commercially develop the results. This culture of mass consumption was already advanced in Europe, Canada, and the United States at the beginning of the twentieth century and was initially enjoyed by the middle classes. The post-1945 increase in prosperity allowed more and more working people to purchase consumer durables.

Designers and manufacturers of the earlier twentieth-century domestic appliances were certainly aware of their potential markets insofar as they wanted their products to sell. Nevertheless, what market research that was carried out was largely unscientific and anecdotal. Initially they relied on the nineteenth-century premise that there were "natural" preexisting markets for a product. The role of promotion and advertising was to make sure that the potential customers were attracted to your particular product. Branding, the process of giving a product an identity, was beginning to develop and was accelerated during the Depression years of the 1930s. Economists and politicians looked to increased consumption as a way out of economic slumps. The late 1920s and 1930s saw the introduction of the marketing methods and psychological selling techniques familiar today. There was a change from "getting commodities to consumers" to "getting consumers to commodities."

This was achieved by advertising techniques that, in the case of domestic appliances, were aimed specifically at women. Advertisements prompted purchase through a combination of guilt and desire. In the United Kingdom and the United States advertisements began to illustrate the housewife, not the servant, using the appliances and exploited rising standards of cleanliness and fears about "household germs." The increasing use of labor-saving appliances may have saved time in some areas, but social and cultural pressures led to increasing standards and more time spent on other areas of housework. The desire to consume was stimulated by aspirational advertisements and planned obsolescence of products.

As Americans were encouraged to become patriotic consumers many of them felt that they needed to make informed choices about the increasing range of products. In 1926 Frederick Schlink, an engineer from White Plains, New York, organized a consumer club that distributed lists of products that were seen as good value and also those "one might well avoid, whether on account of inferior quality, unreasonable price, or of false and misleading advertising." Schlink used these lists to produce a book, *Your Money's Worth,* which led to the founding of Consumers' Research and the *Consumers' Research Bulletin* in 1928.

The Consumers Union was a splinter group from Consumers' Research and was established in 1936, following acrimonious labor relations. Its founding group of professors, labor leaders, journalists, and engineers had a mission to "maintain decent living standards for ultimate consumers," a rhetoric born of the Depression and the strike-breaking tactics of Schlink. It remains independent of both government and industry and depends on membership subscriptions. It first published its magazine *Consumer Reports* in the same year, establishing a tradition of testing and rating products and services. The initial circulation was around 4,000. Appliances were and continue to be tested for performance, energy efficiency, noise, convenience, and safety. Subscriptions had risen to 100,000 by 1946 and continued to grow, even during the McCarthy era when *Consumer Reports* was listed as a subversive magazine. The Consumers Union now has over 4.6 million subscribers, a children's magazine (launched in 1980 as *Penny Power,* now known as *Zillions*) and a web site.

In the United Kingdom, the *Good Housekeeping Magazine* was established in 1922, largely aimed at the servantless middle-class woman. It founded the Good Housekeeping Institute in 1924 to test recipes and "submit all domestic appliances to exhaustive tests and bring those approved to the notice of all housewives," which it continues to do today. The UK Consumers Association, based on the U.S. Consumers Union was founded in 1956 and first published *Which?,* a quarterly magazine of tests and reports in 1957. *Which?* became a monthly magazine in 1959. The UK Consumers Association currently has over a million members. The International Organization of Consumers Unions was established in 1960 and includes consumer associations from the United States, the Netherlands, Belgium, and Australia.

The marketing trends of the 1930s continued after 1945 and in-depth market research developed throughout corporate America in the 1950s. The British Market Research Association was established in 1957, the same year as Vance Packard's critical study of advertising, *The Hidden Persuaders,* was published in the United States. The following quotation from Packard's book illustrates how the advertising industry continued to use the twin themes of guilt and desire in the postwar boom years.

> The cosmetic manufacturers are not selling lanolin, they are selling hope. . . . We no longer buy oranges, we buy vitality, we do not buy just an auto, we buy prestige.
>
> If you tell the housewife that by using your washing machine, drier or dishwasher she can be free to play bridge, you're dead! She is already feeling guilty about the fact that she is not working as hard as her mother. You are just rubbing her up the wrong way when you offer her more freedom. Instead you should emphasize that the appliances free her to have more time with her children and to be a better mother.

Advertisements of the period support this. A Hotpoint ad from *Good Housekeeping* of June 1951 carries the copy "Save 8 Hours Every Week with a Hotpoint All-Electric Kitchen—Gain Extra Time for All Your Extra Duties." The time saved, the advertisement suggests, is "for your family as well as the many added duties you're called on to shoulder these days." Needless to say, the "you" in question was female.

These quotes reflect a set of cultural values that were already in the process of being challenged by the feminist, civil rights, and youth movements of the 1950s and 1960s. *Unsafe at Any Speed,* by the American lawyer and consumer advocate Ralph Nader, was published in 1965 and exposed the lack of safety in the General Motors Corvair automobile. Nader joined the Consumers Union in 1967. Congress passed twenty-five pieces of consumer legislation between 1966 and 1973.

The advertisers and manufacturers varied in their ability to respond to these social and cultural changes. The rise of the affluent teenager created a new market, one that clothing, publishing, and cosmetics companies responded to with vigor. The domestic appliance companies also had to change. By the late 1970s the impact of feminism had been such that the latter comment quoted in Packard was no longer tenable as an advertising concept, even though it was still a reality for many women. A mid-1960s ad for a Nevastik Teflon-coated frying pan from the UK *Good Housekeeping Magazine* had the copy, "Even a Man Can't Go Wrong with Nevastik Pans."

Market research had become more sophisticated, and markets were increasingly divided into socioeconomic groups that could become target markets. This analysis became more sophisticated during the 1980s and 1990s as markets were segmented by postal areas and lifestyles.

It has been assumed that manufacturers and consumers stood in opposition to each other, with the consumer organizations acting as monitors and protectors of the latter's interests. Indeed, the efforts of consumer organizations have led to legislation to improve safety standards and consumers rights after a purchase has been made. But it would be wrong to believe that consumers have been passive recipients of what the producers have given them and that a docile and uncritical public leads to low standards of design. It has been argued that consumers' desires and needs have been created by the producers and, with the aid of their advertisers, have been satisfied by those producers. This implies that consumption is a less authentic and satisfying activity than, for example, working. It also seems to imply that popular forms of culture and material culture are superficial. Given the sophisticated nature of advanced capitalist societies, this attitude can be contested: needs are often no longer natural, but cultural, informed by the many connections and discontinuities within those societies. Many modern objects do not simply—or, indeed, primarily—have "use or exchange" value but more importantly have "identity" value. This can clearly be seen in some of the more fashionable domestic appliances of the 1980s and 1990s. A Dyson vacuum cleaner or a Sony Walkman is a successful piece of technology, but each equally has become a purchase that reinforces its own brand identity and defines the identity of the consumer. The same can be said of older products such as the Aga cooker or the more self-knowing products from the Alessi stable.

The late twentieth century has produced a society where manufacturers, designers, and consumers are linked, knowingly or not. Companies continue to conduct market research but also are quicker to respond to and appropriate ideas that often bubble up from within popular or mass culture. This "circuit of culture" links the identity, production, consumption, regulation, and representation of a commodity within a circular relationship. This model has increasingly applied to domestic appliances over the last twenty years. Many domestic products that were once almost culturally invisible are now recognized as having a meaning. Consumers are now largely more sophisticated and are able to "read" the intended meanings of the manufacturers and to construct or appropriate their own, which will in turn influence the manufacturers and affect how that product is marketed or modified. Nevertheless, the findings of the 1960 UK *Molony Report* on consumer protection remain valid.

> The business of making and selling is highly organized, often in large units, and calls to its aid at every step complex and highly expert skills. The business of buying is conducted by the smallest unit, the individual consumer, relying on the guidance afforded by experience, if he possesses it, and if not, on instinctive but not always rational thought processes.

See also Industrial Design and Designers; Obsolescence

References:

Consumers Association. *Thirty Years of Which? 1957–1987.* London: Consumers Association, 1987.

Consumers Union. *Consumer Reports Best Buys for Your Home, 2000.* New York: Consumers Union, 1999

Cowan, Ruth Schwartz. "'The Industrial Revolution' in the Home—Household Technology and Social Change in the Twentieth Century." In *Material Culture Studies in America,* ed. Thomas J. Schlereth. Nashville, TN: American Association for State and Local History, 1984.

du Gay, Paul, Stuart Hall, Linda Janes, Hugh Mackay, and Keith Negus. *Doing Cultural Studies: The Story of the Sony Walkman.* London: Sage Publications, 1997.

Myers, Kathy. *Understains: The Sense and Seduction of Advertising.* London: Comedia, 1986.

Oakley, Ann. *Housewife.* Harmondsworth, UK: Penguin Books, 1976.

Convenience Foods

Convenience food is very much a twentieth-century concept. In the nineteenth century, the main reason for processing food before sale was to increase its shelf life. This was a matter of increasing concern, given that the growth of the urban population meant that food had to be transported from further and further afield to the place of consumption in order to meet rising demand. Bottling, canning, and drying were methods that assisted food preservation and were amenable to mass-production and distribution. The archetypal canned food is Heinz baked beans, made by the U.S. H. J. Heinz Company, which is now sold all over the world. The disadvantage of canned foods was that the high temperatures at which the food was cooked, in order to kill enzymes and bacteria, also destroyed some vitamins. Canned foods also have a high content of sugar and salt, which are used as flavor enhancers.

Increasing production of preserved foods containing additives led governments to impose legal standards. In Britain, the Sale of Food and Drugs Act of 1875 imposed much stricter guidelines and penalties than earlier legislation. In the United States, the Pure Food and Drugs Act was passed in 1906. A number of minor religious sects stressed the importance of a healthy diet. Notable amongst these were the Seventh-day Adventists, whose headquarters were in the small town of Battle Creek, Michigan. The name of Battle Creek became familiar internationally owing to its emergence as the center of breakfast cereal production. The Adventists championed breakfast cereals because of their nutritional value, but the cereals became popular in the twentieth century because of their convenience. It was the convenience factor that spurred the development of new preservation techniques, including deep-freezing, irradiation, and freeze-drying. These techniques not only extend the life of food, making fewer shopping trips necessary, but also shorten the cooking time, an increasingly important factor as more women went out to work. The convenience of bulk buying led to a shift in food retailing from the local store offering personal service to the self-service supermarket. By 1959, supermarkets accounted for 69 percent of American food sales. In Britain, supermarkets were slower to take hold, but were dominant by the 1970s.

Beverages

Even simple processes like brewing tea and coffee could be simplified by processing. The flavor in coffee beans is a volatile essence, which begins to dissipate when the roasted bean is ground. Hence, traditionally, coffee beans would only be ground immediately before use. In 1878, Chase & Sanbourn of Boston, Massachusetts, packaged ground, roasted coffee in sealed cans to preserve its flavor. In 1901, a Japanese-American chemist, Satori Kato, produced the first soluble instant coffee for the Pan-American Exposition in Buffalo. Eight years later, George Constant Louis Washington of New York produced a soluble coffee powder, which he sold under the George Washington brand name. However, instant coffee was not mass-produced until the late 1930s. The Swiss food company, Nestlé, developed a mass-production method for instant coffee in order to exploit the surplus of Brazilian coffee beans. Nestlé mass-marketed their instant coffee as Nescafé from 1938. The American food giant General Foods produced an instant coffee in 1942 specifically for supply to the United States Army. It was marketed to the public as Maxwell House instant coffee after World War II. However, the American public tended to shun instant coffee, whereas in Britain and Japan, it made up about 90 percent of coffee sales. The standard drying technique involves spraying brewed coffee into a rising column of heated air, which removes the water as steam, leaving a powder residue. Freeze-drying technology improved in the 1950s and was applied to instant coffee in the mid-1960s. Freeze-dried coffee retained more flavor because the volatile oils remained.

Although coffee is the dominant hot beverage in the United States, the British public has always preferred tea. This may explain why the idea of the tea bag originated in the United States, where consumers needed more persuasion to drink tea. In 1904, a New York tea and coffee merchant, Thomas Sullivan, decided to send customers tea samples in muslin pouches. It was in this form that tea bags were first commercially produced in the United States in 1919. At first, manufacturers saw the catering industry, rather than private consumers, as the main market for tea bags, but by the mid-1930s, Tetley, of New York, was mass-marketing tea bags. In Britain, the public at first shunned the tea bag as an inferior product. This was justified insofar as tea bag manufacturers were able to use the fine "sweepings," previously treated as a waste product. These sweepings would have leaked out of the paper cartons used to package loose-leaf tea. Improvements in tea bag technology, giving improved infusion, helped to sell the concept of the tea bag. By 1993, over 80 percent of tea sold in Britain was in the form of tea bags.

Bread

Flour milling and baking became industrialized in the United States in the nineteenth century. Increasingly, bread was not baked at home but purchased from shops and bakeries. One sign of the changing nature of food production and distribution was the formation in the United States in 1898 of the National Biscuit Company (later shortened to Nabisco), which was an amalgamation of 114 bakeries, representing 90 percent of American commercial biscuit production. In Britain, the dominance of national bakery chains is a much more recent phenomenon, with no more than 40 percent of all bread consumed produced in large plant bakeries as late as 1953.

The automation of bread-making began with the introduction of roller milling of flour in the 1870s. Roller mills could produce much finer and whiter flour of a more consistent quality than grindstones. This had two major implications for bread-making: the finer flour could absorb more water, producing a lighter and more malleable dough, and the natural oils in the wheat berry were extracted at an early stage, leaving a flour with a longer life. In the 1920s, the factory bread-making process was accelerated when high-speed dough mixers became available.

The phrase "the best thing since sliced bread" appeared in the United States in the 1930s following the introduction of presliced Wonder Bread. In 1928, after sixteen years of development work, Otto Frederick Rohwedder launched the first practical bread-slicing and wrapping machine in Battle Creek, Michigan. In the same year, the Continental Bakery in New York introduced Wonder Bread, the first nationally distributed wrapped loaf of bread. Two years later, using Rohwedder's machines, it introduced presliced Wonder Bread. Wrapped, presliced bread also appeared in the United Kingdom in 1930. By 1933, 80 percent of the bread sold in the United States was presliced and wrapped.

Sliced bread was convenient and of a standard thickness. Its introduction no doubt helped the sales of electric toasters throughout the 1930s. However, healthfood campaigners argued that the convenience of the presliced white loaf came at the expense of its nutritional value. By the 1940s and 1950s, white bread was routinely enriched by the addition of vitamins and minerals. Stoneground whole wheat flour and unwrapped loaves enjoyed a revival from the late 1950s as a result of the growth of the health foods movement. For example, the American health food guru Gayelord Hauser was a strong advocate of the benefits of wheat germ.

Breakfast Cereals

Prior to the 1860s, breakfast cereal came in one variety—oatmeal porridge. This was not a quick breakfast dish, as it needed to be cooked for a long time. The solution was to cook a large batch and then reheat daily. In 1877, prepacked American Quaker brand rolled oats that had a much faster cooking time than oatmeal were introduced. The first ready-to-eat breakfast cereal, Granula, was invented by James Caleb Jackson of Dansville, New York, in 1863. Jackson found that small cooked granules of graham cracker dough made a suitable cold breakfast cereal, served with cold milk. However, it was not until the 1890s that the idea of ready-to-eat breakfast cereal really took off. John Harvey

Kellogg had become director of the Battle Creek Sanitarium and, with his brother, Will, had begun to develop easily digestible foods for invalids. They developed a baked wheat flake cereal that was marketed in 1895 as Granose, the first flaked breakfast cereal. Soon after, a second breakfast cereal enterprise came into being in Battle Creek, when C. W. Post, founder of the Postum Company, developed Grape Nuts in 1897.

In 1898, Will Kellogg developed Cornflakes, the cereal that became most closely associated with the Kellogg name. Kellogg's became the company name in 1922, replacing the Sanitas Nut Food Company (1898) and the Battle Creek Toasted Flake Company (1906). The first ready-to-eat breakfast cereal to reach the British market was Force Flakes, made in Canada, in 1902.

Although early breakfast cereals followed very healthy formulas, with only small amounts of malt and sugar added for extra flavor, as time went on, sugar content increased dramatically and fiber content fell accordingly. Kellogg's Sugar Smacks, introduced in 1953, had a 56 percent sugar content. In the more health-conscious society of the late 1950s, Kellogg's did introduce healthier cereals, such as Special K in 1955, but the company no longer had a healthy whole-foods image. Muesli, a favorite Swiss breakfast food that contains nuts and dried fruit, has become the epitome of the healthy breakfast cereal.

Frozen Foods

The pioneer of frozen foods was Clarence Birdseye, who based his freezing process on the natural freezing of meat and fish that he had observed in the Arctic zone. He noted that naturally frozen meat and fish seemed fresh when cooked and eaten months later. After returning to the United States, he formed Birdseye Seafoods in 1922 and initially concentrated on chilling fish fillets at a plant in New York. By 1924, he had developed a method of "flash-freezing" by placing cartons of food between metal plates under pressure. He formed the General Seafood Corporation to exploit the flash-freezing technique. In 1929, he sold his company to the Postum Company for $22 million, on the condition that his surname was used as two words, hence the Birds Eye brand name. The expanded company was renamed as the General Foods Corporation. In 1929, cartons of Birds Eye frozen vegetables went on sale in the United States. They were intended to be eaten on the day of purchase, as refrigerators, which were found in only a minority of homes, were only suitable for short-term storage of frozen foods. In 1930, twenty-six varieties of Birds Eye Frosted Foods were test-marketed in Springfield, Massachusetts. The line that was introduced across the United States in 1931 consisted of fish, meat, peas, spinach, loganberries, raspberries, and cherries.

By 1933, 516 stores were stocking Birds Eye Frosted Foods. In 1939, Birds Eye introduced precooked frozen dishes based on chicken, beef, and turkey. As consumption of frozen foods began to increase rapidly in the 1940s, the first specialist self-service frozen-food centers appeared, initially in the New York area in 1945. In Britain, frozen foods became available for the first time in 1946, after a Birds Eye plant was set up in Great Yarmouth. The U.S. company Sara Lee Kitchens produced the first frozen baked foods for the mass market in 1953. A year later, the complete frozen meal appeared when C. A. Swanson & Sons of Omaha, Nebraska, launched TV dinners. In 1957, a new method of cooking frozen foods emerged when the U.S. company Seabrook Farms launched Miracle Pack Prepared Foods, the first boil-in-the-bag frozen foods. The first frozen food to make a major impact in Britain was Birds Eye Fish Fingers, introduced in 1955. These cod sticks coated in breadcrumbs became a favorite children's food.

In the energy-conscious 1980s, a new competitor to frozen foods appeared—chilled foods. The chilling process involves keeping cooked foods at constant temperatures of 0° to 4°C (32°F–40 °F), the recommended temperature range for refrigerators. Although chilled foods have a shorter storage life than frozen foods, they are also quicker to cook and therefore save energy.

Frozen foods have had a profound effect on both the food industry and consumer behavior. For growers of food crops, selling produce to frozen-food companies meant reducing wastage and loss of income through natural decomposition. Some farmers may therefore prefer to sell their whole crop to the frozen food industry. One consequence of this has been that some types of fruit and vegetables are less widely available as fresh produce. The convenience of stocking up on food less frequently is another factor that has reduced the role of fresh food in the diet. An advantage of frozen foods for consumers, however, is that foods are available out of season, thus providing a more varied diet all year round. Calorie-counted, nutritionally balanced frozen or chilled meals may be a boon to the busy consumer, but traditional cooking skills have suffered as a result. Today, for many people, traditional cooking has become a hobby rather than a necessity.

Instant Foods

The term "instant food" covers any dried product that is prepared for cooking simply by adding a measure of liquid, usually water or milk. The first ready-mix food was Aunt Jemima's pancake flour, produced in St. Joseph, Missouri, in 1889. Other instant baking products, such as cake mixes, had their heyday in the 1960s, when the level of female employment rose. These products were marketed on the basis that home baking was a badge of good housewifery, so instant mixes enabled the busy working woman to cheat a little. In 1946, the R. T. French Company of Rochester, New York, introduced the first instant mashed potato product. General Foods introduced Minute Rice, a dried precooked rice, in 1950.

Before the 1950s, all instant products were produced by traditional air-drying, either at ambient temperature or with added heat. By 1940, a new method, freeze-drying, had been developed in Sweden. Food was rapidly frozen and then placed in a vacuum chamber to dry, because, at low pressures, water passes directly from the solid state to the gaseous state, a process known as sublimation. This was particularly effective for any foods with a high water content, as the water is removed rapidly without damaging the structure of the food. The freeze-dried food is sponge-like in texture and therefore absorbs water rapidly. However, the high speed of freezing and drying required for effective results means that the food pieces need to be no more than 2.5 cm (1 inch) thick. The first factory for freeze-drying food opened in Russia in 1954. Freeze-drying is used commercially for drying vegetables and meat, as well as coffee.

Margarine

In its original form, margarine was merely cheap and no more convenient than butter. The first margarine was developed by a French chemist, Hippolyte Mège-Mouriès, in response to a national competition in 1867 to find a cheaper alternative to butter. The name came from the Greek word for pearl, as margarine was whitish rather than yellow. It was based on animal fats such as suet. In Britain, an improved version, developed by the Dutch butter merchants Jan and Anton Jurgens was marketed as "butterine" until 1887, when that was forbidden. In 1903, the development of a process called hydrogenation by the French chemist Paul Sabatier made it possible to use vegetable oils as the main ingredient of margarine. By bubbling hydrogen through liquid oils in the presence of nickel, which acts as a catalyst, the oils are hardened. However, hydrogenation also had the effect of changing unsaturated oils to saturated fats, which are less easily metabolized. The introduction of healthier margarines that were high in polyunsaturated oils, such as sunflower or safflower oil, had the added benefit of producing a softer, more spreadable margarine. The convenience of spreading "straight from the fridge" was used as a major selling point for soft margarines. The more recent spreadable butters are actually a more or less equal mixture of butter and vegetable oils.

See also Breadmakers; Can Openers; Coffeemakers; Cookers; Freezers; General Foods Corporation; Refrigerators; Tea Makers; Toasters.

References:

Burnett, John. *Plenty and Want: A Social History of Diet in England from 1815 to the Present Day.* London and New York: Methuen, 1983.

Chant, Colin, ed. *Science, Technology and Everyday Life, 1870–1950.* London: Open University/Routledge, 1990.

Hillman, David, and David Gibbs. *Century Makers.* London: Seven Dials, 1999.

Origins of Everyday Things, The. London: The Reader's Digest Association, 1998.

Trager, James, ed. *The People's Chronology: A Year-by-Year Record of Human Events from Prehistory to the Present.* London: Heinemann, 1985.

www.affi.com/facts/decades/

www.birdseye.com

www.kraft.com/corporate/about/

www.nabisco.com

www.nestle.com/all_about/history/

Cookers

The character of cooking in the home underwent a dramatic transformation during the twentieth century, partly as a result of technological developments, but also as a result of social changes. In 1900, most households had coal-fired ranges with solid hotplates above small ovens and consumed relatively little preprocessed food. On the whole, processed foods were valued more for their longer shelf lives than for time savings in preparation and cooking. A hundred years later, most households had freestanding or built-in gas or electric cookers (stoves in American parlance) and consumed a wide range of processed foods.

The Evolution of the Kitchen Range

For centuries, cooking arrangements in Europe were based on the system developed by the Romans and diffused throughout Europe in the wake of the military conquests. At its simplest, this involved a raised brick hearth to hold an open fire, set within a wide chimney base. As smoke and hot air rose, they were drawn up the chimney. Different methods of cooking could be achieved by adding devices such as spits, supports for pots and pans, and brick-oven compartments. Cooking on an open fire was slow and inefficient because a lot of heat was absorbed by the chimney walls and by the air in the room. In the mid-eighteenth century, the American statesman and scientist Benjamin Franklin invented a ventilated cast-iron wood-burning stove, through which the hot combustion gases circulated before escaping.

This idea for concentrating the heat source and retaining heat was developed further by Benjamin Thompson, Count von Rumford, in the 1790s. Rumford was born in the United States, in Massachusetts, but his early career as a spy for the British led to his forced departure to Europe. During his employment by the elector of Bavaria in various senior ministerial roles, he developed the solid-fuel range for use in a variety of large-scale catering contexts, including workhouses, army canteens, and hospitals. Perhaps the most innovative feature of Rumford's ranges was the sunken chambers for pans in the range top. The pans were heated by the combustion gases rising up the surrounding flues. Although Rumford produced scaled-down versions of his basic range design, it was another American inventor, Philo Penfield Stewart, who developed the prototype of the nineteenth-century household range. Stewart patented his first range design in 1834 and later moved from Ohio to Troy, New York, where he established himself as a manufacturer.

Early Gas Cookers

In the late eighteenth century, the Reverend John Clayton provided the first account of cooking with gas when he described how eggs and beef had been boiled at a natural gas spring near Wigan, in northwest England. The first person to attempt gas cooking in-

The Beatrice gas cooker is typical of 1930s gas cookers. (The Museum of Science and Industry in Manchester)

side the home was the German gas pioneer Frederick Albert Winsor, who began experimenting at his home in Braunschweig in 1802. However, in the early decades of town gas production, gas was mainly piped to factories, public buildings, and street-lighting installations. In Britain, the first gas cookers were intended for commercial catering use. They bore little resemblance to modern gas cookers, consisting merely of an oven with burners at the bottom and a meat jack suspended from the top. Gas cookers were displayed at the Great Exhibition, held in London in 1851, by which time household gas provision was becoming more widespread. By 1869, gas cookers had evolved into a form closer to the modern gas cooker, with internal shelves and top burners, and were available for hire in Britain. During the late nineteenth century, gas became a relatively cheap fuel in Britain as the industry was by then largely municipally owned, whereas in the United States gas was a more expensive alternative to solid fuel. Therefore, in Britain, gas cookers became slowly but steadily more popular and, by 1900, were found in 2 million homes.

Early Electric Cookers

The first electric oven was installed in the Hotel Bernina, near St. Moritz in Switzerland, in 1889. Electricity was supplied by a hydroelectric power generator.

In Britain and the United States, electric cookers began to feature in public demonstrations and model electrical kitchen displays at major exhibitions in the early 1890s, including the 1891 Crystal Palace Exhibition in London and the 1893 Columbian Exposition in Chicago. The companies that pioneered the commercial production of electric cookers included Crompton & Company in Britain and the Carpenter Company in the United States. The heating elements in these early electric cookers took the form of resistor wires embedded in enameled panels. This heating technology was improved in 1893 by the English electrical engineer H. J. Dowsing, who sandwiched the steel heating wires between two panels, creating a safer and more practical design. Crompton & Company began to manufacture and market cookers to Dowsing's design in 1894. The heating panels were at first placed on the oven sides and later at the top and bottom. The performance of electric cookers benefited from the improvement in heating technology created by the invention of Nichrome (or nickel and chrome) wire by the American Albert L. Marsh in 1905. The boiling plates on the cooktop took the form of radiant coils on fireclay supports, topped by perforated or solid metal plates.

The main problem for electric cooker manufacturers was that there were few electrified homes to sell their products to. Moreover, even fewer homes had a power circuit as well as a lighting circuit. Electric cookers were, and still are, the electric appliances with the highest power rating and, as such, require a dedicated power supply and fuse box. The investment in wiring an electric cooker and the high costs of the heavy electricity consumption were a major disincentive at a time when electric cookers had nothing extra to offer in terms of functionality. Up until World War I, both gas and electric cookers were modeled on the rival solid fuel range. This meant box-shaped ovens with safe-like doors, made of cast iron with a black lead finish. Given the persistence of fears about the safety of gas and electricity, gas and electric cooker manufacturers may have felt that a familiar design would provide a sense of reassurance. Not surprisingly, with such limited sales potential for full-size cookers, manufacturers concentrated their marketing efforts on small, tabletop cooking appliances, such as electric frying pans and chafing dishes. These appliances had the advantage that they could be used in the dining room as well as the kitchen and had no non-electric rivals.

Competition between Gas and Electricity

By 1920, solid fuel ranges were out of general favor, except in rural areas where gas and electricity supplies were absent. They remained so thereafter, although the Aga stove, invented by the Swedish physicist Gustav Dalen in 1924 and marketed commercially from 1929, has sustained a small but devoted customer base. In Britain, the growing importance of gas as a cooking and heating fuel was confirmed by the 1920 Gas Regulation Act, which changed the basis for gas prices from illuminating value to calorific value. The situation was much the same in the United States, where consumption of gas for lighting fell from 75 percent in 1899 to 21 percent in 1919, when consumption as domestic fuel reached 54 percent. World War I had provided an opportunity to demonstrate the convenience of electric cookers, which were adopted for field canteens. In the intensifying competition between gas and electricity, the gas cooker manufacturers had the upper hand, in terms of both price and performance. In 1915, the American Stove Company of Cleveland, Ohio, had introduced the first thermostat for gas ovens, the Lorrain oven regulator. The British equivalent, the Regulo thermostat, was developed by Radiation Ltd. (John Wright & Company) in 1923 and fitted to the Davis Company's New World gas cooker, which also featured a slag wool lagging for better insulation and a base flue. Previously, oven controls, like boiling ring controls, had settings that simply expressed the rate of gas flow, with no reference to the temperature produced. Similarly, electric cookers were fitted with mercury current regulators, and this remained so until the early 1930s. A thermometer attached to the oven door showed the effect of the regulator setting. In Britain, the first automatic temperature controller for electric ovens was the Credastat regulator, introduced in 1931.

Gas boiling rings were also much more efficient than electric ones because the electric elements were slow to heat up, compared to the instant heat of gas. The flat electric plates only provided good heat transmission to pans with similarly flat bottoms that maximized surface contact. Electric boiling rings began to improve in the mid-1920s, when enamel-coated, metal-sheathed elements appeared. This design of boiling ring meant that the pan was in close contact with the heating source without an intervening plate. In the early 1930s, the U.S. company General Electric developed a new type of faster-heating radiant ring, the Calrod strip element, which consisted of resistance coils set in magnesium oxide and sheathed with chromium iron. Combined with bimetallic controls, akin to the automatic oven regulators, the new boiling rings were much more comparable in performance with gas burners.

One of the few inherent advantages of electric cookers at this time was variety of size. Plumbing in a gas outlet was more space-consuming and obtrusive than the electrical equivalent, so gas cookers were invariably full-size cookers. People living alone or families in houses or apartments with small kitchens constituted a ready market for smaller cookers. The British company Belling made particular efforts to exploit this market. In 1919, it introduced the Modernette cooker, a compact, lightweight floor-standing cooker, and in 1929 it launched the Baby Belling, a tabletop cooker.

In Britain, the price differential between gas and electric cookers was largely a result of the non-standardization of electricity supply. This meant that manufacturers needed to produce electric cookers specified to meet the range of voltages in use. The construction of the national grid from 1926 eventually removed this disadvantage. Moreover, in 1930, a group of British electric cooker manufacturers agreed to a common standard that reduced the number of options, thus consolidating production. The electricity utilities introduced cheap rental schemes to overcome the purchase disincentives. An indication of the success of these schemes is that rental of electric cookers was more common than buying until 1938. In the United States, with its standardized electricity supply, electric cookers were much cheaper, but the combined advantages of gas cookers gave them a dominant market position in both Britain and the United States. In Britain, about 75 percent of homes had gas cookers in 1939, compared with about 8 percent of homes that had electric cookers. However, as electric cookers accounted for about a quarter of total cooker production, the balance was shifting in favor of electric cookers. In the United States, gas was less dominant because the larger and more dispersed rural population created a continuing demand for solid fuel and oil stoves. By 1930, gas cookers were the most popular type and were found in 48 percent of homes, while electric cookers were found in just 6 percent of homes.

Cooker Design

While gas-cooker manufacturers tended to be more innovative in design terms than their electric-cooker counterparts during the interwar period, the time lag was much shorter. After 1920, gas and electric cookers gradually evolved their own identity through the use of new materials and surface finishes. Manufacturers began to apply vitreous enamel, which had previously been used sparsely on splashbacks and cooktops, to all surfaces, outside and inside. Although mottled black enamel was used in conjunction with white, mottled grey enamel and white enamel became more common, as a visible break from the traditional black-leaded range. In the 1930s, other colors, such as mottled blue and green, were also popular. Aside from its appearance, the great advantage of the enameled surface was that it was easily cleaned. By 1930, the typical gas or electric cooker stood on four short legs and consisted of an oven, surmounted by a grill compartment, and a cooktop with between two and four boiling rings.

Sheet steel, which was light and more flexible, was available in the 1920s, but was too expensive to be used extensively. The pioneer of the sheet steel cooker was the American designer Norman Bel Geddes, who produced the Oriole cooker design for the Standard Gas Corporation in 1932. Sheet steel was a logical choice for Bel Geddes who, as an advocate of streamlining, sought materials that could provide a seamless profile. The construction process entailed the clipping of bendable sheets to a steel chassis rather than the bolting of rigid plates to a cast-iron frame. The Oriole cooker in white porcelain-enameled steel was notable for its rounded edges, flush front with plinth, and folding splashboard cum tabletop. The plinth served the dual purpose of inhibiting the accumulation of dust and food debris under the cooker and providing storage space. The full-line cooker with a warming drawer below the oven became standard by the 1940s. In Britain, the first white steel gas cooker was the Parkinson Renown, designed for the 1935 George V Jubilee House and produced commercially from 1937. The use of sheet steel encouraged the standardization of core components, which could then be assembled in different combinations, and this standardization lowered production costs.

The Influence of the Fitted Kitchen

While feminists and household economists of the rational school had long espoused the concept of the fitted kitchen, few homes had fitted kitchens until after World War II. The United States was well ahead of Britain in this respect, and U.S. companies began to use the desirability of the fitted kitchen as a marketing vehicle for a range of appliances in the late 1940s. In terms of cookers, this brought an emphasis on ergonomic design and materials. Surplus wartime stocks meant that aluminum became an affordable lightweight option for some cooker parts. Features such as glass doors, introduced in the 1930s, and eye-level grills were heralded as aids to efficiency and economy of movement. The 1950s fitted kitchen also prompted the revival of the split-level cooker with a waist-level or eye-level oven. The term "split level" is used to signify that the cooking units are not integrated vertically, but dispersed horizontally. Split-level electric cookers first appeared in the early years of electric cookers and were initially the more common design in the United States. However, their double width meant that they were too large to fit comfortably in smaller kitchens. In the 1960s, in search of new selling points, manufacturers developed features that extended the potential for producing meals requiring different types of cooking. One option was the cooker with two ovens, allowing simultaneous cooking at different temperatures.

A display of two ranges at a Home Show in Philadelphia, 1951, showing the first Westinghouse electric range (left) from "Grandmother's Day" in 1914; and the "modern" electric range of 1951. (Stady Photo Service)

Oven fans helped to distribute the heat more evenly, facilitating the use of the whole oven, while attachments, such as rotisserie spits, tailored the oven for specialized cooking.

The standard cooktop held four fast radiant rings, and boiling ring technology changed little from the 1930s to 1966, when the ceramic electric hob, or cooktop, appeared. The ceramic hob was the commercial result of an accidental discovery at the Corning Glass Works in the United States in 1952. A malfunctioning furnace produced an opalescent, tough glass with distinctive thermal properties. Heat from bare electric elements placed beneath the glass, and demarcated by patterns on the upper surface, is conducted vertically, but not horizontally. Not only is the ceramic hob extremely efficient, but with its flat surface, it is easy to clean and available as a work surface when not in use for cooking. Manufacturers also gave much attention to the cleanability of ovens. One "self-cleaning" method, introduced in 1969, was the application of a grease-resistant coating to the oven interior. This is known as catalytic cleaning. Another method, introduced in 1978, was pyrolitic cleaning, whereby a short burst of maximum heat after cooking prevents the build-up of hardened grease. Cleanliness was also the motive for the development of the electric cooker hood, which is placed directly above the hob to absorb greasy vapors and cooking smells. Such hoods contain an extractor or exhaust fan and filters. Depending on the hood design, the extracted air may either be recirculated in the kitchen after filtration or vented outdoors.

Since the 1970s, when the fitted kitchen approached its peak of popularity, manufacturers have designed kitchen appliances to fit in with the standard sizes of kitchen units. This also prompted the evolution of the split-level cooker into the modular cooker, whereby the oven, hob, and grill might be completely separate, self-contained units. Modularity has allowed consumers to mix and match gas and electric cooking units to suit their individual needs or preferences. The German manufacturer Neff has been particularly noted for its modular cookers. The Italian company Zanussi has focused more on offering a range of colors, finishes, and style details. For example, the Zanussi ID cooker (1999) could be tailored in terms of types of doors, handles, and knobs, as well as color and finish, to achieve a customized specification. Another trend, associated with a revived interest in cooking as an art rather than a necessity, has created a consumer market for the cooker built to professional catering standards and usually high-tech in design. The latest development in hob technology is the induction hob, which dispenses with heating elements in favor of magnetic heat induction coils. While a British company, the Falkirk Iron Company, experimented with induction cooking in the 1920s, the idea lay dormant until the 1990s. Price, familiarity, and availability of types of energy were the prevailing influences on choice of cooker in 1900, but today the equivalent factors are more likely to be price, performance, convenience, and design. These factors mean that gas and electric cookers are likely to co-exist on more or less equal terms for the foreseeable future.

See also Aga Cookers; Belling Company; Convenience Foods; Electricity; Heating; Microwave Ovens.

References:

Byers, Anthony. *Centenary of Service: A History of Electricity in the Home.* London: Electricity Council, 1981.

Cowan, Ruth Schwartz. *More Work for Mother: The Ironies of Household Technology from the Open Hearth to the Microwave.* London: Free Association Books, 1989.

Giedion, Siegfried. *Mechanization Takes Command.* New York: Oxford University Press, 1948.

Hardyment, Christina. *From Mangle to Microwave: The Mechanisation of Household Work.* Cambridge, UK: Polity Press, 1988.

Lancaster, Maud. *Electrical Cooking, Heating, Cleaning, etc.: Being a Manual of Electricity in the Service of the Home.* London: Constable, 1914.

Cookware

During the nineteenth century the majority of cooking pots had been made of cast iron. A major United Kingdom company manufacturing cooking pots was Kenricks of Birmingham. Lighter wares were of sheet tin. Enameled cast iron was developed in the 1850s. It was easier to clean and more attractive, despite a limited color range. A mottled blue was one of the commonest colors. Germany, Austria, and France were all significant producers. The 1870s saw innovations in the metal industry. Pressed mild steel appeared as a result of the Bessemer and Siemens processes. Aluminum wares were produced in the United States, France, and the United Kingdom from the 1880s. The American West Bend Company of West Bend, Wisconsin, began production of aluminum cookware in 1911. Its main customer was the mail-order company Sears, Roebuck.

The new cooking methods of gas and later electricity had an effect on cookware. Enameled iron saucepans were unpredictable and milk pans boiled over due to the uneven conductivity of iron and enamel. American manufacturers did continue with cast iron, offering bright enameled colors on the outside. British manufacturers moved to steel and aluminum.

The cookware industries of Europe and America went into wartime production between 1914 and 1918 and emerged with improved technologies. The most popular British brands were Tower, Diamond Brand, Swan, and Goat. Steel cookware was often enameled in either green or beige with contrasting rims in a darker shade. In postwar Germany, the once mighty BMW company had to cease producing aircraft engines and move into pots and pans to survive.

Aluminum wares were well suited to electric stoves. In 1934 the Wear Ever Company produced a range with heat resistant plastic handles. Stainless steel, an alloy of steel, nickel, and chromium developed in the 1920s, offered better resistance to rust and acidic corrosion. West Bend introduced their Waterless Cooker in 1921. Based on the suggestion of one of its salesmen, the lid of the cooker was fitted with clamps that prevented the escape of steam during cooking, making the addition of water unnecessary. They sold well and led to the introduction of a range of waterless wares known as the Flavo-Seal line. The U.S. Revere Copper and Brass Company developed a range of stainless steel pans with copper plated bottoms. Marketed in 1939 as Revere Ware, they gave better heat distribution.

In addition to stainless steel materials, chrome plated steel and heatproof glass like Pyrex was becoming popular. Colored aluminum was also fashionable during the late 1940s and 1950s. Anodized aluminum wares have a satin gloss finish and were introduced in 1946 by the Aluminum Utensil Company. West Bend had introduced anodized wares with colored dyes by 1950; these materials improved the aesthetics of cookware during the 1950s when the look of a kitchen became more important to manufacturers and consumers. This trend continued throughout the late twentieth century.

The first nonstick pan was introduced in 1956 by a Frenchman, Marc Gregoire, and his wife, Colette. Gregoire had experimented with the low friction substance PTFE (polytetrafluoroethylene) to produce a smoother running fishing reel. His wife suggested that it would have more commercial success if applied to cookware. The resulting Tefal company became a brand leader as nonstick Teflon-coated pans became popular during the 1960s and 1970s.

Traditional manufacturers also received a boost during the 1970s as a result of public interest in antiques and the positive reassessment of much Victorian taste. A refurbished range or new Aga in a suburban home required the "right" pans. The French firm of Le Creuset benefited enormously. Founded in 1925 at Fresnoy-Le Grand at St. Quentin, its heavy, hand-finished enameled cast-iron pots and pans were just right.

Cultural and social trends also influenced cookware. The success of Asian and Eastern restaurants during the 1960s and 1970s led to a rising interest in cooking such dishes at home. In the United Kingdom, the Habitat stores led the way, selling woks, rice steamers, and chicken bricks. Another trend was toward professional cookware as the Western media began to promote "lifestyle" eating and drinking in the 1980s. For those inspired by celebrity chefs, popular choices were the Elysee Line of 1981 by Cuisinox, a French company established in the 1930s, or Calphalon of 1978, a range of commercial, hard anodized aluminum wares made in Ohio. Another serious choice was Le Pentole designed in 1979 for Industrie Casalinghi Mori in Italy by Nika Sala. This is a stylish modern reworking of a stacking steamer with five pans.

A more recent innovation has been the specialized pan designed for use away from the stove burner (hob). The Tefal Le Saucier is a nonstick saucepan with an integral mixing paddle powered by an electric spindle through its base. It sits on an individual hotplate with electronic controls for heating, timing, and stirring. Equally French is the Tibos electric crepe maker, with a nonstick griddle and spreading device.

Cookware has largely retained its traditional forms throughout the twentieth century; the main advances have been in the better performing and lighter materials used and the aesthetic choices available. Most cookware performs well, dependent on price, and both manufacturers and consumers are equally influenced by popular fashions and tastes.

See also Pyrex; Pressure Cookers; Tefal; Teflon; West Bend

References:

Brears, Peter. *The Kitchen Catalogue.* York, UK: York Castle Museum, 1979.

Feild, Rachael. *Irons in the Fire: A History of Cooking Equipment.* Marlborough, UK: Crowood Press, 1984.

Katz, Sylvia, and Jeremy Myerson. *Kitchenware.* London: Conran Octopus, 1990.

Crate&Barrel

The Chicago couple Gordon and Carole Segal established Crate&Barrel in 1962. Like Terence Conran in the United Kingdom, they realized that others would enjoy the design and quality of kitchen and home products that they had found on journeys within the United States and abroad. The idea came to them while doing the washing up.

They renovated a 1,700-square-foot former elevator factory in Chicago's Old Town district. The decor was, by necessity, very cheap; the walls were lined with crating timber and the products displayed in packing crates and barrels. The first store employed three people and offered gourmet cookware and other contemporary housewares in greater variety and at better prices than elsewhere in Chicago.

The first Crate&Barrel mail-order catalogue was produced in 1967 and the first store outside Chicago opened in 1977. Crate&Barrel opened in San Francisco in 1985 and in New York in 1995. In 1998, it entered a partnership with the world's largest mail-order company, Otto Versand of Hamburg, Germany.

Crate&Barrel has grown into one of the most influential retailers in the United States, with over eighty stores. Its flagship store opened on North Michigan Avenue, Chicago, in 1990. Although the store has been described as resembling a giant food processor, its main forms are a cube with a cylindrical attachment, literally a crate and barrel.

Like Habitat, Crate&Barrel helped create "lifestyle" shopping and influenced both consumers and other retailers. Apart from furniture and linens it sells a wide range of stylish appliances from manufacturers such as KitchenAid and Dualit. The company has a strong philanthropic policy and passes unsold goods on to local charities. It also has financially supported AIDS-related causes.

See also Habitat

References:

www.crateandbarrel.com

Crocker, Betty (U.S.)

In 1921, the Washburn Crosby Company, a forerunner of General Mills, introduced the fictitious "Betty Crocker" as a signature for the advice and information produced by its Home Service Department. This idealized

American housewife was the result of the thousands of baking and cooking inquiries that the company received after organizing a competition. "Betty" sounded friendly and homely, while "Crocker" was the surname of a recently retired company executive.

By 1940, Betty Crocker had become a household name, so it was not surprising that General Mills mechanical engineering division used the name when planning to diversify into domestic electrical goods in 1945. The Betty Crocker cake mixes followed in 1947. *Betty Crocker's Picture Cook Book* was first published in 1950. Written for the growing number of suburban kitchens, it was the first cookbook to have photographs and became a best-seller.

General Mills produced the Tru-Heat electric iron, a toaster, a food mixer, an automatic fry-cooker, a waffle iron, a coffeemaker, and a steam ironer under the name Betty Crocker. Ironically this move to peacetime production was limited by the demands of the Korean War, and General Mills sold the business to McGraw Electric Company in 1954.

The Betty Crocker food brand remains strong, and the name is still used for recipe books and nutritional information. As for the fictional Betty, a series of models and actresses has ensured that her clothes and image keep up to date.

See also McGraw Electric

Curling Irons

See Hair Stylers

D

Deep Fat Fryers

The traditional deep frying pan with a wire basket to hold the food has been in use throughout the twentieth century. Deep fat fryers have different associations in Britain and America. The British saw the stovetop deep fryer as the "chip-pan" used mainly for cooking chips (French fries). Americans tended to associate them more with deep-fried chicken.

The main disadvantage with these models was that left unattended an overheated pan full of fat could cause a dangerous fire. Electric fryers were introduced into the U.S. market in the late 1940s. By the early 1950s rectangular models included Dormeyer's Fri-well and Dulane's Fryryte. Sunbeam produced a circular model that also doubled as a roaster and a casserole. Most were thermostatically controlled so that the oil remained at a constant temperature. Cheaper models did not have this feature and relied on the experience of the cook or a fat thermometer. They could also be used without the frying baskets as electric casseroles and soup cookers.

Like other appliances the fundamentals of the earlier electric models remain but plastics replaced steel or aluminum as the outer casings in the 1970s and additional refinements and safety features have been added. Models in the 1990s featured locking lids, vertical oil drainage, "coolwalls" similar to toasters, replaceable or washable filters in the lids to absorb grease and odors, and controls to raise and lower the baskets without lifting the lids. The main manufacturers are De Longhi, Tefal, Moulinex, and Morphy Richards.

These appliances continue to compete with the traditional fryer, and both now have to meet the challenge of the even-more-convenient ready-cut frozen "oven" or "microwave" fries, which can simply be heated up.

Defrosters

Self-defrosting freezers and refrigerators appeared in the 1960s. Prior to that the excess buildup of ice had to be removed. Most people simply switched off the appliance and left a pan of hot water in the freezer compartment. Once melting was under way, the ice could be scraped off by hand.

The electrical appliance manufacturers spotted the opening for an appliance to assist this process. During the 1950s U.S. companies produced a line of electric defrosters that could be plugged in and then placed in the freezer. Most were small electric heaters encased in aluminum bodies with wooden handles, such as those manufactured by Howell & Company and the Shane Manufacturing Company. Some models had metal and plastic casings. The Osrow Products Company of New York produced an infrared version in the early 1960s.

See also Refrigerators

Design Council

Founded in 1944 as the British Council of Industrial Design, the government-funded Design Council is responsible for encouraging and promoting high standards of industrial design. It was not the first British organization to undertake this mission. In 1914, the Arts and Crafts Exhibition Society, which promoted handcrafted products, turned its attention to industrial design and set up an offshoot, the Design and Industries Association. The Arts and Crafts influence was reflected in its motto, "Fitness for purpose." The government first took an interest in industrial design in 1931, when the Board of Trade first proposed to set up the Council for Art and Industry, whose main purpose from its launch in 1934 was to mount exhibitions showing examples of good design. Ironically, it was during World War II, a period of material shortages and production constraints, that the government began to take a more active role in shaping the design environment. In 1942, the Ministry of Information recruited a number of leading British designers, including Herbert Read, Milner Gray, and Misha Black, to head a new Design Research Unit. The goal was to create an advisory and consultancy network to educate manufacturers to embrace the notion of "total design."

The formation of the British Council of Industrial Design and its Scottish Committee was announced in December 1944 by the president of the Board of Trade. The stated objective of the council was "to promote by all practicable means the improvement of design in the products of British industry." Its motto, "Good design, good business," reflected the government's belief that design was a way of boosting product sales. The council's staff grew from 10 in 1945 to over 1,000 in 1946. With the backing of the Board of Trade, the council mounted its first exhibition, "Britain Can Make It," at the Victoria and Albert Museum in 1946. Featuring 5,000 exhibits produced by 1,300 companies, the exhibition was visited by 1.4 million people. The products were carefully selected to convey the council's notions of good taste in design to both consumers and manufacturers.

This 1968 Murphy television model was shown in the Design Centre.(The Museum of Science and Industry in Manchester)

Gordon Russell, an influential British furniture designer whose career spanned the Arts and Crafts Movement and Modernism, took over as director in 1947. The council began to promote its ideas through publications (such as *Design* magazine, which was introduced in 1949), film strips, and displays. In 1951, the council undertook a national survey of British design, which generated the Design Index, a stock list of approved products, and a pictorial reference library. The most high-profile venture of the council in its early years was its contribution to the 1951 Festival of Britain. The festival was the brainchild of the Royal Society of Arts as a means of celebrating the centenary of the Great Exhibition. The role of the Council of Industrial Design was to choose 10,000 products of British manufacture for exhibition. Eight and a half million visitors attended the festival, which had its main site on the South Bank of the Thames in London.

In spite of the success of the Festival of Britain, the council felt that it was not getting its message across to the public. The pattern of sales of consumer products suggested that the majority of the public were attracted by the "cheap and cheerful" rather than the quiet good taste that the council espoused. The council felt that it needed a permanent showcase for good design. The result was the Design Centre for British Industries, a "shopping guide to well-designed British goods," which opened in London in 1956. The Scottish Design Centre opened in Glasgow in 1957, which was also the year that the council introduced an annual design award. The Design Centre Awards, later renamed the Design Council Awards and then the British Design Awards, were originally restricted to consumer goods. They were later extended to capital goods, such as engineering products and car components, in 1974, to medical products in 1975, and computer software in 1986. In order to bring good design to the attention of a wider audience, the council introduced a black and white triangular label as a symbol to identify goods that met its design criteria. The label scheme was discontinued in 1988.

The council's name was shortened to the Design Council in 1972. The 1970s was a period of consolidation and regional expansion. New offices were opened in Cardiff in 1974 and Belfast in 1978. In the 1980s, the council began to adapt to a changing political and business environment. In 1983, it introduced the Funded Consultancy Scheme, whereby eligible companies could claim free design consultancy services. New facilities were opened at the London Design Centre, including an Innovation Centre in 1985, a Materials Information Centre in 1988, and a Young Designers Centre in 1989.

Further rethinks in the 1990s brought several changes of premises and a slimming down of the council's functions. In 1990, the Design Council Scotland moved to a different location in Glasgow. A year later, the Design Council Northern Ireland also relocated to new offices in Belfast, while a new regional English office was opened in Leeds. The last of these relocations saw the council move its London headquarters from Haymarket to Covent Garden in 1998. In 1993, the government's industry minister announced a comprehensive review of the Design Council in consultation with manufacturers, designers, and educators. The outcome of the review was that the council ceased to provide direct design consultancy services, transferring its consultancy database to the Chartered Society of Designers.

The end of the second millennium provided the opportunity for the council to relaunch itself. The Millennium Products initiative, managed by the Design Council, was launched in September 1997 by the prime minister. After three rounds of applications, a total of just over 1,000 Millennium Products were selected by the end of 1999. These design achievements were celebrated in the Spiral of Innovation artwork commissioned by the Design Council and installed in the Millennium Dome at Greenwich in London.

References:

Sparke, Penny. *Design in Context.* London: Bloomsbury Publishing, 1991.

Stewart, Richard. *Design and British Industry.* London: John Murray, 1987.

www.design-council.org.uk/dib/history/

Detergents

In the broadest sense, a detergent is a cleansing agent of any kind. The term has become commonly used in a narrower sense to describe particular chemical cleansing agents developed in the twentieth century. These detergents replaced soap and washing soda as the preferred agent for washing clothes.

Soap is made by mixing animal or vegetable fats with sodium hydroxide (caustic soda) or potassium hydroxide (caustic potash). This causes the fatty acids to form sodium salts. In its normal bar (tablet) form, soap is not ideal for washing large loads of clothes. For centuries, the usual practice was

Examples of the kind of general and specialized detergents available in the 1930s (The Museum of Science and Industry in Manchester)

to presoak clothes in water to which lye, an alkali derived from wood or plant ash, or urine had been added. This had a mild bleaching effect and loosened dirt, which could then be removed by rubbing with soap. In 1791, a French chemist, Nicholas Leblanc, developed a process for making washing soda (sodium carbonate), which performed the same function as lye or urine. In Britain, most people used soap very sparingly until 1853, when the government repealed a soap tax imposed in 1712. The first powdered soap, Babbitt's Best Soap, went on sale in New York in 1843. The first British soap powder was Hudson's Soap Extract, introduced in 1863. By 1900, soap was also available in flake form. In 1918, the British company Lever Brothers (now Unilever) introduced Rinso, the world's first granulated laundry soap.

The German company Henkel took the first step toward modern detergent technology with its Persil washing powder, introduced in 1907. The name was derived from two of its constituents, perborate and silicate. Perborate releases oxygen from the water molecules so that it becomes available to act on stains. Persil was an improved soap powder, rather than a modern detergent, but its self-activation process was the key to the successful laundry detergents that followed. The U.S. company Procter & Gamble developed an equivalent product, Oxydol. After electric washing machines became available in 1907, the disadvantages of soap-based laundry products became more evident. When soap is used in water containing magnesium and calcium (hard water), insoluble salts are created, which form scum on the surface. The first nonsoap detergents were developed in Germany in the late nineteenth century, with coal tar as the base and sulfuric acid as the reagent. Nekal, introduced in Germany in 1907, was the first detergent to be marketed commercially.

However, although early detergents eradicated the scum problem, they were not as effective at washing clothes as the improved soap powders. In 1933, Procter & Gamble marketed its first detergent, Dreft, as a cleanser for washing dishes in hard water. It launched the first synthetic laundry detergent, Tide, in 1946 as the "washing miracle." Tide became the leading American laundry detergent in the 1950s. The British equivalent of Tide was Surf, introduced by Unilever in 1949. These new detergents contained optical brighteners, which enhanced the ap-

Child holds a box of Tide detergent at the "Food Fair" in Fresh Meadows, New York, 1956. (Bettmann/Corbis)

pearance of white clothes. Previously, a similar effect had been achieved by adding a blue powder to rinsing water. From the 1950s, powder detergents were based on alkyl benzene, a light, clear oil derived from coal tar or petroleum. In 1968, the first biological detergents appeared. These contain enzymes, which are natural catalysts with specific properties. The digestive enzymes used in detergents break down biological stains: lipase acts on fats, protease on proteins, and amylase on starches. Another advantage of enzymes is that they are effective at low temperatures. The increasing popularity of front-loading automatic washing machines created a need for low-lather detergents. Silicone is one of the ingredients that can be added to reduce foaming. In the mid-1980s, liquid detergents were developed.

While advances in detergent technology improved their washing performance, it became apparent that a number of the key ingredients had damaging environmental consequences. Phosphates and surfactants, the surface-active agents that improve wetting (the penetration of water into fabrics), evade the biological processes used to purify sewage. Phosphates, which may also be found in fertilizers, enrich rivers and lakes, leading to a buildup of algae and bacteria that use up oxygen needed to sustain fish. Surfactants result in the release of effluent foam to rivers and lakes. In the 1980s, concerns about these adverse effects led to the introduction of biodegradable detergents that were free of bleaches and phosphates. However, it seems that the substitute ingredients, zeolites and polycarboxylic acids, may not themselves be neutral, as a new kind of unpleasant foam has appeared since their use.

See also Clothes Dryers; Dishwashers; Washing Machines

References:

www.proctorgamble.com

Whirlpool's "Energy Star" dishwasher, April 2000 (Courtesy, Whirlpool)

Dishwashers

In 1900, the domestic dishwasher did not exist in either mechanical or electric form. Nineteenth-century patents for dishwashers were conceived in terms of commercial or institutional use. The most influential of these has been L. A. Alexander's U.S. patent of 1865. Alexander's dishwasher consisted of a tub with a rotor at the bottom attached by a crankshaft to a handle in the lid. A rack with tangential slots for dishes was placed above the rotor. As the user turned the handle, the rotor threw water outward against the dishes. While Alexander's dishwasher was conceived for commercial catering use, a similar but smaller-scale dishwasher was patented by American housewife Josephine Cockran in 1886.

The first commercial hand-cranked dishwasher intended for home use was shown by the Walker Company of Syracuse, New York, at the New York State Fair in 1910. Walker produced an electric version in 1918, with a small electric motor at the base to drive the agitator. Hand-cranked models continued to be sold in the 1920s. A British example was the Polliwashup machine, effusively advertised as the "greatest household labour-saver of all Time." General Electric purchased the Walker Company in 1930 and began to remodel the dishwasher. It brought out the first square-tub model in 1932. The first electric dishwasher sold in Britain, in 1937, was a U.S. product, a Thor model made by the Hurley Machine Company.

In the late 1930s and again ten years later, a combined washing machine and dishwasher appeared on the market in the United States. The two functions were served by having a long agitator for clothes washing that was interchangeable with a short agitator plus dish rack. This model was a top loader but from the late 1940s, as with automatic washing machines, manufacturers began to favor the front-loading design. The price of dishwashers in Britain fell when Hoover began to manufacture them there. The introduction of plastic sink-top models by companies such as Electrolux was another way of making dishwashers more affordable and at the same time addressing the problem of space constraints in the typical British kitchen.

The performance of dishwashers benefited from improvements in soap and detergent technology in the 1950s. By the mid-1960s, special dishwashing powders and rinse aids were available. Considering the very limited commercial success of dishwashers even in the United States, it is surprising that Kelvinator demonstrated a high-tech concept of dishwashing in Seattle in 1962. This was the water-free and detergent-free ultrasonic dishwasher. The concept never got beyond the prototype stage, so its reliability is untested.

Dishwashers were still a luxury item in the United States until the late 1960s when annual sales reached 2 million units. In Britain, only 2 percent of homes had dishwashers by 1973. This slow pattern of growth has baffled those historians of domestic technology who have argued that dishwashing, as a frequent and tedious chore, is an obvious candidate for automation. Recently, manufacturers have applied the same technical innovations to dishwashers as to washing machines and for the same reasons. The ability of the fuzzy logic chip to optimize water and detergent use allows manufacturers to promote dishwashers as environmentally friendly, setting aside the issue of electricity consumption.

See also Washing Machines; Waste Disposal Units

References:

Byers, Anthony. *Centenary of Service: A History of Electricity in the Home.* London: Electricity Council, 1981.

Giedion, Siegfried. *Mechanization Takes Command.* New York: Oxford University Press, 1948.

Do-It-Yourself

The term "do-it-yourself" or DIY now refers to one of the most popular, if enforced, leisure occupations within the Western world. It is supported by an equally large manufacturing and retail sector. A largely post-1945 phenomenon it covers a huge range of practical household tasks that are now carried out by the resident of the home rather than a specialized professional. This is especially so in countries like the United Kingdom, where private property ownership is high, and primarily appeals to people on low or middle incomes, the wealthy being able to afford interior designers and professional craftspeople.

In the United Kingdom, 70 percent of families lived in rented accommodation before 1939. The repairs were the responsibility of the landlord and there was little incentive to redecorate or improve someone else's property. Also, labor was cheaper and culturally such manual work was not seen as the preserve of the middle-class office worker. Couples often worked together on their gardens and the "man of the house" may have hung the occasional picture, but to attempt the work of the professional painter, plumber, or electrician would have appeared eccentric.

Even if one had wanted to attempt such things the odds were stacked in favor of the professional. There were few detailed manuals and household guides were rudimentary in this area. To take decorating as an example, the wallpaper still came with a selvedge and had to be trimmed, either by hand or with a machine fitted with circular blades. The adhesive paste had to be mixed in a bucket.

World War II brought hardships and shortages and a "make-do-and-mend" attitude, which was in turn followed by a desire to improve the domestic environment. The two attitudes were not necessarily polarities; those who had to mend or make things often found that they enjoyed the creative or technical challenge and translated former tasks into hobbies.

The slowly rising affluence of the postwar period led to rising expectations and the need to replace bomb damaged houses with new towns and suburbs. Despite this, many young married couples still had to live with one set of their parents for a few years before they could find a place of their own, which was often rented. Buying a house was expensive and could require a 30–40 percent deposit. Local governments built many houses and usually decorated them in standard colors. The London County Council used green, cream, and battleship gray; Stevenage new town in Hertfordshire used white walls and gray paintwork. Some local governments let tenants decorate their own houses, others did not, and local government workmen appeared to redecorate two rooms every four or five years. Many people aspired to better brighter homes, inspired by events such as the 1951 Festival of Britain, but often the materials (wallpapers, for example) were scarce in the early 1950s. The influence of the "American dream home," deluxe modern interiors featured in magazines and Hollywood films, was also important.

By 1957, the industry was big enough to stage a Do-It-Yourself show at London's Olympia Exhibition Center. It ran for three weeks and attracted over a quarter of a million people. Magazines such as *Practical Householder, Handyman,* and *Do-It-Yourself* appeared in the late 1950s, and companies began to respond to this changing mood. They offered step-by-step guides to projects such as converting attics, boxing in stair rails, and removing old fireplaces. They recommended the right tools and helped popularize new materials such as Formica and Melamine. Black & Decker, for example, began to produce power tools aimed at the home market.

These materials were able to transform previously dull rooms, especially kitchens, where a laminate could be glued to old furniture and worktops. Not everyone approved; Richard Hoggart, writing in his influential *The Uses of Literacy,* glumly commentated on the changes in the working-class home of 1957. "Chain-store modernismus, all bad veneer and sprayed on varnish stain, is replacing the old mahogany; multicolored plastic and chrome biscuit barrels and bird cages have come in."

Despite the fears of some cultural critics, the market continued to grow. The rising affordability of television provided a mass audience for televised DIY programs such as the BBC show hosted by Barry Bucknell. In 1962, the BBC bought a derelict house in West London, which Bucknell "did up" over a period of months, removing old fireplaces and covering paneled doors and stairs. The project inspired many terraced house dwellers to transform their homes into brighter more modern spaces. Over 2,000 people came to view the house before it was sold. *The Daily Mail Ideal Home Show* began to introduce DIY displays.

By the 1960s the pioneering days of DIY were over and it had become an accepted part of modern life. The range of paints, tools, and equipment continued to expand as the stigma of doing it yourself fell away. Paint companies such as Crown and Dulux produced "fashion" colors aimed at the homeowner rather than the professional. Dulux is the paints division of ICI and was a pioneer in promoting DIY paint. It remains one of the major brands within the United Kingdom. Its success illustrates how the company aimed itself at the new DIY consumers. In 1961 it introduced an Old English sheepdog into its advertising, both on television and in magazines. Much of the advertising was directed at women and the dog achieved remarkable brand recognition, so much so that Old English sheepdogs are often referred to as "Dulux Dogs."

The lead content of paints was reduced and then removed. Nondrip paints were introduced in the 1970s, followed by one-coat gloss paints in the 1980s. The large DIY stores began to introduce their own lines of paints in the 1980s. Wallpapers became washable and ready pasted. DIY was seen as an important pastime and many households were proud that they had "done it themselves."

Culturally, DIY has tended to follow marriage or a long-term commitment. One suburban male interviewed in 1998 probably speaks for many:

> I never used to do any DIY before I was married. When I lived at home, I just used to play football every weekend and that was it. And quite honestly when we got married it was

just a question of you went out and, if you could find somewhere to live and you could afford it, then you bought it, and if there was anything to be done then you did it yourself. You couldn't afford to pay anyone to do it for you. I've just learned as I've gone along.

DIY was given a further boost by the economic uncertainties of the early 1970s, following global recession and the increase in oil prices in 1973. Many people could not afford to buy a new house and invested in improving what they had. Even so home ownership did rise in this decade, and a British survey carried out in 1979 showed that 51 percent of the male respondents claimed that DIY was one of their main leisure pursuits.

DIY continues to be a major pastime, covering home painting and decorating, electrical work, gardening, car maintenance, and small-scale building work. America has had a different experience of DIY due to its size and diversity. For many established eastern farmers and the homesteaders attracted to the prairies in the 1910s and 1920s, doing it yourself was a way of life. Many urban Americans continued to rent their apartments. The rise of suburbia in the United States led to similar trends as those in the United Kingsom, although professional decorators and repair companies have survived longer in the U.S. labor market. Nevertheless U.S. companies such as Black & Decker and a host of smaller companies produce tools, fittings, and accessories. Large supermarkets have DIY sections. In the United Kingdom, the Sainsbury group introduced Homebase stores in the early 1980s and the Woolworth's group their B&Q stores. The U.S. market has Black & Decker's own stores and also large suburban outlets such as Builders Square, Home Depot, Ace Hardware, and Sears.

The domestic area continues to be a center for much of modern life as computers and the Internet offer home-based services. There is a growing trend toward working at home in some sectors. As a result, the desire for attractive affordable surroundings remains high.

See also Black & Decker; Consumers; Drills, Electric; Ideal Home Show; Wallpaper Stripper

References:

Hoggart, Richard. *The Uses of Literacy.* London: Chatto & Windus, 1957.

MacDonald, Sally, and Julia Porter. *Putting on the Style: Setting up Home in the 1950s.* London: Geffrye Museum, 1990.

Meadows, Daniel. *Nattering in Paradise—Suburbia in the 1980s.* London: Simon & Schuster, 1988.

Ryan, Deborah S. *The Ideal Home through the Twentieth Century.* London: Hazar Publishing, 1997.

A. F. Dormeyer

A. F. Dormeyer developed the electric Household Beater in 1927. It was designed so that the motor could be easily detached from the bracket that held the beater blades, a forerunner of detachable beaters. It was originally manufactured by the MacLeod Manufacturing company, which later changed its name to the A. F. Dormeyer Manufacturing Company.

The company continued into the 1960s producing beaters, deep fat fryers, and coffee percolators, including the magnificently named Automatic Electric Hurri-Hot Electri-Cup.

Drills, Electric

The first electric hand drill is a good example of an appliance becoming truly popular almost sixty years after its invention. At the beginning of the century household tool kits could include either a hand drill or an Archimedean drill for making small holes. The first electric model was invented by the German Wilhelm Fein of Stuttgart in 1895. The American Black & Decker Company introduced its first model in 1915 followed by the first on/off trigger switch for a drill in 1917.

Housed in metal cases, these drills were used by workpeople in factories and workshops. They were not seen as a domestic product until the rise of the do-it-yourself (DIY) phenomenon of the postwar period. Black & Decker produced affordable electric drills for the DIY enthusiast, retailing at around seventeen dollars in the 1940s. This and other models had metal casings, and their styling appeared to be influenced by "ray-guns" seen in magazines like *Amazing Stories* and science fiction film serials such as *Buck Rogers* and *Flash Gordon*. Some companies were developing ergonomic designs. Atlas Copco commissioned the Swedish designer Rune Zernell for its LBB 33 model of 1955; he used ergonomic models and noise reduction chambers to make it slimmer and easier to use.

Bosch, the giant German engineering company, also began to make drills for the domestic market. Founded in Stuttgart in 1886, Bosch began producing industrial equipment and has diversified into appliances, power tools, and communications technology.

The rise of DIY and the confidence of homeowners to tackle bigger jobs as well as a more competitive market led to increasing product specifications and refinements. Variable speeds and hammer adjustments were introduced in the 1960s and 1970s, along with sanding attachments. The hammer action makes the drill bit vibrate backward and forward, pounding the material at the tip of the drill. This is especially useful for masonry, especially combined with two speed gearboxes that increased the torque (twisting force) of the drill at the lower speed. Increasingly models also featured a reverse action for use with a screwdriver attachment. Plastic replaced metal casings. Battery-operated cordless drills appeared on the DIY market in the 1980s. Convenient for light tasks they are not as powerful as drills that plug in to power outlets. That decade also saw the introduction of electro-pneumatic drills with a powerful hammer action and torque control to limit the twisting force at the chuck, ensuring that screws are not driven in too tight.

Today the electric drill is an essential part of the home tool kit and continues to be improved. The main manufacturers are Black & Decker, Bosch, Hitachi, and Makita.

See also Black & Decker; Do-It-Yourself; Rawlplugs

Reference:

Heskett, John. *Industrial Design*. New York: Oxford University Press, 1980.

Dyson Appliances

James Dyson, the founder of Dyson Appliances, is a British industrial designer and entrepreneur who studied at the Royal College of Art in London. His early product designs include the Ballbarrow, a wheelbarrow made more maneuverable by substituting a ball for the front wheel. He is best known for the Dual Cyclone line of bagless vacuum cleaners that he designs and manufactures.

Based on his observations that traditional vacuum cleaners become less effective as the pores of their dust bags become clogged with dust, he set about designing a new kind of vacuum cleaner. Dyson's dual cyclone system uses centrifugal force to separate the heavier dust particles from the air, thereby maintaining a clean airstream and full suction. In 1979, he made an industrial version of the cyclone cleaner for the Ballbarrow factory. It took five years of further development and more than 5,000 discarded prototypes to create the world's first bagless cyclonic vacuum cleaner, the G-Force, which Dyson patented in 1984. Another nine years were to pass before Dyson began to reap the commercial potential of his invention.

Dyson tried to interest existing vacuum cleaner manufacturers in his patent with little success. He managed to negotiate licenses with a Japanese company and an American one. Production went ahead in Japan, but the American company, Amway, pulled out. Dyson then discovered that Amway was producing a cyclone cleaner, purportedly of its own design. In 1987, he sued Amway for patent infringement, a case that took five years to be settled in Dyson's favor. During that period, he relied on the income from the Japanese license to meet his legal costs and pay the patent renewal fees. With the court case won, Dyson was able to set up Dyson Appliances at a factory in Wiltshire, in the south of England, to mass-produce his designs.

The Dual Cyclone line, more effective at removing finer dust particles than the G-Force, began with an upright model in 1993, soon followed by a cylinder (canister) version. All Dyson cleaners feature a transparent plastic dust container and cylinder models have a unique "stair hugging" shape. The classic Dyson colors are gray and yellow, but variations have been introduced to differentiate new models—for example, the use of purple casing for a high-efficiency filter. The de Stijl models are a stylistic tribute to the Dutch design movement of that name. Another Dyson first is the Recyclone, the first vacuum cleaner to be made by taking plastic waste from the manufacturing process and recycling it. At the end of the twentieth century, Dyson began working on the development of a robotic vacuum cleaner.

Dyson Appliances quickly achieved great commercial success in Britain, in spite of its products being at the higher end of the price range. In 1996, Dyson sold 400,000 vacuum cleaners, taking cumulative sales income above the £1 billion mark. The company now accounts for half of the British vacuum cleaner market by value. As exports only represent about 15 percent of sales, there is considerable foreign growth potential.

See also Vacuum Cleaners

E

Eastman Kodak

George Eastman, the founder of Eastman Kodak, is the person who almost single-handedly popularized photography, beginning in 1900 with his cheap "point and press" Brownie cameras. The commercial success of its cameras and photographic film made Kodak one of the most widely recognized brand names in the world.

Eastman Plates and Film

Eastman became a keen amateur photographer while he was employed as a junior clerk at the Rochester Savings Bank, in New York State. Through reading British photographic magazines, he learned about a new dry gelatin emulsion coating for photographic plates and began to work on his own formula. After experimenting for three years, Eastman patented his dry gelatin photographic plate in 1879 and an emulsion-coating machine for making them. In 1880, he leased premises and began to manufacture dry gelatin plates. A year later, he founded the Eastman Dry Plate Company in partnership with Henry A. Strong. By 1883, the company needed larger premises and moved to a four-story building. It was reconstituted as the Eastman Dry Plate and Film Company, a corporation with $200,000 stock held by fourteen shareholders, in 1884.

While dry gelatin plates were much easier to use than their wet collodion predecessors, Eastman was still not satisfied because glass plates were a heavy burden for the field photographer. His goal was to find a lighter and more flexible support than glass. In 1884, he brought out Eastman Negative Paper on rolls and a roll-holder attachment. However, the resulting images were inferior to those from glass plates, as the grain of the paper was visible. The next development was Eastman American Film, described as a transparent "stripping" film. Introduced in 1885, it was a paper strip coated with two layers of gelatin. The base layer of gelatin masked the paper grain and, after processing, the paper backing was stripped away.

Early Kodak Cameras

The Kodak brand name made its debut in 1888 to launch Eastman's first camera. The Kodak box camera, priced at $25, was marketed with the slogan "You push the button, we do the rest" and came preloaded with a 100-exposure roll of stripping film. The completed film was returned inside the camera to Eastman Kodak for processing and reloading at a cost of $10. In 1889, the company was reconstituted as the Eastman Company and introduced a celluloid roll film, the first commercial transparent roll film. The first daylight-loading film and camera followed in 1891.

As business flourished in 1891, the company opened new factories in Rochester and in Harrow, near London, where it had established a British office in 1885. It was renamed the Eastman Kodak Company in 1892 and opened a separate camera factory in Rochester in 1893. The Pocket Kodak camera of 1895 brought a new feature, the exposure window, while the 1898 folding pocket Kodak camera established the 2.25-in-by-3.25-in negative as standard. Eastman Kodak began the twentieth century by creating a mass market for snapshot photography with the launch of the No. 1 Brownie box camera. Designed by Frank Brownell and made from wood and cardboard, the No. 1 Brownie cost just $1, with 6-frame roll films priced at 15 cents each. In 1908, the company brought out the world's first commercial cellulose acetate film, described as safety film to distinguish it from highly flammable cellulose nitrate film. Eastman Kodak continued to expand its U.S. and overseas operations in the years prior to World War I. An Australian subsidiary was added to its British and French ones, and a research laboratory and new headquarters opened in Rochester.

Movie Photography, Color Films, and Precision Cameras

During World War I, Eastman Kodak turned to producing aerial cameras for military use. After the war, the company began to bring movie photography within the reach of the amateur. In 1923, it produced its first 16-mm Cine-Kodak camera and the Kodascope movie projector. Moreover, 16 mm movie-film was the format of the first Kodak color film, introduced in 1928. A line of 8 mm movie format Kodak film, cameras, and projectors came out in 1932. In 1931, Eastman Kodak gained a foothold in the precision camera market by acquiring the Nagel Camera Company of Stuttgart, Germany, which became Kodak A.G. and produced the first of its 35-mm precision Kodak retina cameras in 1933. George Eastman committed suicide in 1932, when the spinal problems that he had suffered for years became too painful to bear. Aside from his technical and business achievements, he made his mark as a progressive employer who fostered employee loyalty through social benefit, saving-and-loan, and profit-sharing schemes. In 1919, he gave a third of his stock holdings to employees. His philanthropy extended to the wider community through charitable donations, particularly in the areas of health and education.

The Kodachrome line of color film was launched in 1935 with a 16 mm movie format, followed by 35 mm transparency and 8 mm movie film in 1936. In 1938, the company introduced the Super Kodak Six-20 Camera, its first camera with a built-in photoelectric exposure control. During World War II, Eastman Kodak contributed to the war effort by introducing the Airgraph or V-Mail service for microfilming letters. It was also during the war, in 1942, that the company produced its first color negative film, under the Kodacolor brand. After the war, Eastman Kodak produced its first camera for the revived television industry, the Eastman Television Recording camera of 1947. During the 1950s, the company continued to build on and expand its successful camera, projector, and photographic film lines. In 1955, it began to sell color films without the cost of processing included, giving customers the choice, for the first time, of using other processing laboratories. In 1958, Eastman Kodak produced its first single-lens reflex camera, the German-made Kodak Retina Reflex Camera, and a fully automatic slide projector, the Kodak Cavalcade. The highly successful line of Kodak Carousel slide projectors first appeared in 1961.

Cartridge and Disc Formats

Eastman Kodak's most successful camera model of the 1950s was the Brownie Starmatic camera, introduced in 1957, 10 million of which were sold within five years. In

1962, the company's U.S. sales hit the $1 billion mark. A year later, Eastman Kodak's commitment to convenience photography yielded its first Kodak Instamatic camera, which used a new film format—cartridge film. As the film was completely enclosed in the plastic cartridge, there was no chance of accidental exposure to daylight. A simple cube flash attachment for Instamatic cameras was available from 1965. By 1970, 50 million Instamatic cameras had been sold. The cartridge format was also applied to movie cameras in the 1965 Super 8 movie line. Eastman Kodak continued to develop equipment and film for high-end specialist uses. For example, it supplied cameras, film, and other equipment for the U.S. space program. In 1972, a more compact version of the Instamatic camera, the pocket Instamatic, with its 110 film cartridge, was launched. It was an immediate success, generating sales of 25 million units in less than three years. By 1981, annual sales revenue had risen to $10 billion. Eastman Kodak introduced yet another convenient and compact film format, the film disc, in 1982.

In 1988, the Fun Saver line of single-use, disposable cameras was introduced. The first camera in the line was the Fun Saver 35 camera, followed in 1989 by the Fun Saver Stretch 35 panoramic and Weekend 35 all-weather cameras. Like the early Kodak cameras, the Fun Saver cameras were returned intact to the company for processing after use. Eastman Kodak began recycling disposable cameras in 1990 and had recycled its 50-millionth disposable camera by mid-1995. Recycling recovers about 86 percent of materials by weight either for direct reuse or reprocessing. The company was rewarded for its commitment to recycling by being awarded of the World Environment Center's Gold Medal for International Corporate Achievement in 1998.

The Video and Digital Age

The company entered the video and digital markets in 1984, when it began to manufacture floppy disks and videocassettes in 8 mm, Beta, and VHS formats. In the same year the company underwent major restructuring. Two years later, it developed the first mega-pixel electronic image sensor. The 1.4 million pixel capacity of the sensor was equivalent to the content of a 5-by-7 in photographic image. Other new products included a 14-in optical disc, an image printer, Supralife batteries, and Ultralife lithium batteries, the first 9-volt lithium batteries for consumer use.

The Kodak PhotoCD system, launched in 1990, was based on the Kodak Color Management System, proposed as a worldwide digital color standard and later licensed to other companies. The company began to produce writable CDs (compact discs) in 1992 and brought out a portable CD player in 1993. The growing market for writable compact discs led Eastman Kodak to open a dedicated production plant in Ireland in 1997. The Kodak Professional Digital Camera System, based on Kodak's mega-pixel sensors, was introduced in 1991. The first camera was a Nikon camera fitted with a Kodak sensor, but Eastman Kodak subsequently worked on the digital range in partnership with Chinon Industries, in which it later acquired a majority stake, and Canon. The Kodak Professional DCS 520 Digital Camera of 1998, developed with Canon, was designed specifically to meet the needs of professional photographers. As the price of digital cameras fell, the company developed consumer models, such as the point-and-shoot Kodak Digital Science DC120 Zoom model of 1997 and the pocket-sized Kodak Digital Science DC200 model of 1998.

APS—Convenience, Choice, and Quality

In 1992, Eastman Kodak began a joint research and development project in partnership with the Japanese companies Canon, Fuji, Minolta, and Nikon to develop the Advanced Photographic System (APS). APS offers consumer convenience—including a choice of three print shapes—allied with high technology. The Information Exchange (IX) film technology means that the film, held in a compact 24 mm cartridge, has a magnetic coating that stores information about date and time, exposure length, and size settings. Eastman Kodak holds 60 percent of the APS patents, and the company launched its Advantix line of APS cameras and film in 1996.

Kodak Today

Today, Eastman Kodak is a world leader in both consumer and specialist photographic products. It has won eight Oscars for its technical contributions to the movie industry and is dominant in the field of medical laser imaging. The company's involvement in the development of plastics for photographic film, film containers, and camera bodies resulted in sideways expansion into the fields of synthetic fibers, coal-based industrial chemicals, and general plastics. In the 1990s, Eastman Kodak decided to concentrate on its core businesses and divested itself of peripheral interests by selling off the Eastman Chemical Company and its pharmaceutical businesses. In spite of growing competition from Japanese companies since the 1950s, Eastman Kodak has been able to maintain a healthy share of the camera and film markets in the United States and overseas. The one notable exception has been Japan. In 1995, the company petitioned the U.S. government to take action against anticompetitive trade practices in the Japanese photographic film market. Two years later, the U.S. government responded by filing the case with the World Trade Organization, as yet unresolved.

See also Camcorders; Cameras; Television

References:

Coe, Brian. *Cameras from Daguerreotypes to Instant Pictures.* London: Marshall Cavendish, 1978.

www.kodak.com

Ekco

Eric Kirkham Cole (1901–1965) began making wireless sets at home in the spare time his house-wiring and rechargeable battery business afforded him. He was helped by a young man, Stan Clements, and an interested customer, William Streatfield Verrells, who provided £50 of capital. They set up business with five employees above a sweetshop at 505 London Road, Southend, England, in 1926. The company's evocative name derives from that of Eric Cole. The business prospered and by 1930 they had taken possession of a large factory at Prittlewell, Southend, with a payroll of one thousand.

The company is perhaps best known for its innovative use of phenolic moldings for its radios. Its early products were high-quality radios in wooden cases. Their 1930–1931 models 312 and 313 were housed in phenolic cases made by AEG (Allgemeine Elektrizitats Gesellschaft) in Germany. Due to tightening import controls, Ekco went into plastics production themselves and developed a large plastics division within the company, paying fees and royalties to AEG. The results were the Consolette models of 1931 that featured art deco styling. The SH 25 and RS 3 models, designed by J. K. White, had distinctive "tree and river" moldings on the speaker grille. Both also featured a new tuning scale that was the first in the United Kingdom to have the station names on it.

The early 1930s were bad for Ekco, as they were paying royalties to AEG and also suffered a fire at the works. Sales fell, staff had to be laid off, and the company was in crisis. The company response was to invest in innovation. In 1932 it approached the architects Serge Chermayeff and Raymond McGrath to design some new models. Each was paid for six sketches. Those of the Russian born Chermayeff (1900–1996) were chosen for production, resulting in the AC 64 and AC 74. These were high-quality, attractively priced models that featured truly modern cabinets in black phenolic. They did not attempt to ape wooden cabinets and had simple but elegant curved lines. The speaker grille had a bayonet fitting and could be removed so that the speaker silk could be cleaned or replaced. Likewise, the tuning scale was also removable and could be replaced by a new one as more stations became live.

These models sold well in 1933 and helped Ekco to recover, and they returned to Chermayeff and McGrath to design new models for 1934. They also invited the modernist architect Wells Coates (1895–1958) to supply designs. Coates's solution was inspired, resulting in the round AD 65 of 1934, AC 76 and AC 36 of 1935, and the A 22 of 1945. These models exploited and celebrated the molding qualities of plastics and produced results that would have been impossible and uneconomic in wood. They were probably the first authentic expression of modern design in Britain applied to a domestic appliance. The circular speaker grille was in the center with the tuning scale in a 180-degree arc above it. The later A 22 featured an even more elegant 360-degree scale. Most were manufactured in brown and black Bakelite, but a number of more expensive models were in Onyx Green and Pearl Ivory. Overall, these models sold very well. In design terms its nearest rivals were the simple veneered cases designed by Gordon Russell Ltd. for the Murphy Radio Company and the plastic cabinets of the Ferranti Company.

Coates also designed the Ekco Thermovent electric heater and a television case with a lift-up top. Designed in 1934 for the transatlantic liner *Queen Mary*, the Thermovent was housed in an elegant Bakelite panel. It was placed on the domestic market in 1937.

Ekco continued to use Chermayeff and Coates as well as Misha Black (1910–1977). Black designed the 78 series of 1937–1938. Despite this success Ekco was aware that popular taste still liked wood veneered cabinets and even before the outbreak of World War II was beginning to produce more wooden than plastic cabinets. Postwar products were well made but rather dull. An exception was Coates's Princess-Handbag portable radio of 1948. It featured contrasting light and dark blue plastic panels and a transparent carrying strap. Aimed at housewives it was designed to look equally at home in the living room as the kitchen. The company was eventually merged with Pye of Cambridge.

See also The Modern Movement and Appliance Design; Plastics; Radio

References:

Bayley, Stephen. *In Good Shape: Style in Industrial Products, 1900–1960.* London: Design Council, 1979.

Going, Tom. "A Resounding Ekco." *Plastiquarian* 6 (1990).

Hogben, Carol. *The Wireless Show!* London: Victoria and Albert Museum, 1977.

Mossman, Susan, ed. *Early Plastics: Perspectives, 1850–1950.* London and Washington, DC: Science Museum/Leicester University Press, 1997.

Sparke, Penny. *A Century of Design: Design Pioneers of the Twentieth Century.* London: Mitchell Beazley, 1998.

Electric Fans

The electric fan was one of the first electric appliances. Schuyler Skaats Wheeler, chief engineer at the New York-based Crocker & Curtis Motor Company, patented the first electric fan in 1882. However, this was in the days of the large and unwieldy direct current (DC) motor. The electric fan for office or home use requires a compact and reliable electric motor. In 1888, the Croatian-born engineer Nikola Tesla was granted an American patent for a multiple-phase motor, which was the first commercially viable electric motor. Tesla sold the patent rights to the Westinghouse Company and went on to develop the first small-scale AC (alternating current) motor. Westinghouse applied the Tesla AC motor to a three-blade fan, which went on sale in 1889. This was a small desk or table model. The Westinghouse electric fan was improved by the incorporation of a speed regulator in 1891.

Early table fans were purely functional, with little effort expended on appearance. In 1911, Peter Behrens, the artistic director of AEG (Allgemeine Elektrizitats Gesellschaft), remodeled his 1908 electric fan design.

Behrens, a member of the Deutsche Werkbund movement, believed that design should be expressed through form, not through unnecessary ornament. The wavy radial bars of the guard of his electric fan were a stylistic interpretation of a functional element. This kind of modernistic styling of electrical appliances was rare before the 1920s.

In the 1920s, small electrical appliances were promoted as desirable gifts. To increase the appeal of the electric fan, makers created dual-purpose models, such as fans combined with fruit bowls or table lamps. The use of Bakelite instead of cast iron for stands and rubber instead of brass for fan blades reduced the weight and increased the convenience of the electric fan. Lighter colored mottled plastics were used for decorative effect and the fan stem was elongated to produce a more streamlined look. A taller stem turned the table fan into a floor-standing fan. Table fans with fabric bands instead of rigid blades, such as the Singer/Diehl Bandolero model made for the U.S. retail company Sears, Roebuck, were marketed as suitable for nursery use. The table fan has changed little since the 1930s. Cheaper models use modern lightweight plastics, while retro fans with classic metal stands and blades are more expensive. Battery-operated portable fans are intended for outdoor use in hot weather.

In 1953, the German engineer, Bruno Eck, invented a new type of fan, the cylindrical or tangential fan. It produced a narrower stream of air than the conventional oscillating fan, but this was an advantage when the fan was combined with a heating element to create the fan heater. Ceiling fans, typically with four blades, usually share the electricity supply to a central pendant light fitting, but as permanent fixtures, these had limited sales potential in temperate climates. However, one type of fan that is a permanent fixture has become widely used in homes. This is the exhaust fan, installed in kitchens and bathrooms to prevent problems associated with the condensation of water vapor. It is fitted into an opening, either in a window or outside wall, with the blades on the outer face so that air is drawn out of the room. Shutters behind the blades prevent heat loss when the fan is not operating. Early models included the compact Vent-Axia Silent Six, designed by J. C. Akester for De La Rue in 1935. Modern exhaust fans may have features such as variable speed and reversing action. Extractor fans may also be fitted to cooker (range) hoods (with added filters to absorb grease) for more efficient extraction of moist air and cooking smells direct from their source.

See also Hair Dryers; Heating; Pifco; Vacuum Cleaners; Westinghouse Electric

Electrical and Musical Industries (EMI)

The prominent British record company EMI began life in 1897 as the British subsidiary of Emile Berliner's Gramophone Company. The change of name to Electrical and Musical Industries came in 1931 as the result of a corporate merger. EMI is probably still best known as the record company that signed the Beatles in 1962, but its most significant contribution to recording history was the development of the stereophonic recording technique in 1931.

In 1888, Berliner, a German immigrant to the United States, invented the gramophone and founded the Gramophone Company to market his new invention. The London offices of the Gramophone Company Ltd. opened in 1897, and the first British recordings took place in 1898. Concurrently, another American inventor, Eldridge R. Johnson, filed a patent for an improved gramophone with the addition of a spring motor. The Johnson and Berliner companies merged in 1901 to form the Victor Talking Machine Company. Johnson's Consolidated Talking Machine Company had adopted the "His Master's Voice" trademark for its record label. The trademark name came from a painting by Francis Barraud, featuring a fox terrier called Nipper; the commercial rights to the image and name were purchased in

1899, and the distinctive trademark was retained by the Victor Company and its associates. In Britain, the HMV name was later used for the company's retail operations.

The British Gramophone Company soon embarked on its own overseas expansion. In 1898, it opened branches in Germany, France, and Italy, followed between 1900 and 1903 by a branch in Russia with a record pressing plant, a branch in India, and sales offices in Australia, Japan, and China. By 1906, 60 percent of the company's profits were generated outside Britain. In 1902, the company signed its first major recording artist, the Italian operatic tenor Enrico Caruso. A year later, 12-inch shellac discs, which became the standard format for 78-rpm discs, were introduced. The Australian operatic soprano Nellie Melba opened the company's new recording studios at Hayes, Middlesex, in 1907. By 1913, a third of British homes had gramophones, so business was thriving. World War I not only temporarily disrupted the record trade, but also brought the loss of the company's German and Russian branches.

The first HMV store opened on Oxford Street, London, in 1921. In 1925, the new electric recording process was introduced, and the Gramophone Company set up a new company, Electrola, in Germany. The year 1931 was the most eventful in the company's history. First, two years after the Radio Corporation of America acquired the Victor Talking Machine Company in 1929, the British Gramophone Company merged with the Columbia Graphophone Company to create Electrical and Musical Industries. Second, the company opened its world-famous recording studios at Abbey Road, London. Third, EMI gave the world's first demonstration of stereophonic recording. EMI's stereophonic technique was developed by the British sound engineer Alan D. Blumlein. While stereo recording was largely restricted to professional environments, such as radio broadcasting, for more than twenty years it was the Blumlein system that became the world stereophonic standard in 1958. Last but not least, EMI began experimenting with electronic television.

EMI engineers began their television research and development efforts by improving the resolution of the mechanical scanning technique to produce a 243-line picture. However, EMI soon concluded that electronic scanning held greater advantages. Led by Isaac Schoenberg, a Russian engineer who had emigrated to Britain in 1914, the EMI research team produced the Emitron television camera in 1933. It delivered a 405-line picture at a rate of 25 frames per second. In 1934, EMI formed a consortium with Marconi, which had already been supplying transmitters to EMI. In February 1937, following four months of trial broadcasts by Marconi-EMI and Baird Television, the British Broadcasting Corporation decided to adopt the Marconi-EMI 405-line system. Television sets were manufactured under the HMV brand name.

In 1940, EMI appointed A&R (artists and repertoire) managers for the first time in order to develop new talent. One of these new managers was George Martin, who later signed and produced the Beatles. EMI also extended its repertoire by using licensing agreements and acquisitions to capture American recording artists. In 1948, it signed licensing agreements with the Paramount Record Corporation and the Mercury Record Corporation. The acquisition of Capitol Records in 1955 netted major recording stars, including Frank Sinatra and Peggy Lee. EMI also expanded its overseas interests by setting up ventures in Mexico in 1957 and in Japan, in partnership with Toshiba, in 1961.

EMI's golden era coincided with the blossoming of British pop music in the early 1960s, represented by the signing of Cliff Richard in 1959 and the Beatles in 1962. In 1964, EMI artists dominated the British singles chart, with eight EMI records totaling forty-one weeks at number one. In 1966, the company marketed its first prerecorded cassettes. Buoyant record sales encouraged retail expansion, and by 1970 HMV had fifteen stores in London and southeast England. Further expansion on a nationwide basis to thirty-five stores in 1976 made the HMV chain one of the leading music retailers. In 1973, a new subsidiary, EMI Music Publishing, was formed and soon strengthened by several acquisitions in the music publishing field. Consolidation through acquisition and retail expansion was the business pattern for the next twenty years. Notable acquisitions included the Liberty/United Artists record company in 1979, Chrysalis Records in 1990, and the Virgin Music Group in 1992. In 1979, EMI merged with Thorn Electrical Industries.

The introduction of the compact disc format in 1982 provided an impetus to recorded music sales. HMV Music Retailing became a separate division in 1986 and opened stores in Australia, Japan, the United States, and Hong Kong in 1990. The HMV division diversified its retail line in 1995 through the acquisition of Dillons bookstores. By 1996, HMV had 300 retail outlets worldwide. The EMI Group emerged in 1996 from the reversal of the Thorn-EMI merger. EMI continued to reap the financial benefits of signing key recording artists. In 1996, EMI released the Spice Girls' first album and in 1997, the American country and western singer Garth Brooks became the biggest-selling male artist ever in the United States with the release of his latest album for EMI. In 1997, EMI acquired a half-share in the rights to the valuable 1959–1991 Tamla Motown catalog when it acquired 50 percent of the Jobete music companies. A further separation of business interests occurred in 1998, when EMI sold off HMV and Dillons as the HMV Media Group, which then acquired the Waterstones bookstore chain. HMV celebrated its independent status by opening a new flagship store on Oxford Street, London, in October 1998 that ranked as the world's largest record store.

See also RCA; Record Players; Tape Recorders; Television.

References:

Martland, Peter. *Since Records Began: EMI First One Hundred Years.* London: Batsford, 1997.

Pandit, S. A. *From Making to Music: The History of Thorn-EMI.* London: Hodder & Stoughton, 1996.

www.emigroup.com/world/

www.hmv.co.uk/hmv/Customer_Service/about_hmv.html

Electrical Association for Women (EAW)

The EAW was formed in 1924 to fulfill two main objectives: "to give the woman in the home useful knowledge of electrical apparatus" and "to provide a platform for the woman's point of view on all matters relating to electricity which may affect her private or public interests." It was conceived by Mrs. Matthews, a member of the Women's Engineering Society. She had tried unsuccessfully to interest the Institution of Electrical Engineers and the British Electrical Development Association in supporting her proposal. The meeting that launched the EAW was and held at the home of the society's president, Lady Parsons, in November 1924.

The first director of the EAW was Caroline Haslett, who had completed an engineering apprenticeship with a firm of boiler makers during World War I. As secretary of the Women's Engineering Society, she had been an early advocate of Mrs. Matthews's proposal, and she proved a committed and inspirational director who saw electricity both as a means of liberating women from "soul-destroying drudgery" in the home and a source of "interesting and well-paid jobs" for women. She epitomized the missionary fervor that characterized the EAW's approach. Other leading members of the EAW were also influential women, such as Mrs. Hammer, mayor of Hackney and the first woman to be elected as chairman of a municipal electricity committee.

The EAW carried out its educational mission through publications and courses tailored to the needs of different categories of

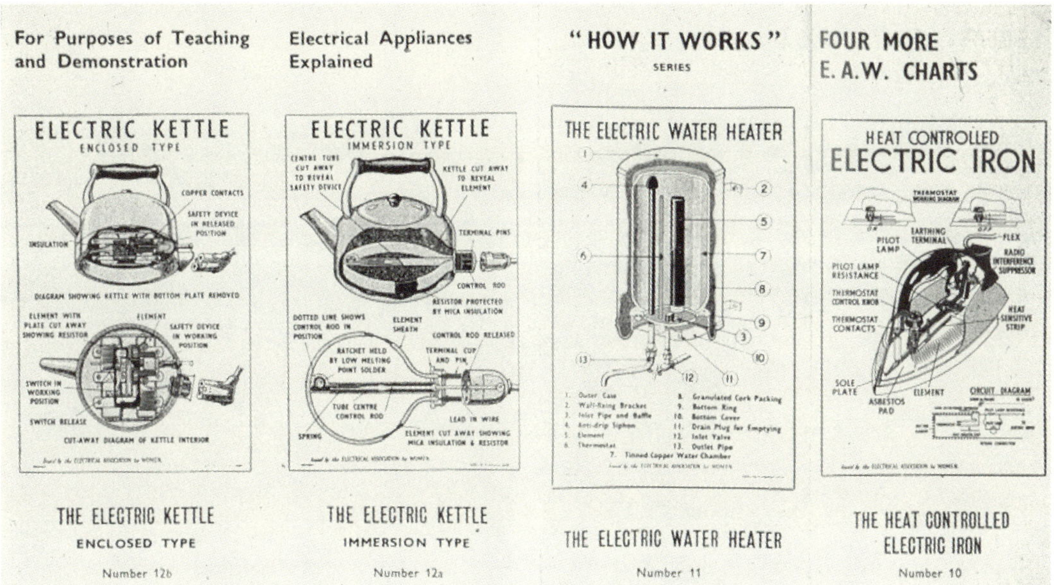

One of the EAW's instructive "How It Works" leaflets (The Museum of Science and Industry in Manchester)

women. It ran a course for appliance demonstrators leading to a qualification in electrical housecraft (homemaking). Other courses were directed at schoolgirls, domestic science teachers, and housewives. Well-illustrated leaflets and handbooks demystified electricity by explaining "how it works," in the case of appliances, and "how to do" tasks such as changing a plug. From 1926, the EAW produced a quarterly magazine, *The Electrical Age for Women,* which contained wide-ranging articles on electrical matters and reports from the regional branches. The network of regional branches began with Glasgow, Birmingham, and Manchester, grew to forty-seven branches in 1934, and more than eighty by 1945.

The EAW adopted a very practical and constructive approach to its advocacy and lobbying role. It conducted surveys of its members on the ergonomics of electric socket placement. Another survey resulted in its report "The Design and Performance of Domestic Electrical Appliances," which was submitted to manufacturers and engineers. Annual conferences featured special sessions on subjects such as kitchen planning. It also sought to raise awareness among women of different backgrounds through exhibition showpieces, including the All-Electric Flat at the 1930 Bachelor Girls Exhibition in London and its All-Electric House in Bristol, which opened in 1935.

By 1986, when the EAW went out of existence, declining membership was just one of the factors in its demise. Electricity and electrical appliances had become familiar fixtures of everyday life, making the EAW more or less redundant, although recognized as a pioneering and effective consumer watchdog.

References:

Randell, Wilfred. *Electricity and Woman.* London: Electrical Association for Women, 1945.

Scott, Peggy. *An Electrical Adventure.* London: Electrical Association for Women, 1934.

Worden, Suzanne. "Powerful Women: Electricity in the Home, 1919–40." In *A View from the Interior: Feminism, Women, and Design,* edited by Judy Attfield and P. Kirkham. London: The Women's Press, 1989.

Electricity

Electricity utility companies came into existence in the last quarter of the nineteenth

"The electric light in houses—laying the electrical tubes." Engraving in Harper's Weekly, *June 24, 1882 (Library of Congress)*

century. Their role was to generate electricity and supply it to both public and private users. Generating plants were located in or near to the urban areas in Europe, Canada, and the United States where potential demand was greatest. At first, electricity was mainly used for public lighting, particularly of streets and theaters, and then, as efficient electric motors became available, for operating industrial machinery. However, the high cost of wiring, appliances, and electric current inhibited the use of electricity in the household until the 1920s and 1930s, when electricity utilities began to target the household market.

Electricity has had a more profound impact on the household than any other form of energy because of its versatility. Fuels such as coal, oil, and gas produce heat and light, but this energy can only be converted to motive power by driving an engine. The engine can then supply power to machines by turning drive shafts or belts. Electricity provides an intermediary form of energy transfer that can be more easily scaled to a range of applications. In a power station, the mechanical energy of the engine is converted to electric energy in a generator.

The Basic Principles and Properties of Electricity

Early scientists and philosophers observed the existence of electricity in its static form without being able to explain it. A sixteenth-century English scientist, William Gilbert, coined the term "electric." It is derived from the Greek word for amber (elektron), because amber was one of the first substances observed to generate an electrostatic charge when rubbed. In the early eighteenth century, scientists recognized that the electrostatic charge could be positive or negative, depending on the properties of the substance. By 1900, scientists and engineers had developed devices for storing and generating electricity. However, it was only when scientists were able to explain the structure of the atom in the twentieth century that the nature of electricity was fully explained.

Friction is one of three methods of generating an electric charge; the other two methods are induction and chemical change. Electromagnetic induction is the principle on which electric generators, motors, and transformers are based. Chemical change is the basis for the storage of electric charge in cells and batteries. Each method relies on the same fundamental property—the availability of free electrons in the outer shell of an atom. The atoms of all elements consist of a dense nucleus that is positively charged, surrounded by a number of negatively charged electrons that orbit the nucleus. The outermost electrons of the atoms of some elements may be transferred to the atoms of other elements, causing the recipient substance to become negatively charged and the donor substance to become positively charged. The availability of free electrons is one of the characteristics of metals, making them good conductors of electricity.

In 1745, Ewald Georg von Kleist discovered that an electrostatic charge could be stored in a water-filled glass bottle with conductive metal sheet around the inside and outside of the bottle. This type of capacitor became known as a Leyden jar. In the 1780s, the Italian physiologist Luigi Galvani concluded that the contractions that he observed when he touched the legs of a dead frog with a scalpel were the result of the release of a charge induced by static electricity in the air. Galvani's compatriot Alessandro Volta subsequently supplied the real explanation for the phenomenon, which was a chemical reaction between the frog's body fluids (the electrolyte) and two conductors (the electrodes). Volta, a professor of physics, had previously experimented with the accumulation of electric charge by repeatedly transferring the electrostatic charge from a resin disc to a plate covered in tin foil. Inspired by Galvani's observations, Volta developed the first true electric battery, known as the Voltaic pile, which consisted of a stack of alternate zinc and silver plates separated by paper discs soaked in brine. Scientists went on to develop more efficient types of battery.

The Danish scientist Hans Christian Ørsted was the first person to predict and demonstrate the electromagnetic effect. In 1820, he placed a magnetic needle below a wire connected to a battery and found that the needle moved in response to the magnetic field around the wire. In the same year, the French scientist André-Marie Ampère built on Ørsted's work by showing that whether the direction of current in two parallel wires was the same or opposed determined their magnetic repulsion or attraction. He also established that a coil of wire (a solenoid) acts like a permanent magnet when current flows. It was left to the English scientist Michael Faraday to demonstrate the practical applications of electromagnetism. Faraday invented the electric motor in 1821 and the transformer and simple generator in 1831. All of these inventions were based on the electromagnet, consisting of a bar of iron and a coil of wire.

Commercial Generation of Electricity

After 1831, generators were developed to power carbon arc lamps and limelights. Batteries were used to start or excite these generators. In 1866, a number of scientists and engineers realized independently that residual magnetism in the magnetic core of the electromagnet was sufficient to self-excite the generator. The term "dynamo," which had been used for any generator, thereafter became exclusively applied to generators that were self-exciting. The 1870s was the pivotal decade for the electricity industry. The Belgian engineer Zénobe Théophile Gramme invented the first practical electric generator for commercial use in 1871. Gramme's generator was known as the Gramme Ring because of its distinctive ring-shape armature, which produced a sustained current. Prior to this, generators were only capable of powering electric lighting on a small scale, sufficient, for example, to light a theater. In 1872, the German company Siemens and Halske introduced another improvement, the drum armature. These developments were complemented by the invention in 1876 of an improved form of carbon arc lamp called the Jablochkoff candle. The Paris Electrical Exhibition of 1878 was one manifestation of the sudden surge of interest in electricity.

Trials had revealed problems with maintaining continuous high-speed generation until the American engineer C. F. Brush introduced the open coil dynamo in 1878. Brush then set up companies to provide public lighting by carbon arc lamp. His first company was the California Electric Light Company, set up in San Francisco. In 1880, he formed the Anglo-American Brush Lighting Company to provide the same services in Britain. In 1881, the American inventor Thomas Alva Edison decided that it was crucial to increase the availability of electricity in order to stimulate demand for his incandescent filament lamp. He developed the "jumbo" dynamo and opened his first public generating plant, the Pearl Street Power Station, in New York in 1882.

In Britain, the first electricity venture that provided electricity for both public and private use was set up in Godalming in 1881. It proved to be a commercial failure and closed in 1884 owing to insufficient demand. The 1882 Electric Lighting Act empowered the Board of Trade to authorize local authorities or private ventures to set up as electricity suppliers. For private companies, the disincentive was that local authorities had the right to take over these operations after twenty-one years. For example, in early 1882, Edison's company set up a power station (driven by reciprocating steam engines) at Holborn Viaduct in London, but closed it four years later when it had proved to be unprofitable. In 1884, the British engineer Charles Algernon Parsons invented the steam turbine, which was more efficient at providing rotary force. The reciprocating engine relied on the alternate action of two pistons, whereas the turbine relied on the continuous rotation of a shaft with vanes. A number of small hydroelectric plants were built in the 1880s, but the first major hydroelectric scheme opened in 1895 at Niagara Falls, New York. It was primarily intended to power an aluminum smelter, but also provided electricity to the major city of Buffalo, 22 miles away.

Direct Current versus Alternating Current

Before the late 1880s, all generators supplied direct current (DC), current that flowed in one direction. Batteries also produce direct current. In 1884, the Croatian engineer Nikola Tesla immigrated to the United States and began to develop equipment for generating and using alternating current (AC). Alternating current involved the periodic reversal of the current direction. The disadvantage of DC generators was that they could only transmit electricity efficiently over a distance of a few miles. The alternator (AC generator) produced current that could be transmitted long distances without significant energy loss, particularly at high voltages. Another advantage was that the voltage of alternating current could be easily "stepped" (raised or lowered) by a transformer so that the high voltage current produced by the generating plant was delivered to the end user as safer, lower voltage current. In the United States, the AC versus DC debate became a high-profile battle between the industrialist George Westinghouse, pioneer of AC, and Thomas Edison, advocate of DC. Westinghouse gained tremendous publicity for AC through the Niagara Falls hydroelectric project, particularly as a result of the 1901 Pan-American Exposition held in Buffalo, which had electricity as its central theme. Similar rivalries existed in Britain. The electrical engineer Sebastian Ziani de Ferranti was a leading advocate of AC, while the physicist William Thomson, Baron Kelvin, was one of the notable proponents of DC.

Electric Motors

The electric motor is an essential part of any electric appliance that performs work, rather than just heating, lighting, or converting electric signals to sound or pictures. Therefore, the development of small, practical electric motors was a necessary precondition for the development of appliances such as vacuum cleaners and washing machines. An electric motor is essentially the opposite of a generator or dynamo. Both work on the same principles of electromagnetic induction. In the case of the motor, the input of electric current creates magnetic forces that supply mechanical energy, rather than the reverse. The first electric motors were DC motors. The first practical AC induction motor was invented by Nikola Tesla in 1888. Westinghouse acquired Tesla's patents and manufactured an electric fan in 1889, the first appliance based on the compact AC motor. Given that DC and AC transmission systems coexisted, the development of a motor that could be driven by either, termed a universal motor, was a significant step forward. Chester Beach, an employee of the Arnold Electric Company in Racine, Wisconsin, prototyped the first fractional horsepower universal motor in 1905. As this motor was lightweight and high-speed, it was ideal for domestic appliances. Beach and coworker L. H. Hamilton formed the Hamilton Beach Manufacturing Company in 1911 to produce a range of small appliances based on the universal motor.

Business Structure and Standardization

In Britain, electricity supply existed in both direct and alternating forms with a variety of voltages until the creation of the National Grid in the 1920s and early 1930s. In contrast, American electricity utilities and manufacturers soon recognized that standardizing electricity supply would encourage faster growth of demand and mass-production of equipment and appliances. The small number of key players in the industry simplified the process of agreeing on standards. Two companies, General Electric and Westinghouse, held most of the important key patents and also held stock in many of the electricity utilities. In 1885, the utility companies formed the National Electric Light Association (NELA) as a forum to coordinate functions such as marketing and advertising. The Electric Club, founded in 1886 and based in New York, provided a more informal forum for engineers and managers to discussion issues of common interest. In 1895, General Electric and Westinghouse agreed to pool their patents through a cross-licensing arrangement. As a result, by 1910, the United States had a standard electricity supply in the form of alternating current delivered to the user at 120 volts. Standardization yielded immediate benefits. As the price of electricity was far lower in the United States than in other countries, demand grew much faster.

In both the United States and Britain, utilities and manufacturers concentrated on meeting the needs of public services, such as lighting and transport, and the manufacturing industry. Creating infrastructure for supplying electricity to public and industrial customers was cost-effective because demand was predictable. Extending the supply infrastructure to residential areas was more risky because demand was unknown. The domestic customer also presented a different set of challenges for manufacturers, who tailored their plant and equipment for individual contracts. Initially, the larger manufacturers had little interest in mass-producing off-the-shelf appliances for domestic consumers. Until the 1920s, domestic appliance manufacture was a specialist area, served by smaller companies. By 1907, 8 percent of American homes were wired for electricity, and this percentage had doubled by 1912. In Britain, only 2 percent of homes were wired for electricity by 1910. Most of these homes were wired for lighting alone, as there was little expectation that consumers would want electric appliances.

Wiring and Connecting

The problem with only providing a lighting circuit was that when consumers did start to acquire electric appliances, usually smaller items such as irons, the lighting circuit did not provide a convenient means of connection. By the mid-1880s, specific fittings for securely connecting filament lamps to light sockets had become standard. In the United States, the screw fitting became the norm, whereas the bayonet fitting was preferred in Britain. Manufacturers began to fit these kinds of connector to smaller appliances that could be safely operated from the lighting circuit. However, light fittings were usually suspended from the center of the ceiling, out of easy reach for frequent changes of connection. Double, or even triple, light sockets meant that there was no need to disconnect the light bulb, but trailing flexes remained a hazard. In spite of the inconvenience, it was not uncommon for light circuits still to be used for appliances in the 1940s. For those homes that had appliances with high power ratings, particularly cookers (stoves), a separate, exclusive circuit was necessary in order to prevent the overloading of circuits, and this is still true today. Wall and floor sockets for appliance power circuits appeared in the 1890s, but were rarely found in homes until after World War I. In 1917, six leading U.S. manufacturers agreed on a standard power socket in the form of a double (or duplex) socket with T-shape slots to take plugs with two blades in line or in parallel. The parallel blade plug eventually supplanted the tandem blade plug and, from 1962, a third pin was added for the earth (ground) wire on higher rated appliances. In Britain, the standard power socket in the 1920s and 1930s was a single socket with two holes to accommodate a plug with two round pins. A three-pin plug and socket for 13 amp ring mains was introduced in Britain in 1947, but older wiring schemes often remained in use for many years.

The Economics of Supply and Demand

After World War I, the electricity utilities in both Britain and the United States faced the same conundrum: in order to stimulate increased demand, they needed to offer competitive prices, but in order to reduce unit operating costs, power stations needed to balance output more efficiently. By relying mainly on industrial demand for electricity, electricity utilities suffered from an imbalance in the power load, with a heavy load demand during the day and light demand at night. As generating capacity had to match the peak load, capacity was wasted during the off-peak period. The obvious market for off-peak electricity was the household. As only 6 percent of British homes and 35 percent of American homes were wired for electricity in 1920, there was great growth potential.

In Britain, the government began to direct the electricity industry toward better coordination and greater efficiency. The 1919 Electricity Supply Act created a body of Electricity Commissioners to coordinate the supply industry. In the same year, supply engineers and appliance manufacturers took an independent initiative and set up the British Electrical Development Association (BEDA). The role of the BEDA was to promote both wings of the industry to all types of consumer. In 1926, the government created the Central Electricity Board to rationalize power stations and connect them through a standardized national grid. The electricity grid was fully operational by 1933. While the grid provided standardized mains transmission of AC current at a uniform voltage and frequency, household supply was not completely standardized until the early 1970s. The standard (3-phase, 50-cycle, 240-volt AC) was prescribed in 1947, but took twenty-five years to implement.

Silhouetted 230,000-volt powerlines feed the San Fernando Valley with electricity. (Roger Ressmeyer/Corbis)

The electricity industry in the United States had a head start over the British industry in attracting the domestic consumer, not only because of the higher penetration level that existed, but because supply was already standardized. The NELA began to mount nation-wide campaigns targeting the domestic consumer. As the household market became more promising, General Electric and Westinghouse began to acquire small appliance companies in order to gain a foothold. Recognizing the need to foster consumer confidence, manufacturers formed a new alliance, the National Electrical Manufacturers Association (NEMA), in 1926. Its role was to develop technical standards by defining products and processes.

The electricity industries in Britain and the United States used similar approaches to stimulate consumer demand. The starting point was to increase the number of homes that were wired for electricity. In the 1920s, wiring up new homes became common practice, but the big challenge was to persuade landlords to wire older rented homes. The utilities offered incentive schemes, including deferred payment through monthly installments. In Britain, where electric domestic appliances were significantly more expensive than in the United States, the electricity utilities introduced low-cost appliance rental schemes. As the main objective was to increase electricity consumption, renting appliances was regarded as a loss leader. The main competition came from gas, which could claim to be more efficient and reliable for heating and cooking. In Britain, it was also a significantly cheaper option. The main areas where electricity had a competitive advantage were in its cleanliness and labor-saving potential. The latter was particularly appealing to middle-class consumers, as demand for domestic servants exceeded supply. The electricity industry promoted electricity as the "silent servant," with the added attractions of being more available, dependable, and tireless than the human servant. Advertisements tended to exaggerate these qualities, giving the impression that electric appliances took all of the drudgery out of housework, with the housewife just managing the operation.

In Britain, the electricity industry had an independent ally in its quest to convince women of the joys of the electric home. The Electrical Association for Women (EAW) was launched at a meeting of the Women's Engineering Society in 1924. Its objectives were to inform women about "electrical apparatus" and to provide a platform for women's views on all matters relating to electricity. It was particularly useful to the industry because of its educational emphasis. The industry restricted itself to providing electrical housecraft (homemaking) courses and employing trained female demonstrators to assist customers. The EAW ran a broader educational program that was more empowering. Teaching women basic maintenance skills, such as how to change a plug, helped to improve understanding and overcome fears about safety. The United States did not have a precise counterpart of the EAW, as the nearest equivalent was the NELA Women's Committee, which had a similar mission but was tied to the industry.

By 1930, the proportion of homes wired for electricity had risen to more than two-thirds in the United States and to one-third in Britain. However, appliance ownership remained low, and the onset of a worldwide economic depression, in the wake of the 1929 Wall Street Crash, meant that private consumers either could not afford new appliances or were more reluctant to spend freely. As a consequence, the U.S. government took a more active role in regulating the electricity industry in the national interest. The National Power Policy Committee closely scrutinized electricity pricing policies, which had the effect of prompting the NEMA to set up its own National Recovery Administration code to monitor production and prices. In Britain, the industry was already subject to tighter regulation and was largely publicly owned through local authorities. The degree of state control was taken to its ultimate limit after World War II when the Labour government came to power with a manifesto of nationalizing key industries. This was accomplished by the 1947 Electricity Act.

Fossil Fuels and the Environment

After World War II, electricity consumption increased sharply, partly as a result of rising levels of ownership of domestic appliances. In most countries, electricity generation was heavily reliant on fossil fuels, particularly coal and oil. This became a matter of concern for three reasons—depletion of resources, rising fuel costs, and adverse impact on the environment. These concerns came to a head in the 1970s. An energy crisis hit Britain, the United States, and other developed countries when the Organization of Petroleum Exporting Countries (OPEC) imposed huge price increases and production cuts. The crisis was precipitated by U.S. involvement in the Arab-Israeli war, but it was sustained by OPEC's strategic desire to dictate the terms of supply and demand. Over the decade, the price of a barrel of oil increased tenfold. In Britain, the government responded by introducing the 1976 Energy Act in order to control and conserve fuel and significantly increasing taxes on oil in 1979. In the 1980s and 1990s, oil prices fluctuated, falling when OPEC members sought to increase sales and rising in response to events such as the Gulf War. Oil-consuming countries began to exploit other oil sources, such as oil and gas deposits below the North Sea and, where possible, to seek alternatives to oil.

Coal posed a different set of problems. Industrialized countries had been exploiting their coal resources for much longer than they had been using petroleum. Coal deposits in smaller countries, such as Britain, were being rapidly depleted and mining costs were rising, while in the United States, with its more extensive coal resources, coal was still relatively cheap. In Britain, as household consumption of coal decreased, hundreds of mines closed in the 1950s. Demand from the electricity industry sustained the British coal industry until the 1970s. Thereafter, competition from North Sea gas hastened the decline of coal mining. The percentage of electricity generated by coal-fired power stations fell from 74 percent in 1992 to 50 percent in 1995. While the motives for this change were primarily economic, the consequences were also environmentally beneficial.

The third argument for reducing the use of fossil fuels, particularly coal, in electricity generation was to cut down environmentally harmful emissions. Power stations that burn fossil fuels emit carbon dioxide, one of the major "greenhouse gases" responsible for global warming. Greenhouse gases collect in the upper atmosphere where they absorb reflected solar radiation, preventing the release of heat to space. As a result of escalating fossil-fuel consumption during the twentieth century, there has been a significant rise in global temperatures, resulting in the increased incidence of drought and a rise in sea level caused by the melting of polar ice caps. While carbon dioxide can be drawn off and stored under pressure, this process itself consumes considerable energy. Specific commitments to reduce greenhouse gas emissions were made at the 1992 United Nations Conference on Environment and Development in Rio de Janeiro, familiarly known as the first Earth Summit. The failure to achieve scheduled reductions led to the 1997 United Nations Kyoto Protocol, which set a new target for industrialized countries of a 5 percent reduction in emissions by 2012. However, the protocol has not been ratified by the required number of signatories, and Britain and Germany are among the few nations to be on schedule for compliance. In Britain, this has been largely achieved by switching from coal to gas-fired power stations, which produce less carbon dioxide.

The destructive effects of the emission of sulfur dioxide from coal-fired power stations are more localized, but devastating. When rain absorbs sulfur dioxide, it forms a dilute sulfuric acid known as acid rain. In North America and northern Europe, acid rain has damaged building masonry, destroyed forests, and polluted lakes and rivers. Sulfur dioxide emissions can be reduced by desulfurizing the flue gases from power stations. This process adds about 10 percent to the cost of generating electricity, but this is considerably less expensive than the equivalent process for removing carbon dioxide.

Nuclear Electricity and Safety Concerns

During World War II, scientists in the United States succeeded in creating the first nuclear chain reaction. State funding of nuclear research was motivated by its potential military applications. However, in the wake of the 1945 Hiroshima and Nagasaki nuclear bombings, antinuclear pressure groups were formed to oppose the proliferation of atomic weapons. Governments turned to civil nuclear applications as a means of defusing criticism and offsetting the costs of nuclear research and development. After the world's first commercial nuclear power plant opened at Calder Hall in Britain in 1957, there was a steady growth in nuclear power generation. The 1970s energy crisis and environmentalist criticism of fossil fuels made the nuclear option still more attractive to the electricity industry. Nuclear reactors consumed only tiny amounts of radioactive materials in relation to the amount of energy produced, and nuclear power plants did not emit greenhouse gases or cause acid rain. However, fears about the safety of nuclear power plants were heightened in 1979 when a pressurized water reactor at the Three Mile Island nuclear power plant in Pennsylvania leaked radioactive material.

Seven years later, the Three Mile Island accident paled in comparison with the impact of the disaster at the Chernobyl power plant in the Ukraine, when two explosions destroyed a reactor. Suddenly, arguments about the environmental friendliness of nuclear power were swept aside by the consequences of the release of 9 tons of radioactive material, ninety times the amount released by the Hiroshima bomb. Five million people were exposed to radiation and thirty-one people in the immediate vicinity died. The resulting cloud of radioactive isotopes spread as far west as Ireland and as far south as the Mediterranean. In Britain, concerns were exacerbated by studies into the increased risks of childhood leukemia in the vicinity of the Sellafield nuclear reprocessing plant, which has the dubious distinction of being the world's biggest discharger of radioactive waste.

Moreover, from the early 1990s, the electricity industry could no longer claim that nuclear power stations were low cost after the politically embarrassing removal of nuclear power plants from the privatization schedule of the British electricity industry. The financial prospectus had revealed that, when waste disposal costs and the decommissioning costs were included, nuclear electricity was not an attractive investment. Nevertheless, nuclear reaction remains a significant source of energy production. In Britain and the United States, just over a fifth of electricity was generated by nuclear power plants in 1992. The proportion rose to more than a third in Britain after the opening of Sizewell B, the world's most advanced pressurized-water reactor power plant, in 1995. All current reactors are based on nuclear fission, but the vision for the future is nuclear fusion, hailed as a safer and more efficient form of nuclear energy production.

Alternative Energy

The 1970s energy crisis encouraged electricity utilities to explore renewable energy alternatives to fossil fuels. As alternative energy, by its very nature, depends on factors outside human control, the potential for exploitation varies greatly from one country to another. The main alternative energy options depend on river flow, tidal effects, wind patterns, and hours of sunshine. While hydroelectricity was a long-established renewable energy technology, it is specific to locations where there are fast-flowing rivers or that are suitable for constructing large dams. These locations are often far from the main areas of electricity consumption, so the costs of the transmission infrastructure may be high. Hydroelectricity supplies about 20 percent of the world's electricity, but the percentage varies widely from country to country, depending on physical geography. France is the only country that has made a major effort to harness tidal power.

Solar energy has the greatest potential in the tropical and subtropical zone, where sunshine is abundant throughout the year, and in the subpolar regions, where there are very long days in summer. It has limited potential in the cloudy, temperate belt of Europe, Canada, and the United States. Two solar power plants were built in California's Mojave Desert in the 1990s. However, even where long hours of sunshine do exist, the high cost of solar power installations may be a strong disincentive. An alternative is to exploit solar energy at the consumer level. Solar panels can provide central heating by using solar energy to heat pipes through which air or water is circulated by pump. While wind-powered electricity generation has increased steadily since the 1970s, by the mid-1990s the total global production was only equivalent to the output of a single nuclear power plant. Environmentalists have been ambivalent about wind power because while it is a clean and renewable source of energy, wind farms have a dramatic visual impact on the natural landscape. There is also concern that the noise from tall, highly aerodynamic turbines affects the ecology of the local area by discouraging wildlife.

Energy Conservation

Because there were no easy solutions to the challenge of safer and more environmentally responsible electricity generation and no sign of demand declining, energy-efficiency became the key to controlling consumption and reducing environmental damage. In the 1970s, manufacturers were slow to respond, except in the field of electric lighting. The Dutch company Philips, the world's leading manufacturer of electric light bulbs, introduced new energy-saving light bulbs that were small, coiled fluorescent tubes with standard light socket fittings. In the 1980s, manufacturers began to improve the energy-efficiency of larger appliances, such as washing machines. By reducing water intake, less electricity was needed to heat water. Improved detergents that were effective at lower temperatures and increasing use of synthetic fabrics that demanded lower temperature water also encouraged reduced use of electricity. The microwave oven was promoted as an energy-efficient, and therefore cheap, method of cooking. While the application of valve and transistor electronic technologies had been restricted to audio and audio-visual appliances, microelectronics began to pervade a wide range of appliances. Microelectronic systems not only enabled appliances to be automatically controlled to very precise parameters, but also required little power themselves.

Governments, which had become more proactive in promoting energy conservation as a result of the 1970s energy crisis, played a complementary role. Advisory agencies provided information about energy choices and encouraged passive energy conservation through sustainable building design. In Britain, grants became available for improving home insulation and thus reducing energy loss. For most consumers, reducing energy costs was the main incentive for saving energy. However, as energy-efficient appliances were more technologically sophisticated and thus more expensive to manufacture, the popularity of such appliances hinged on the willingness of consumers to take a long-term financial view of the choices. In practice, many consumers take a short-term view, so there is still much untapped potential for energy saving.

See also British Electrical Development Association (BEDA); Electrical Association for Women (EAW); General Electric; National Electric Light Association (NELA); Westinghouse Electric

References:

Electricity Supply in the United Kingdom: A Chronology. London: The Electricity Council, 1987.

Hannah, Leslie. *Electricity before Nationalisation: A Study of the Development of the Electricity Supply Industry in Britain to 1948.* London: Macmillan, 1979.

Hannah, Leslie. *Engineers, Managers and Politicians.* London: Macmillan, 1982.

Nye, David E. *Electrifying America: Social Meanings of a New Technology.* Cambridge, MA: MIT Press, 1990.

Schroeder, Fred E. H. "More 'Small Things Forgotten:' Domestic Electric Plugs and Receptacles, 1881–1931." In *Technology and Choice: Readings from Technology and Culture,* edited by Marcel C. LaFollette and Jeffrey K. Strine. Chicago and London: University of Chicago Press, 1991.

Electrolux

With annual unit sales exceeding 55 million, the Swedish company Electrolux is the one of the world's leading domestic appliance manufacturers. It is the world market leader in floor care appliances and is ranked first in Europe and third in the United States in the production of white goods (stoves, washing machines, and refrigerators).

Electrolux was formed in 1919 by the merger of Elektromekaniska and Lux, both based in Stockholm. Formed in 1901, Lux started as a manufacturer of kerosene lamps and in 1912 launched its first vacuum cleaner, a cylinder model known as the Lux 1. The Electrolux Model V vacuum cleaner of 1921 introduced a new feature—toboggan-like runners for ease of movement. In 1925, Electrolux extended its product range by moving into the manufacture of refrigerators. The first Electrolux refrigerator was the

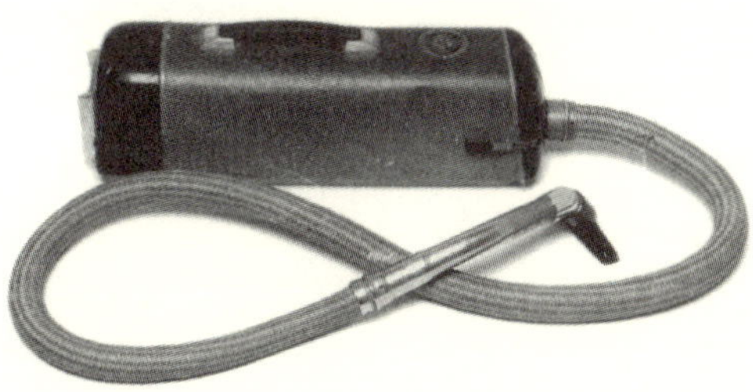

1950s Electrolux model ZA55 cylinder vacuum cleaner (The Museum of Science and Industry in Manchester)

D-fridge, based on the absorption method of refrigeration developed in 1922 by Swedish engineering students Baltzar von Platen and Carl Munters.

During the late 1920s, the success of both its vacuum cleaners and refrigerators in Europe led Electrolux to set up manufacturing plants in a number of countries, including Germany, France, and Britain. In 1931, Electrolux began manufacturing vacuum cleaners in the United States at a plant in Connecticut. During World War II, while production in Europe was stifled, Electrolux opened new factories in Australasia. Expansion continued after the war. In 1951, Electrolux produced its first domestic washing machine.

The absorption refrigerator pioneered by Electrolux had never matched the market share of the compression refrigerator, which was favored by the major American refrigerator companies. In 1956, Electrolux launched its first compression refrigerator and its first chest freezer, in recognition of the growing consumption of frozen food. It brought the refrigerator and freezer together as the combined fridge-freezer in 1959, the same year that it introduced its first dishwasher, a cylindrical model designed to sit on a kitchen worktop or draining board.

Electrolux extended its white goods range in 1962 by acquiring the Swedish company ElecktroHelios, whose products included cookers. It continued to consolidate and expand its operations in the 1960s and 1970s with further acquisitions and new sales subsidiaries. Particularly significant was the acquisition of two American companies. In 1974, the purchase of National Union Electric/Eureka elevated Electrolux to the position of world leader in vacuum cleaner production, and the purchase of Tappan in 1979 provided a foothold in the American white goods market.

The company's expansion strategy followed the same pattern in the 1980s. Electrolux attained leadership of the European household appliance sector in 1984 by acquiring the Italian company Zanussi. Other major acquisitions in Europe included the white goods division of the British company Thorn-EMI (1987) and the white goods division of Germany's Buderus group (1989). In the United States, Electrolux picked up the prestigious Kelvinator and Frigidaire brands in 1987 by taking over White Consolidated.

In the 1990s, Electrolux made fewer high-profile acquisitions. It worked on its corporate image, particularly in the area of environmental awareness. A low-energy refrigerator, which halved electricity consumption, was launched in 1991, and CFC-free refrigerators were produced from 1993. Electrolux also developed a system for recycling refrigerators by extracting and sealing CFC refrigerants. Since 1989, Electrolux has had an agreement with the Sharp Corporation for marketing its white goods in Japan. This relationship has also assisted the development of Electrolux's latest innovation, the Screenfridge, a refrigerator that doubles as a household communications device. With its built-in display screen and communication ports, the Screenfridge combines Sharp's expertise in LCD (liquid crystal display) and computer technology with core Electrolux refrigerator technology.

See also Refrigerators; Vacuum Cleaners; Zanussi

References:
www.electrolux.com/node418.asp

Electronic Games

There are three types of electronic games equipment for home use: the games console connected to a television set; the home or personal computer; the hand-held game set with built-in controls and a small display screen. Every potential move in a game has been identified and mapped electronically, ready to be activated by use of the controls. In 1962, Steve Russell, a computer programmer working at the Massachusetts Institute of Technology, developed a game called Spacewar for the PDP-1 computer. Other early electronic games followed, but were just intended to amuse fellow computer programmers.

In the early 1970s, the first commercial electronic games were those designed for use in amusement arcades. The American Nolan Bushnell founded Atari in 1972 and introduced Pong, an electronic version of tennis. Shortly afterward, another American company, Magnavox, introduced a similar electronic ball game, Odyssey, for home use. Two years later, Atari produced a Pong console for home use. To play Pong, two players used the controls to move their "bats" up and down to direct the "ball," presented on the television screen in simple black and white graphics. The limitation of this first generation of home video games was that the games were coded into the console circuitry ("hard-wired"). Some consoles provided a small selection of games, but the user could not add new games.

The next generation of video games arrived in 1976 when the American electronics company Fairchild introduced a console with the game, a version of tic-tac-toe, supplied on a cartridge. In 1977, Atari launched its Video Computer System, a console incorporating a programmable microchip designed to work with interchangeable games cartridges. The video game craze entered a new phase with the arrival of more sophisticated arcade games, beginning with Space Invaders in 1978. This in turn stimulated the home games market, with Atari's Pac-Man generating sales of 7 million units. Concerns began to arise, and still remain, about the addictive nature of electronic games. Major toy and games manufacturers entered the market, including America's Mattel, which brought out its Intellivision system in 1981.

Competition increased as more games were developed for use with home computers. The video games market became oversaturated and by the mid-1980s, analysts were predicting its demise. These predictions proved premature. With a blend of imaginative creativity and improved technology, two Japanese companies, Nintendo and Sega, took video games into the next phase. Nintendo's first console, the Family Computer, based on an 8-bit chip, appeared in Japan in 1983. Two years later, the 8-bit Nintendo Entertainment System arrived on the world market, followed by the Master System. In 1987, the Nintendo console was the best-selling toy/game in Canada and the United States. The popularity of Nintendo's Mario and Luigi and Sega's Sonic the Hedgehog marked the rise of more character-based, story-driven games. Nintendo's Super Mario Brothers 3 game, launched in 1990, amassed sales revenue of $500 million. In 1989, Nintendo introduced the Game Boy, the first hand-held electronic game system, which ran a game called Tetris.

In the early 1990s, 16-bit chips came into use in games consoles, improving speed, graphics, and general performance. The Sega Megadrive was the market leader at this time, aided by the fact that it was able to draw on the company's catalogue of arcade games, such as Sega Rally. In 1994, Nintendo launched the game Donkey Kong Country, which was heralded as a major advance in graphic realization. Meanwhile, computer-based games took a major step forward with the move from floppy disk to CD-ROM (compact disc with read-only memory) as a storage medium. As CD-ROMs can hold much more information than floppy disks, it became possible to incorporate high quality 3-D graphics and sound, creating a degree of virtual reality (VR). However, even by the mid-1990s, increased home computer ownership had discouraged sales of games consoles.

In 1996, the Japanese electronics company Sony made a late but dramatic entry into the market with the launch of its PlayStation, the first games console to use a 32-bit chip. Sega also introduced its 32-bit Saturn model. Atari had bypassed the 32-bit chip and moved from 16-bit consoles to the 64-bit Jaguar console in 1993. However, the Jaguar failed to achieve its sales potential because its games software lagged behind the hardware. The PlayStation cost $500 million to develop. However, it was soon challenged by the Nintendo 64, the first commercially successful 64-bit chip games console. An astonishing half-million Nintendo 64 systems were sold in Japan on the first day of release in June 1996. Competition between Sony, Nintendo, and Sega has continued to flourish, with development efforts split more or less evenly between hardware and software. By 1998, Sony had won the battle of the CD models, and the Saturn virtually disappeared. One of the key reasons for Sony's success was its willingness to court an older market of sixteen–twenty-five-year-olds. It benefited from the highly successful Tomb Raider series of games. Tomb Raider's main character is Lara Croft, a cross between Indiana Jones and a sexually aware Barbie doll.

By the end of the 1990s, the games had evolved into a number of types—racing games based on cars, skiing, and so on; fighting or "beat'em-up" games based on either wrestling or science fantasy; milder "token collecting" obstacle games; and more thoughtful construction games like SimCity, where players can build their own virtual city. Game development is a key part of the industry, and some games have production budgets of $4 million.

Nintendo upgraded the Game Boy with a smaller pocket version and the introduction of a color screen (Game Boy Color). It also introduced optional extras in the form of a camera and a printer. Nintendo also launched the Pokémon game that has generated an unprecedented range of spin-offs and become a franchise in its own right. The last games console to be launched in the twentieth century was Sega's Dreamcast, the first Internet-enabled console. With its 128-bit chip, Dreamcast was heralded as the fastest console, with the best quality 3-D graphics. Sony was due to launch the Playstation2 in 2000. It is also a 128-bit machine that can play DVDs. Like the Dreamcast, it has been designed with the Internet in mind and players will be able to buy add-on modems. Sony announced that Playstation2 would be for sale on their web site. It duly crashed after receiving 600,000 hits in one minute! The next major step could be when more sophisticated VR technology, which already exists in the form of VR headsets, remote controls, and gloves, becomes available at mass-market prices.

The success of electronic games has been worldwide and, like movies and pop music, has given children and adults a new form of global language. They have also led to the now familiar merchandise that spins off from the original product in the form of magazines, toys, souvenirs, models, comics, TV programs, movies, and web sites. Their popularity continues, despite the continuing concerns about a generation of home-based children who may prefer virtual to actual experiences.

See also Atari, Inc.; Computers; Nintendo; Sony Corporation; Television; Virtual Pets

References:

Herz, J. C. *Joystick Nation.* London: Abacus, 1997.

Ericsson

The Swedish company Telefon AB L. M. Ericsson is one of the world's leading telecommunications companies. Three-quarters of the company's business lies in telephone network infrastructure and a quarter in consumer products. Two-fifths of the world's mobile phone users are served by an infrastructure based on Ericsson equipment. With a world market share of 16 percent, Ericsson ranks third in mobile phone sales overall and second in digital mobile phone sales. The company has 100,000 employees worldwide and is responsible for 15 percent of Sweden's exports. Ericsson products are sold in 140 countries, of which the most important, in order of revenue generated, are the United States, China, and the United Kingdom.

The company's founder, Lars Magnus Ericsson, was raised on a small farm. After serving an apprenticeship with a blacksmith, in 1867 he got a job in a telegraph factory in Stockholm. Ericsson was a diligent employee and became a skilful instrument maker. Keen to further his ambitions by studying abroad, he prepared by learning German and English

Ericsson skeleton telephone supplied to the U.K. National Telephone Service (The Museum of Science and Industry in Manchester)

in his spare time. His dedication was rewarded when he received two grants from Swedish government to study in Germany. When he returned to Sweden, Ericsson opened a telegraph equipment repair shop at his home in Stockholm in 1876. He soon began designing and producing improved telegraph equipment.

The invention of the telephone created new opportunities. In 1878, Ericsson designed his first telephone, a "magnet telephone with signal trumpet," and sold twenty-two. In 1880–1881, the Swedish government decided to create a national telephone network. After comparing Ericsson telephones with American Bell telephones, the government awarded the contract to Ericsson on the basis of better ringer and microphone design. In 1881, Ericsson received its first overseas order from Norway and produced its first wall telephone. The growing success of the company warranted the construction of a new factory in Stockholm in 1883. With larger premises, Ericsson began manufacturing switchboards and telephone exchanges. By 1889, Sweden had the highest telephone density in the world. In 1892, Ericsson introduced the first desktop telephone with a single handset. This was the innovative "skeleton" design, in which the exposed magnet bars formed the support for the handset.

By 1894, exports exceeded domestic sales. In 1896, the company was reconstituted as a new corporation, Aktiebolaget L. M. Ericsson & Co. By 1899, Ericsson had 1,000 employees and a new factory in St. Petersburg. The steady growth of the business was disrupted by World War I, which led to the factory being requisitioned for military production. In the aftermath of the war, the market was constrained by high raw material prices and wage inflation. Ericsson continued to be an innovator in telephone design. In 1930, Ericsson commissioned the Norwegian engineer Johann Christian Bjerknes and the Norwegian artist Jean Heiberg to develop a new handset phone in bakelite. The resulting DHB 1001 model went into production in 1932 and proved to be one of the most long-lived and influential telephone designs. One of Ericsson's most adventurous telephone designs was the one-piece Ericofon, also known as the Cobra, developed by the Swedish design team of Ralph Lysell, Hugo Blomberg, and Gösta Thames. The Ericofon was the first one-piece telephone, with the dial placed on the underside of the base and the mouthpiece above. Commissioned in 1949, it was available in a range of colors from 1956. The design was licensed to other companies, including American Telephone & Telegraph. By this time, Ericsson had reached the production landmark of 6 million telephones. The Dialog telephone of 1966 was Ericsson's first telephone with a keypad instead of a dial.

In 1956, Ericsson developed its first mobile phone system. Operating in the Stockholm area, this was the world's first automatic mobile phone system, allowing direct dialing. The mobile telephone handset weighed 40 kg (88 lb). It was designed to run off a car battery and was as expensive as a car. Ericsson continued to develop and improve its mobile phone technology, but even by the mid-1970s, its mobile phones weighed a hefty 9 kg (20 lb). Ericsson was also at the forefront of the cellular phone revolution of the late 1970s and early 1980s. In 1981, the Scandinavian countries (Sweden, Norway, Denmark, and Finland) agreed on a common cross-border cellular system, the Nordic Mobile Telephone system (NMT) developed by Ericsson. In 1982, Ericsson was awarded forty contracts for cellular phone infrastructure in the United States. The cellular phone boom encouraged Ericsson to set up a mobile telephone laboratory in Lund in 1983. By 1984, the Scandinavian NMT network had more than 100,000 subscribers, making it the world's largest cellular phone service. The NMT standard was also adopted by a number of other European countries and Saudi Arabia.

Ericsson launched its first handheld mobile phone, the Hotline, in 1986. It weighed 700 g (1.5 lb). Meanwhile, standards were being developed for the next phase of mobile phone technology, digital cellular phone networks. The first standard, GSM (originally an acronym for the Groupe Spécial Mobile, but later reinterpreted as the Global System for Mobile Communications), was developed in Europe between 1982 and 1987 with the backing of the Conference of European Posts and Telecommunications. Ericsson won large orders for GSM equipment from Britain and Germany as well as Scandinavia in the late 1980s. These first GSM systems became operational in 1991. In 1992, Ericsson introduced the world's first GSM digital handheld mobile phone, the GH197, and produced 30,000 for the German market.

In the last years of the twentieth century, Ericsson continued to consolidate its strong position in the mobile telephony field. It installed a digital mobile phone network in Japan in 1996, and total production of handheld mobile phones reached 10 million units. In 1997, the Ericsson GF 788, designed by Yeo Chung Sun in Singapore, weighed only

135 g (5 oz), making it the smallest mobile phone to date at that time. In the same year, Ericsson introduced a digital cordless phone for home use. New product lines for Ericsson, pagers and personal organizers, were launched in 1998. The latest Ericsson mobile phones incorporate PC cards and infrared modems to enable a range of voice and data communications.

See also British Telecom; Mobile Phones; Telephones; Vodafone.

References:

www.ericssonus.com/us/consumer/
www.ericsson.co.uk/UK/corporate/history.shtml

F

S. W. Farber, Inc.

The S. W. Farber company was established in 1900 when S. W. Farber, a tinsmith, set up a shop in Manhattan's Lower East Side. The initial products were copper and brass bowls and vases. In 1910, the company introduced Farberware, the line of kitchenware for which it is best known, with chrome and silver-plated table accessories. It produced kitchenware, pots, and pans in the 1920s and moved into the appliance field in 1930 when it developed the first Farberware coffee percolator. Farber improved this with the Coffee Robot coffeemaker of 1937 that could keep coffee warm long after it had brewed. The Broiler Robot, which followed in 1938, was advertised as "the first broiler that could cook at the dinner table."

Farberware stainless steel and aluminum pans were introduced in 1949 after the company had moved away from wartime demands and invested in stainless steel production. It produced the first stainless steel electric frying pan in 1954. Sold to the multinational Hanson Industries in 1987, it is now a part of Salton Inc. and continues to produce coffeemakers, juicers, and waffle makers.

See also Coffeemakers
References: www.farber.com

Fax Machines

A facsimile (fax) machine is a machine that copies the image of a document, sends it to another location, and reproduces it exactly. Early facsimile machines were developed in the nineteenth century, but in its modern form, compact and cheap enough for household use, the fax machine is a product of the late twentieth century.

Nineteenth century facsimile machines, such as those developed by Alexander Bain and Frederick Bakewell in England and Giovanni Caselli in Italy, used mechanical methods of copying and reproduction. The contents of the document were traced by a pen and the movement of the pen was recreated at the other end. This technique was slow and laborious. In 1873, Willoughby Smith showed that pulses of light converted to pulses of electric current and sent by wire could be reconverted by a selenium photoelectric cell. This technique was used to send newspaper photos. A breakthrough occurred in Germany in 1902, when Arthur Korn invented a method for transmitting photographs by electric wire. This process, telephotography, was used to send the first intercity fax, between Munich and Berlin in 1907. The first machine to employ document scanning, as in the modern fax machine, was the Belinograph, invented by the French-

Executive using a fax machine (Walter Hodges/Corbis)

man Edouard Belin, in 1913. A strong light beam scanned the document and a photoelectric cell converted the light readings into electrical impulses for transmission. From the 1920s, Telecopiers based on Belin's design became widely used in the newspaper industry for "wiring" photographs.

Until the 1960s, facsimile machines were too expensive and unwieldy to reach a mass market. In 1966, America's Xerox Corporation produced the Magnafax Telecopier, which was much smaller, easier to use, and transmitted a letter-size document in about six minutes. The relative slowness made such machines expensive to use. In the 1970s, Japanese companies began to develop faster, smaller, cheaper, and more efficient fax machines. Japan was also the country with the highest level of ownership. By 1985, the number of fax machines in use in the United States had grown to 550,000, whereas Japan had 850,000 and Europe only 120,000. From the mid-1980s, the availability of combined telephone and fax machines made the fax machine more appealing for home use. By 1989, there were 4 million fax machines in the United States.

Fax technology has continued to improve. Early fax machines required expensive thermal paper impregnated with carbon. In such machines, the thermal paper, held on a drum, is heated when current flows through the stylus, releasing the carbon to the paper surface to recreate the transmitted document. To improve print quality and enable the use of ordinary office paper, manufacturers developed "plain paper" fax machines that use either ink film or ink-jets. The Japanese company Canon launched the first plain paper fax machine in 1987. Ink film still relies on thermal technology but the ink is transferred from a thermal film to plain paper, whereas ink-jets spray ink through tiny perforations in the printer head directly onto plain paper. Early fax/phones had to be manually set to act as phone or fax, which could be frustrating. This was solved by the development of automatic fax/phones that could detect whether the incoming message originated from a telephone or fax machine and respond accordingly. Today, the top range fax machines are fully compatible with personal computers, so that they can fulfill a variety of scanning and printing functions.

See also Brother Industries; Computer Printers; Computers; Matsushita; Modems; Telephones

References:

Goodman, Robert. *How Electronic Things Work . . . and What to Do When They Don't.* New York: McGraw-Hill, 1999.

www.hffax.de

Food Mixers

The mixing of drinks, the whisking of batters, and the beating of dough had traditionally been done by hand, using a wooden or metal whisk and a bowl. By 1900, a mechanical version existed for mixing cake batter. It had a cylindrical tinplate body and hand-powered beaters driven by cogwheels.

The arrival of the small electric motor signaled change. Its first use was in scientific laboratories for mixing liquids. Such models were on sale in London in 1904. In the same year George Schmidt and Fred Osius of Racine, Wisconsin, manufactured the first electric milk-shake mixer. The Hamilton Beach Manufacturing Company of Washington, North Carolina, used their "universal fractional horsepower motor" for a commercial food mixer in 1910 and a drink mixer in the following year. About the same time, Troy Manufacturing Products, a subsidiary of the Hobart Company, produced a large electric bread-dough-mixing machine. Designed by Herbert Johnson, this machine was aimed at the commercial market and met with success. The first truly domestic model was the Universal mixer/beater of 1918. Produced by Landers, Frary and Clark of Connecticut, it utilized a small electric motor attached to a stand. Troy Manufacturing Company followed with the H-5, KitchenAid model, launched in 1919. This model was the first domestic mixer with its own fixed stand and bowl and featured the innovative "planetary" system where the mixer rotates the beater in one direction while moving the bowl in the opposite one. Further refinement came in 1926 when Air-O-Mix produced the first combination base-mounted mixer that could also be hand-held. This set the standard for all the other models that followed.

The next landmark for mixers came in 1930 with the introduction of the Sunbeam Mixmaster. Its design was improved by the in-house designer Ivar Jepson, and it became the most successful appliance produced by Sunbeam. With the benefit of more reliable motors, quality construction, and its "mixfinder dial" the Mixmaster set the market standard. It enjoyed a long and successful life and was still a top seller in 1959.

By the end of the 1930s all models had a more unified look, with the motor encased in a pressed steel body shell. These were usually painted white, although some were chrome plated. Bowls were increasingly made of Pyrex. The KitchenAid mixer was redesigned by the streamlining-designer Egmont Arens, giving it the curved form that it retains today. Two trends continued into the 1940s and 1950s: an increasing range of attachments and smaller hand-held mixers.

The idea of a multiple-use machine was current in America by the late 1940s. Attachments were fitted to a "nose" connection at the front of the machine. The A. F. Dormeyer Meal-Mixer of 1949 featured an add-on mincer and juicer. Hand-held machines such as the Dormey and Sunbeam Jr were introduced in a range of fashionable colors. Metal casings were beginning to be replaced by plastics.

Europe continued to follow America. The British Kenwood A 700 Chef mixer of 1948 resembled models like the American KitchenAid. The Kenwood line also included a large commercial model and a smaller portable model. Like their American cousins, they were available in a range of "gay" colors. Despite these changes and the superficial styling of handles and controls, most mixers still betrayed their industrial origins. A change in style and substance came in 1957 with the Braun KM 321 Kitchen Machine, designed under the guidance of Dieter Rams. The motor became almost invisible, encased in a carefully proportioned white plastic case. *Design* magazine commented that unlike other mixers it "appeared to have been designed with a woman, a kitchen sink, and a crowded shelf in mind."

The Kitchen Machine inspired changes in other models. Kenwood employed the designer Kenneth Grange to remodel the A 700. The resulting A 701 of 1960 maintained a metal body but employed cleaner, simpler lines. It was a great success.

The 1960s and 1970s saw mixers develop more attachments, especially blenders. Changing tastes and lifestyles have diminished the popularity of mixers, as fewer people bake their own bread and cakes. This led to the rise of the food processor in the 1970s and 1980s. Nevertheless mixers remain popular with serious cooks, and the 1990s witnessed a return to more industrial, retro styled models. The KitchenAid Kitchen Machine maintains its curved 1930s look with a heavy-duty metal body, stainless steel bowl, and dough hooks. It has attachments for a food mincer, juicer, disc shredder, pasta maker, and sausage stuffer and comes in a range of fashionable colors.

See also Braun; A. F. Dormeyer; Hamilton Beach; Kenwood; KitchenAid; Pyrex; Sunbeam

Reference:

Goldberg, Michael J. *Groovy Kitchen Designs for Collectors, 1935–1965.* Atglen, PA: Schiffer Publishing, 1996.

Sparke, Penny. *Electrical Appliances.* London and Sydney: Unwin Hyman, 1987.

Food Processors

During the 1950s food mixers came with a wider and wider range of accessories of varying power and utility. A Frenchman, Pierre Verdun, developed the food processor in 1963. Verdun was a salesman for a catering equipment company, and his access to professional kitchens led him to realize that a simple, multipurpose machine would be popular. His model, the Robot Coup, was initially marketed to professional cooks. Later known as the Magimix, it has become one of the most popular kitchen appliances. It is compact, versatile, and free of complicated attachments as well as being strong, fast, and easy to use. One problem of the larger mixers was that the attachments were fiddly and difficult to clean. The Magimix arrived in the United Kingdom in 1974 and sold well. This model established the standard design; a clear plastic bowl sits on the base of the machine

The Cuisinart Power Prep Plus food processor DLC214 (Courtesy, Cuisinart)

that houses an electric motor that drives a spindle, which fits through the base of the bowl. The spindle can house a range of cutting blades, the standard being a double-edged cutter. Other blades include dough hooks and shredding discs. The lid of the bowl is a "locking" lid with a "chimney" into which the food is placed. Once switched on, the blades rotate at high speeds capable of chopping meat, fish, fruit, nuts, and vegetables within seconds. The basic design of the Magimix has remained unchanged. A simple addition has been a handle for the bowl. Because the original Magimix was based on the commercial model and was expensive, other manufacturers began to produce cheaper versions. Magimix responded with the Robot Chef. The first American model was the Food Processor by Cuisinart, introduced at the 1973 National Household Exposition in Chicago.

Braun's Multipractic of 1983 changed the basic configuration by placing the bowl and the motor side by side. This led to the introduction of "combination" machines that had both a processor and a juicer/blender. These are currently produced by Braun, Hamilton Beach/Proctor-Silex, Kenwood, Morphy Richards, Moulinex, and Sunbeam.

The food writers Prue Leith and Susan Campbell have noted that food processors helped to develop the Nouvelle Cuisine of the 1980s, thanks to their ability to puree vegetables that could be used as a roux to thicken dishes and produce mousses.

See also Blenders/Juicers; Braun; Food Mixers

References:

Campbell, Susan. *The Cook's Companion.* London: Chancellor Press, 1980.

Furnival, Jane. *Suck, Don't Blow.* London: Michael O'Mare, 1998.

Food Warmers

Most of us eat hot food soon after we have cooked it, but there are occasions when the food has to be kept warm for longer periods. This problem is by no means new, and food warmers have been in use for centuries. They keep the food warm before it arrives at the table and also allow second helpings to be served hot.

Large country houses often had coal-fired warmers to keep the food hot prior to serving. These were essential if the kitchen was a long way from the dining room, as many of them were. In smaller houses the kitchen range, and later the Aga, had ovens that could keep food hot without burning it.

Table food warmers can be very simple. Many restaurants use heated metal trays with insulated handles. A simpler domestic variant is the candle warmer. This is "powered" by night-lights, small candles in white metal cases, that sit in a rectangular tray below a perforated stainless steel shelf on which the food sits. Single-candle models were also popular in the 1950s for keeping coffeepots warm. These were often of metal wire, occasionally decorated with ceramic or copper inserts.

Some of the earliest electrical appliances of the 1890s were simple hot plates. Al-

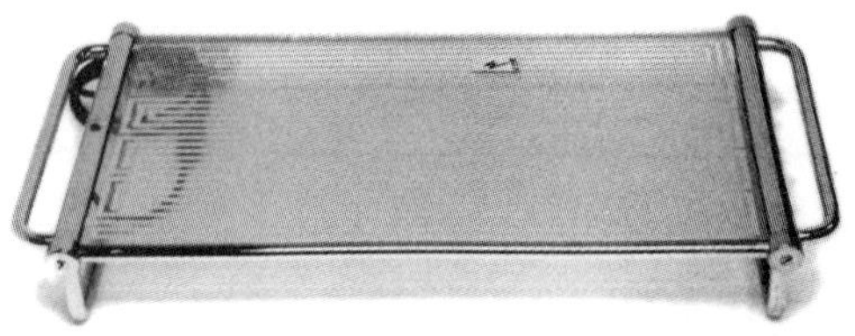

1960s electric warming tray, with the conductive element "printed" on the glass surface (The Museum of Science and Industry in Manchester)

though lacking thermostatic controls, these were modified into heating stands for the table. Such a model was featured in a photograph entitled "Electricity's Aids to the Tea Table," which appeared in the *Daily Mail* in 1920. The development of thermostatic controls allowed safer and more efficient models to be developed. Electric trays of the 1970s and early 1980s usually had a glass heating surface fitted inside aluminum or chrome-plated frames with wooden handles. The trays could be heated to a maximum of 93°C (200°F). Some, such as the Salton Hotray, had "hot-spots" that reached up to 120°C (248°F) for keeping coffeepots warm.

A development of the late 1950s was the electrically heated serving trolley, which could keep a whole meal warm. The largest models, cabinets on wheels, featured a sliding lid that opened to reveal glass and metal compartments for different vegetables or courses. Below was a larger warming cupboard with a tray for meats and casserole dishes. They also came with Pyrex dishes that could be carried straight to the table. Most were finished in a foil wood-grain finish in an attempt to make the product blend in with the furniture of the dining room. Smaller, sideboard versions were also produced. The most popular British model was the Ekco Hostess line, which gave the product the generic name of the "hostess trolley." (At this time Ekco was a division of Pye; Philips now produces the Hostess line).

A domestic adaptation of commercial food warmers, the serving trolley was popular for the emerging middle-class custom of giving dinner parties. It allowed a host or hostess to cook a complete meal in good time and leave it until the guests arrived. Their popularity declined with the rise of more informal eating habits and the arrival of the microwave oven. They were often seen as a symbol of middle-class domesticity.

Serving trolleys were one of the first large appliances to venture out of the kitchen into the more conservative areas of the house where the use of plastics and other new materials was culturally proscribed. Like the television hidden inside a mock regency cabinet they have been criticized for trying to look like something they were not. Serving trolleys are still in production, and a twenty-first-century model has a more confident brushed stainless steel finish.

See also Cookware; Ekco; Philips Electronics; Pyrex

References:

Boxshall, Jan. *Every Home Should Have One: Seventy Five Years of Change in the Home.* London: Ebury Press/Good Housekeeping, 1997.

Campbell, Susan. *The Cook's Companion.* London: Chancellor Press, 1980.

Freezers

The freezer (or deep-freezer) is an artificially cooled cabinet used to freeze fresh foods and store prefrozen foods. While frozen foods will stay frozen at any temperature below freezing point, temperatures have to be well below freezing point in order to freeze fresh foods sufficiently quickly to maintain their cell structure. Most domestic freezers are built to operate at between –20°C and –40°C (–4°F and –40°F). Domestic freezers did not become commonplace until the 1970s.

In 1842, H. Benjamin took out a British patent for a method of freezing by immersion in ice and brine, but the potential of this technology for food preservation was not recognized. Similarly, Ferdinand Carré developed an ice-making machine in 1857, but this is more significant as the inspiration for the absorption type of domestic refrigerator than as a prototype freezer. In 1908, H. S. Baker froze fruit in Colorado, but the first person to fully appreciate the value of freezing food was Clarence Birdseye. Birdseye was employed as a naturalist by the U.S. government and was sent on an expedition to Labrador, Canada, in 1912. Birdseye observed that meat and fish that froze naturally in the Arctic conditions seemed fresh when cooked and eaten months later. He returned to the United States and in the early 1920s developed a method of rapid freezing with food placed between metal plates. He formed a company to exploit his technology and filed a U.S. patent, granted in 1930.

The first commercial frozen foods produced by the Birdseye process, cartons of frozen vegetables, were sold in the United States in 1929. These were marketed on the assumption that they would be eaten on the day of purchase, given that few homes had refrigerators for short-term storage. Therefore, for consumers, the only advantage of frozen foods initially was the availability of vegetables out-of-season. In 1930, a wider range of frozen foods—fish, meat, and vegetables—were test-marketed in Springfield, Massachusetts. For growers, there was the advantage of reducing wastage through natural decomposition and loss of income. Consumption of frozen foods in the United States increased rapidly in the 1940s. Specialist self-service, frozen-food centers appeared in the New York area in 1945. Frozen foods did not reach Britain until 1946.

The American company Frigidaire brought out a domestic freezer in 1930, but production soon ceased owing to lack of demand. The first commercially successful domestic freezer was the Deepfreeze, launched in 1950 by the American company of that name. This was a small chest freezer, designed to maintain temperatures no higher than –18°C. The first domestic chest freezer capable of maintaining low enough temperatures to freeze fresh food was available in the United States in 1955. In Britain, the Swedish company Electrolux brought out the City Box chest freezer in 1956. Within a few years, it became standard for refrigerators to have small frozen food compartments.

Frozen meals went on sale in 1954, and became known in the United States as TV dinners. As the range of frozen foods increased, the domestic freezer became a desirable appliance. In the early 1960s, manufacturers tailored the freezer to the requirements of the modular fitted kitchen by developing smaller front-opening freezers, with matching refrigerators. The front-opening dual fridge-freezer, with the two units stacked in one tall cabinet, appeared in the same period. Its verticality was out of alignment with the horizontality of the fitted kitchen, but the advantage for the consumer was that the combined fridge-freezer with one mechanism was cheaper than separate units. The only major development in freezer technology since then was the introduction of frost-free freezing by the Dutch company Philips in the 1980s.

Even in the United States, where the majority of homes had a refrigerator by the early 1940s, the freezer remained a luxury appliance until the 1970s. In Britain, home ownership of freezers grew from 4 percent in 1970 to 41 percent in 1978. Freezer sales were boosted in the 1980s by the growing ownership of microwave ovens, both reflecting changing lifestyles. Households without an adult based full-time in the home appreciated the convenience and immediacy of the frozen or chilled ready meal cooked rapidly in the microwave oven.

Together, the freezer and commercial frozen foods had a dramatic effect on eating and shopping habits and diet in the last quarter of the twentieth century. Preserving food by freezing does not require the addition of preservative substances, such as salt. Frozen food retains both its natural flavor and its nutritious value. Freezing has increased the availability of highly perishable foods, such as soft fruits, and made our diets less restricted by seasonal factors. However, the dominant purchasing power of frozen-food manufacturers has reduced the availability of some fresh fruits and vegetables in season, thereby constraining consumer choice.

See also Convenience Foods; Electrolux; General Foods Corporation; Refrigerators; Whirlpool Corporation

References:

Byers, Anthony. *Centenary of Service: A History of Electricity in the Home.* London: Electricity Council, 1981.

Giedion, Siegfried. *Mechanization Takes Command.* New York: Oxford University Press, 1948.

Frying Pans

The frying pan remained little changed for over 300 years. Whether made of tinned copper or cast iron the frying pan had a broad, shallow body and a long handle to keep the cook's hand out of the fire. A close relative was the chafing dish, which by the late nineteenth century was a pot or pan that sat in a lower pan of hot water. Both were supported by a stand over a flame below. The heat maintained the water at a simmer, which allowed for the slow cooking of foods like soups and fondues.

The common frying pan was amongst the first objects to be electrified in the 1890s. A British example dates from 1898. It had an element fitted below the pan and socket at the end of the wooden handle. Due to the cost of electricity it was a luxury item. It never gained popularity when electricity became more widespread, as the increasing efficiency of gas and electric hot plates meant that the traditional pan was just as effective and easier to use.

In 1911, Westinghouse introduced an electric chafing dish. Made of sheet steel, it could be turned over and used as a hot plate. Little development followed. The main setback was developing a dependable and easily variable heat control that could compete with a traditional hotplate. In 1953, Sunbeam introduced the Automatic Frypan. It was a square cast-aluminum pan with a built-in element. The black plastic handle featured a heat control and "fry-guide" reminiscent of the "mix-finder" of the Sunbeam Mixmaster. The traditional frying pan gained a new lease of life with the introduction of nonstick coatings in 1956 that made it even easier to use and clean. The quality of the coatings varied. The UK Good Housekeeping Institute received calls from worried users who enquired if it was safe to eat the black bits of coating that were coming off the pans! By the late 1960s the quality of the coatings had improved.

The electric fry pan could also stew, braise, and bake. With the lid on, it could also be used for roasts and casseroles. By the 1970s it was also known as a multicooker. This versatility was limited by its size and was soon challenged by the microwave. Although still in production, the electric frying pan never gained mass acceptance as a replacement for its traditional rival.

See also Sunbeam; Tefal; Teflon

G

General Electric

General Electric (GE) was founded in 1892 with the merger of the Thomson-Houston Electrical Company and the Edison General Electric Company. Both had concentrated on electric lighting and the associated heavy electrical engineering to supply the power for industry, public utilities, railroads, and street lighting across the United States and in Britain.

GE had begun to produce some domestic appliances before 1920. Irons, toasters, and heaters had been produced before 1914, but it was in 1922 that a separate appliance and merchandise department was formed. GE saw that the American economy was expanding, and it realized that there was a huge potential market, which would follow in the wake of electrification. For a giant company like GE, it was easier to expand in the appliance market by acquiring smaller manufacturers like the Hughes Stove and Hotpoint companies than establishing its own divisions. The Hotpoint Company had begun by making electric irons whose soles were as hot at the tip as in the middle. Between 1925 and 1930, GE introduced larger appliances such as stoves, refrigerators, washing machines, and vacuum cleaners. GE also introduced the Calrod metal surface unit for electric stoves. It also began to spend more on advertising, concentrating on creating its own GE brand in addition to its subsidiaries. Between 1922 and 1930, the annual advertising budget rose from $2 million to $12 million.

GE is well known for its larger appliances. The launch of its refrigerators, such as the Monitor Top of 1926, was accompanied by a huge advertising campaign. GE was later to embrace industrial design when they engaged Henry Dreyfuss to design a refrigerator in 1935. This replaced the utilitarian Monitor Top.

GE continued to take advantage of the increasing prosperity of America in the 1950s and 1960s, as refrigerators became necessities rather than luxuries. GE remains one of the largest appliance manufacturers in the United States and is particularly strong in stoves and refrigerators, with a number of brands, including GE, Hotpoint, RCA, and GE Monogram. GE operates in over 100 countries, with 250 manufacturing plants in 26 countries. It employs 340,000 people worldwide, including 197,000 in the United States. GE sold its home electronics business to the French Thomson Electronics in 1987, which now produces answering machines, camcorders, CD players, televisions, and audio and video equipment under the GE brand.

See also Cookers; Irons; Industrial Design and Designers; RCA; Refrigerators; Toasters; Washing Machines

References:

Nye, David E. *Electrifying America: Social Meanings of a New Technology.* Cambridge, MA: MIT Press, 1990.

www.ge.com

General Electric Company

From modest beginnings, the General Electric Company (GEC) became the largest British electrical company with diverse interests in the fields of heavy electrical engineering, appliance manufacture, telecommunications, and advanced defense technology. Following a series of mergers, acquisitions, and divestitures in the 1980s and 1990s, the company was renamed the Marconi Corporation in December 1999.

The company began life in 1880, when a German immigrant, Gustav Binswanger, began an electrical wholesale business in London. Binswanger changed his surname to Byng and in 1886, Byng formed the General Electric Apparatus Company in partnership with another German immigrant, Hugo Hirst. The company produced its first catalogue of electrical goods in 1887. It made the transition from selling to manufacturing electrical goods in 1888 by acquiring a factory in Manchester, where it began to make telephones, electric bells, overhead light fixtures, and switches. The company was reconstituted as a private limited company, the General Electric Company in 1889. GEC began in

1920s GEC travel iron, shown here in its alternative mode as a hot plate for heating water for shaving or a hot drink (The Museum of Science and Industry in Manchester)

1893 to manufacture electric incandescent lamps, which became a highly successful line of business. In 1909, GEC acquired the rights to use the Osram trademark, which had been adopted by the German company Auer Gesellschaft in 1906 to mark the addition of tungsten filament lamps to its existing line of osmium filament lamps.

In 1900, GEC was incorporated as a public limited company. Two years later, it began production at its first purpose-built factory, the Witton Engineering Works, near Birmingham. It manufactured a range of electrical domestic appliances, including irons and stoves, under the Magnet brand name. By 1914, GEC had branches in Europe, Japan, Australia, South Africa, and India. During World War I, the company was required to direct its efforts toward military production. Its output included radios, signaling lamps, and arc-lamp carbons. Wartime profits enabled GEC to acquire the heavy engineering company, Fraser and Chalmers, in 1918. This placed it in a position to play a major role in supplying plant and equipment for the development of the national electricity grid from 1926. It moved to a new head office in 1921 and set up industrial research laboratories in London two years later. The company also expanded by diversifying into new areas, such as radio production. By the outbreak of World War II, GEC was well positioned to contribute to many aspects of military defense and communications technology, whilst maintaining production of essential civilian goods such as lamps. The development of the cavity magnetron, which improved radar systems, was one of GEC's major wartime achievements.

The postwar years were a period of relative stagnation for GEC, which faced increasing competition both in Britain and abroad. The turning point came in 1961, when GEC acquired Radio and Allied Industries. This takeover brought Arnold Weinstock (later Lord Weinstock) into the GEC fold. Weinstock became managing director of GEC in 1963 and soon established himself as one of the most dynamic figures in British industry. He began to strengthen the company through a process of internal streamlining and restructuring, combined with an aggressive approach to mergers and acquisitions. In 1967, GEC acquired one of its major British competitors, Associated Electrical Industries (AEI), for £120 million. Formed in 1929 as the financial holding company for a group of electrical manufacturing companies, of which the leading constituents were British Thomson-Houston (BTH), Metropolitan-Vickers (Metrovick), and Edison Swan. AEI had suffered financially from internal duplication and competition. One example of this was that in the 1930s BTH and Metrovick independently developed jet engines, thus becoming the first companies in the world to do so. AEI had a strong position in the British electric lamp market, particularly through BTH's Mazda brand and the Ediswan brand, and its Hotpoint Electric Appliance Company was successful in the buoyant domestic appliance market of the 1950s. By the mid-1960s, this combination of strengths and weaknesses made it ripe for takeover.

Weinstock followed up his AEI coup by orchestrating a merger with another major rival, English Electric, in 1968. English Electric was the result of the merger of a number of British heavy electrical engineering companies in 1918. It embarked on a program of diversification in the late 1930s by beginning production of diesel-electric locomotives and aircraft. In 1946, English Electric took over Marconi's Wireless Telegraph Company, thus acquiring interests in telecommunications and radar.

In the 1970s, GEC continued to expand and diversify, ending the decade as Britain's largest private employer. In 1974, it took over Yarrow Shipbuilders, and in 1979 it acquired Avery, manufacturers of weighing equipment. During the 1980s, GEC's business became increasingly oriented toward the high-technology telecommunications and defense sectors, which offered the greatest growth potential. The upgrading of telephone networks following the deregulation of telephone services prompted GEC and Plessey to form GEC-Plessey Telecommunications (GPT) in 1988. Subsequently, GEC and Siemens jointly acquired Plessey. Another joint venture was launched in 1989, when GEC and the French company, Compagnie General D'Electricitie, became equal partners in GEC-Alsthom, which operated in the power generation and transport sectors. The final acquisitions during Weinstock's reign as managing director were the takeover of parts of Ferranti in 1990 and of Vickers Shipbuilding and Engineering in 1995.

Lord Weinstock retired in 1996, and his successor began another phase of restructuring to consolidate business in the high-technology areas. The new strategy hinged on the sale of peripheral businesses and the dissolution of joint ventures. In 1998, Alsthom was floated as an independent company, generating the funding for GEC to buy out Siemens' stake in GPT. GEC then strengthened its international position in the telecommunications and information technology markets by purchasing the U.S. companies Reltec and Fore Systems in 1999. Marconi Electronic Systems, formerly GEC-Marconi, was divested in 1999 and merged with British Aerospace to form British Aerospace Systems. This left GEC with 45,000 employees and business interests in 100 countries. The process was completed in December 1999, when GEC was renamed the Marconi Corporation.

See also Electricity; General Electric; Lighting; Telephones; Westinghouse Electric

References:

Jones, R., and O. Marriot. *Anatomy of a Merger: A History of GEC, AEI, and English Electric.* London: Jonathan Cape, 1970.

www.marconi.com/about_marconi/heritage/

General Foods Corporation

The General Foods Corporation was the name chosen for the expanded U.S. food manufacturing company created by the Postum Company's acquisition of the General Seafood Corporation and the Salt Company in 1929. The founder of the Postum Company was C. W. Post, who developed a range of cereal products at his base in the small town of Battle Creek, Michigan. Post's first product was Postum, a cereal beverage that he devised in 1895. Battle Creek was the headquarters of the Seventh-Day Adventists, a religious sect that placed great value on healthy living, including a wholesome diet. It became famous as the center of breakfast-cereal production after Post and the Kellogg brothers set up factories there. In 1897, two years after the Kellogg brothers had introduced Granose, the first flaked breakfast cereal, Post marketed his Grape Nuts cereal. Post seemed destined to follow in the wake of the Kelloggs, introducing his Post Toasties corn flakes in 1908, ten years after the Kellogg equivalent. In 1927, the Postum Company began to diversify, acquiring the rights to market Sanka decaffeinated coffee.

Meanwhile, naturalist Clarence Birdseye had developed a method for replicating the natural fast freezing of meat and fish that he had seen in the Arctic. He had observed that this rapid freezing meant that the food seemed fresh when cooked and eaten months later. In 1924, he perfected a method of "flash-freezing" by placing cartons of food between metal plates under pressure, and formed the General Seafood Corporation. Birdseye sold his company to the Postum Company for $22 million, on the condition that his surname be used as two words, hence the Birds Eye brand name. The first cartons of Birds Eye frozen vegetables went on sale in the United States in 1929. Following test marketing in Springfield, Massachusetts, an expanded line of Birds Eye frozen foods went on sale across the United States in 1931. In response to the growing market for instant meals, Birds Eye precooked frozen chicken, beef, and turkey meals were introduced in 1939.

In 1942, General Foods produced an instant coffee for the United States Army that it marketed to the public as Maxwell House instant coffee after World War II. The brand name originated in 1892, when it was applied to a special coffee blend developed for the Maxwell House Hotel in Nashville, Tennessee. General Foods continued to bring out new products that reflected changing lifestyles and dietary tastes. In 1957, it introduced Tang beverage crystals as a new breakfast drink. In 1966, it launched Shake 'n' Bake coating mix in chicken and fish flavors and Cool Whip, a nondairy whipped topping. General Foods latched onto the new designer coffee trend of the 1970s, introducing its International Coffees line of flavored coffees in 1973. It acquired the meat products company Oscar Mayer & Company in 1981. Oscar Mayer had been founded in 1883, when three brothers leased a meat market in Chicago.

In 1985, General Foods was itself acquired by the Philip Morris Group, which then acquired Kraft in 1987. General Foods and Kraft made a complementary combination, as Kraft was primarily a dairy products company. Founded in 1903, Kraft began as a wholesale cheese business in Chicago. It developed a line of processed foods, including Velveeta processed cheese (introduced in 1928), Miracle Whip salad dressing (1933), and Cheez Whiz processed cheese spread (1952). In 1989, Philip Morris amalgamated the two food companies to create Kraft General Foods, the largest U.S. food manufacturer. The acquisition of Nabisco brand cold cereals from the R. J. R. Nabisco Group in 1993 expanded the company's share of the breakfast cereal market. Known from 1995 as Kraft Foods, the company continues to modify existing lines and develop new lines. The Birds Eye name is now the property of Agrilink Foods, which acquired it through purchasing Dean Foods in 1998.

See also Convenience Foods; Freezers

References:

www.birdseye.com/about_body.html
www.kraft.com/corporate/about/

Glen Dimplex

The Glen Dimplex Group is one of the largest manufacturers of electric heating and domestic appliances in the world. It is a privately owned Irish group with an annual turnover of over IR£600 million. The company currently employs around 6,000 people and manufactures in Ireland (North and South), England, France, Germany, and Canada.

It has grown from the Glen Electric Company, which was established in Newry, Northern Ireland, in 1973 with 34 employees. The company expanded throughout the 1970s and 1980s and has acquired a number of very well-known British brands, including Dimplex in 1977, Burco and Morphy Richards in 1985, Belling in 1992, Roberts Radio in 1994, and the Goblin and Aquavac vacuum cleaner brands in 1998. It has also expanded into the European heating market with a significant holding in the French heating group Muller and the acquisition of the German portable electric heater and small appliance company AKO ISMET Elektrogerate GmbH. Although the parent group is virtually anonymous, it has astutely built upon a growing stable of long established brands within the heating, floor-care, and small appliance market.

See also Belling Company; Cookers; Heating; Irons; Morphy Richards; Vacuum Cleaners

References:

www.glendimplex.co.uk

Glisdomes

Introduced in 1993, these are small plastic discs with a nonstick Teflon coating on one side. They are designed for moving furniture and large appliances such as washing machines and refrigerators. They are placed underneath at each corner and the appliance slides rather than having to be "walked." They normally work well on carpets and vinyl floors but less so on concrete.

See also Teflon

Goblin

The floor-care appliance company Goblin was founded by the British inventor of the electric vacuum cleaner, Hubert Cecil Booth. Born in Gloucester in 1871, Booth

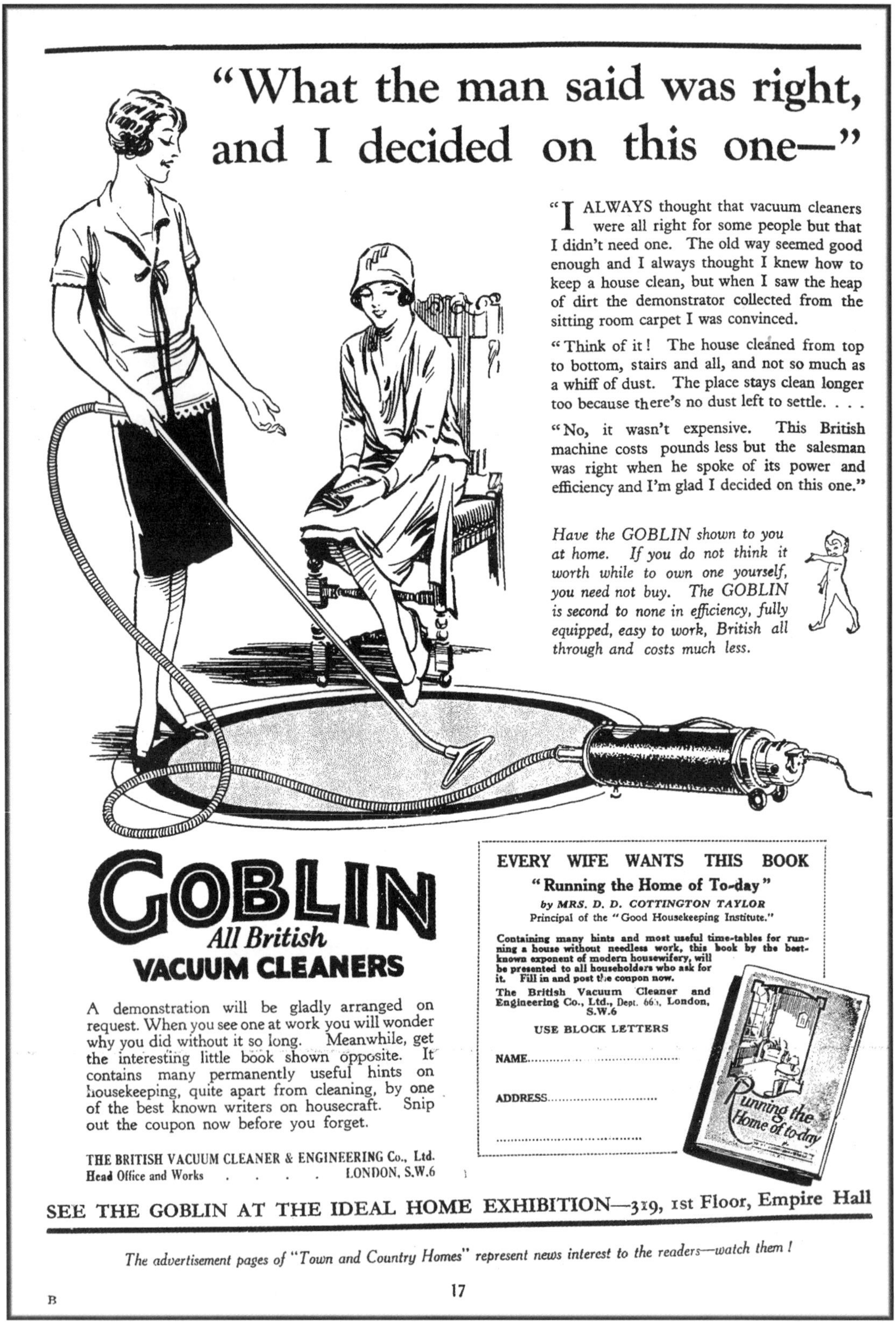

Goblin advertisement, Town and Country Homes *(April 1930) (The Museum of Science and Industry in Manchester)*

completed a three-year course in mechanical and civil engineering at the City and Guilds Central Institution in 1892. His early engineering career was in the fields of marine engineering and Ferris wheel construction. In 1901, he set up his own engineering consultancy and worked on a variety of civil and mechanical engineering projects. His interest in vacuum cleaners was awakened by a demonstration of an American cleaning system that involved blowing compressed air at carpets or upholstery to drive dust out. Booth quickly concluded that reversing the blowing action to produce suction would be more effective, as the dust would be collected rather than dispersed.

In 1901, Booth patented his vacuum cleaner, which in its original form became known as Puffing Billy. He offered a commercial vacuum cleaning service from 1902, and one of his early commissions was to clean blue pile carpets in Westminster Abbey for the coronation of Edward VII. His early vacuum cleaners consisted of a large electric motor or petrol engine, fan, and dust-collecting box mounted on a pushcart or horse-drawn cart. The cart stayed outside the building to be cleaned and long hoses were fed through doors or windows to extract the dust. He marketed his services by carrying out demonstrations using transparent suction tubes so that viewers could see the amount of dust being extracted. In 1903, he registered his company as the British Vacuum Cleaner Company. Recognizing that the market for his vacuum cleaning services was likely to be limited to public buildings and large houses, Booth developed a smaller version of his vacuum cleaner, the Trolley Vac, in 1904. However, this model was still heavy and expensive and thus unsuitable or unaffordable for smaller households.

After the introduction of the lighter, cheaper, and more mobile Hoover upright vacuum cleaner in the United States in 1908, European vacuum cleaner manufacturers began to develop more compact cylinder vacuum cleaners. Booth's company was the most successful of the British manufacturers. The company was renamed Goblin (BVC) Ltd. in 1924. Hubert Booth was chairman of Goblin until 1952. Today, Goblin is based in Tralee, Ireland, and is part of the Glen Dimplex group. In addition to the cylinder vacuum cleaner, its core product line, it also produces hand-held vacuum cleaners, wet and dry vacuum cleaners, and steam cleaners.

See also Vacuum Cleaners

References:
www.glendimplex.co.uk

H

Habitat

The Habitat group of stores was a pioneer in the field of retailing in the United Kingdom and had a profound influence on middle-class taste throughout the 1960s, 1970s, and 1980s. It was also a major influence on retailing and "lifestyle" shopping in Europe and the United States.

Habitat was the brainchild of the designer Terence Conran. Born in 1931, Conran trained as a designer and worked in furniture, fabrics, and ceramics in the 1950s. He traveled in Europe and became an advocate for the modernist influenced designs he saw in Italy and Scandinavia. He was also aware of the simplicity of rural styles that survived in France, along with the delights of French food, which was also being promoted by the food writer Elizabeth David. His earliest foray into the public arena was the Soup Kitchen cafe in London in 1954, followed by the Orrery restaurant in 1956. Both these ventures were well received, and Conran then began to turn his interests and energies toward furniture and household retailing. Apart from having a mission to promote good design, he was equally frustrated with the conservatism of the British retailers.

The first Habitat shop, a reflection of Conran's tastes, opened on 11 May 1964 in London, selling a variety of furnishings, furniture, and kitchenware. The interior was very simple, and the goods were stacked in bulk on shelves and on the floors. The store was a great success and four more branches had opened by 1968. The stock was a mixture of Bauhaus modernism in the form of Braun electrical appliances and traditional French Sabatier cooks' knives and Le Creuset cookware. Further expansion followed in the 1970s, with branches opening in the larger British cities and also in Paris (1975) and New York (1977). It also established a mail-order service based on its annual catalogues. By the early 1980s, it had forty-seven branches in Belgium, France, Iceland, and Japan.

The Habitat group continued to expand during the 1980s. In 1981, it merged with the children's clothes and equipment retailer Mothercare, another 1960s success story in lifestyle shopping. Conran also took over the long-established Heals furniture store in London in 1983 and later took over Richard Shops (a British clothes retailer) and the French FNAC chain. In 1986, the Conran Group became a major retail presence by combining with the British Home Stores, forming the Storehouse Group. This rapid corporate growth took Conran away from the Habitat chain, and it began to be accused of losing its way and becoming rather conservative. Following falling sales across the Storehouse Group Conran resigned as chairman in 1990, keeping only his Conran Shop, which had opened in 1986.

Ironically Habitat was bought by a competitor with very similar ideals, the Swedish giant IKEA, in 1996. Since then it has trimmed the number of stores and returned to its roots as a purveyor of "good design at sensible prices." In 1998, it appointed Tom Dixon as head of design, who has introduced a line of classic products from the 1950s and 1960s and has invested in young designers.

Terence Conran continues to be a force in design and popular taste through the Conran Shop (an upmarket Habitat) and a growing number of restaurants such as Quaglino's and the Bluebird. He is also active in publishing, and his Conran foundation funded the Boilerhouse Project at the Victoria and Albert Museum in 1982 and the establishment of the Design Museum in London in 1989.

Habitat has been a huge influence on late twentieth-century tastes and influenced the way many people think about how the things they buy and use look. This has led to a more visually literate group of consumers who have in turn influenced the designers, manufacturers, and retailers of appliances. By combining the traditional and the contemporary under the banner of "good design for the mass market" it educated an important and affluent market. Its influence is now carried on in Spain (Vincon), France (Prisunic), and America (Crate&Barrel, Williams Sonoma) and is best summed up by the words of Maurice Libby, one of the earliest staff members: "We were very ignorant in those days and didn't even know what a Mouli Grinder was. Terence would drag us, almost unwillingly, to his flat for meals and there we saw a lifestyle we'd never expected."

It is probably Conran's greatest achievement that this lifestyle is now shared by millions of "ordinary" people.

See also Braun; Cookware; Crate&Barrel; IKEA; The Modern Movement and Appliance Design

References:

Habitat Annual Catalogs, 1979 –1988.

Philips, Barty. *Conran and the Habitat Story.* London: Weidenfeld and Nicolson, 1984.

Hair Clippers

The hair clipper was the first electrical haircare appliance designed for use by men rather than women. Leo J. Wahl invented the first practical electric hair clipper in 1919 and set up the Wahl Clipper Corporation in Sterling, Illinois. The hair clipper basically consists of an electrically driven blade with a comb guide, and the design has changed little since its introduction. Today's hair clippers are supplied with a range of combs for different trimming grades and are also available as cordless models with rechargeable batteries.

See also Hair Dryers; Hair Stylers

References:

Matranga, Victoria Kasuba, and Karen Kohn. *America At Home: A Celebration of Twentieth-Century Housewares.* Rosemont, IL: National Housewares Manufacturers Association, 1997.

Hair Dryers

A hair dryer uses a fan powered by an electric motor to drive air over a heating element, producing a flow of heated air. In 1889, the fan became the first appliance to be fitted with an electric motor. A suitable heating element became available when Albert L. Marsh patented an element made from nonrusting nickel-chrome (or Nichrome) wire in the United States in 1906.

The classic pistol shape of the hand-held hair dryer was established by 1914. The motor and fan were housed at the back, above the handle, with the heating element in front of the fan in a cylindrical nozzle for channeling hot air. Early hair dryers had wooden handles and casings of lightweight metal, such as chromium, stainless steel, or aluminum. Perforations in the casing allowed air to be drawn onto the fan. In the 1920s, an

Hair dryer with chrome-plated metal shell and wooden handle, c. 1930 (The Museum of Science and Industry in Manchester)

alternative to the hair dryer was the hair dryer attachment for a vacuum cleaner. A number of companies, including Air-Way and Eureka in the United States, produced vacuum cleaners that, by means of the appropriate attachment, could be used for other functions such as hair drying and paint spraying. In the 1930s, the use of Bakelite in models such as the Supreme, made by the British company L. G. Hawkins, enabled the shape of the hair dryer to become more streamlined. The Bakelite shell was formed as two pieces that were screwed together, giving easy access for repair.

Short, permanently waved hairstyles, which became increasingly popular in the 1920s and 1930s, did not require styling as the hair dried, so floor-standing hooded hair dryers were developed for hairdressing salons. These hair dryers were too bulky and expensive to have much potential for home use, even in a more compact, less robust form.

After World War II, lighter-weight plastics, such as melamine, replaced Bakelite. The basic shape of the hand-held hair dryer remained the same, but new pastel shades changed the look. Functionality was enhanced by the provision of variable speed and heat settings and optional attachments. These attachments included soft plastic hoods with flexible hose, brushes, combs, and stands, which left both hands free for styling hair. A typical British model of the 1960s was the Pifco Princess, supplied with a vanity case to hold the hair dryer and attachments. Compact travel hair dryers became available. By 1970, 33 percent of British homes had a hair dryer.

The 1960s established two trends in professional hair care that influenced appliances and products for use in the home: unisex hairdressing salons and the celebrity hairdresser. The blurring of the distinction between barber shops for men and hair salons for women was reflected in changes in hairdryer design. No longer was the hair dryer targeted predominantly at the female consumer. The pistol shape became more exaggerated and pastel shades were replaced by black, chrome, white, and strong colors. In the 1960s, the British hairdresser Vidal Sassoon became as famous as some of his celebrity clients and used his high public profile to create a hairdressing empire, comprising a chain of salons and a line of brand-name products. That trend has continued and hair dryers bearing the names of leading hairdressers compete with those bearing the names of longer established makers, such as Germany's Braun and America's Remington. Hair dryers for home use also share the same design and performance features as those for professional use, including an array of styling attachments such as volumizers, diffusers, straighteners, and concentrator nozzles. There are also "intelligent" hair dryers, such as the Natura by the Dutch company Philips, which has a sensor to monitor hair temperature and adjust heat output accordingly.

See also Hair Clippers; Hair Stylers; Vacuum Cleaners

References:

Hillman, David, and David Gibbs. *Century Makers.* London: Seven Dials, 1999.

Hair Stylers

Hair stylers (curling irons) perform the dual function of drying and styling. The nature of the styling function has been determined by the desired hairstyles of the period. Hence, the first type of electric hair styler, which appeared at the beginning of the twentieth century, was an electric heater for curling tongs. However, this was only a novelty item, and it was half a century before an electric hair styler with real commercial potential appeared. During the intervening period, hair styling relied on manual aids such as metal curl pins. The spiked cylindrical roller, developed in the late 1950s, was suitable for being electrically heated. The heated roller unit consisted of a tray in which metal pins supplied heat to the rollers before use. The range of hair stylers available today, which tend to take a wand shape, began to be developed in the 1970s. The styling head may take the form of a single tong (for curling), dual tongs (for straightening), a cylindrical or flat brush, or paired crimping plates. Similar stylers are available as plug-in appliances or in cordless form, using either rechargeable batteries or disposable gas cartridges.

See also Hair Clippers; Hair Dryers

References:

Matranga, Victoria Kasuba, and Karen Kohn. *America at Home: A Celebration of Twentieth-Century Housewares.* Rosemont, IL: National Housewares Manufacturers Association, 1997.

Hamilton Beach

In 1904, L. H. Hamilton and Chester Beach joined the Arnold Electric Company of Racine, Wisconsin. Beach was a former farm boy and Hamilton a former cashier. Here Beach improved a small lightweight "universal" electric motor capable of running on either AC or DC currents. Known as the fractional horsepower motor, this was an important advance in a country that used both currents and significant for the appliance industry. Beach and Hamilton joined together to produce their first product, an electric hand-held massager, in 1910. They founded their own company, the Hamilton Beach Manufacturing Company, in Racine in 1911. Its main aim was to develop more universal motor-driven appliances.

Hamilton Beach began with producing a commercial drinks mixer, The Cyclone. Wisconsin was a dairy state, and Racine was also home to the Horlick Malted Milk Company. Beach developed an electric drinks mixer for use in their factory. Malted milk was frequently prescribed by doctors and also promoted by the manufacturers as a health drink. As a result, the Hamilton Beach drinks mixer was purchased by drug stores and soda fountains. Its first domestic product was an electric sewing machine. This featured ingenious attachments, including one for mixing cake batter! Not surprisingly, blenders became one of its major lines.

The company was acquired by Scovill in 1923. Scovill had been established in 1802 as a brass foundry and grown to become a large corporation. Hamilton Beach became one of its eleven divisions manufacturing domestic and cooking appliances, producing an ever-expanding range of appliances around its core products of mixers and blenders. It introduced hair dryers in the 1930s, coffeemakers in the 1940s, and steam irons in the 1950s. In 1986 Scovill Inc. sold Hamilton Beach to Glen Dimplex Ltd, which later sold it to NACCO Industries of Cleveland in 1990. NACCO had already purchased Proctor-Silex, and in 1990 the two companies merged to become Hamilton Beach/Proctor Silex Inc.

Proctor-Silex began in 1920 as the Proctor and Schwartz Company. Having bought the Liberty Gauge and Instrument Company of Cleveland, which pioneered the thermostatic control in 1926, Proctor produced a variety of thermostatically controlled appliances, such as irons. Proctor Electric merged with Silex, who manufactured coffeemakers and irons, in 1960.

Hamilton Beach/Proctor-Silex is currently the largest manufacturer of small electric appliances in the United States, employing over 4,000 people worldwide. It sells around 32.5 million appliances every year, accounting for around one in four sales in the market. Both companies have had a number of different owners throughout their histories, but their strong brand identities (established in the 1930s and 1940s) have always remained.

See also Blenders/Juicers; Irons; Proctor-Silex

Heating

For centuries, the open fire, built on a hearth beneath a chimney, was the focal point of the home and a symbol of home comfort. It was the dominant form of domestic heating in 1900, although gas fires had been available for fifty years. While solid fuel, particularly coal, remained cheap, there was little incentive to break with tradition, even though the disadvantages of open fires were all too evident. Carrying coal or wood to the fireplace was heavy work, and laying the fire was time-consuming, as were clearing away the ashes and cleaning residues from surfaces. The open fire was also slow to reach its full heating strength. Not surprisingly, convenience and cleanliness were the major selling points cited in favor of gas and electric fires. While open fires continued to be widely popular until the middle of the twentieth century, they eventually fell victim to a combination of changing lifestyles and environmental concerns. In the 1950s, the creation of smokeless zones were among the measures taken to improve air quality in increasingly smoky cities. This prompted the development of self-contained heaters to burn smokeless fuels, such as coke and coalite (a lightweight, smokeless coal formed into pellets), but many people had come to want instant heating that did not require laborious preparation. For this reason, most homes in the late twentieth century, whatever their main form of heating, possessed at least one supplementary electric fire (or heater) as a standby.

Improving the open fire was a desirable goal for eighteenth-century inventors. In 1740, the American statesman and scientist Benjamin Franklin invented a ventilated cast-iron wood-burning stove, which retained and threw off more heat than an open fire. Sixty years later, the American-born political advisor and inventor Benjamin Thompson, Count von Rumford, redesigned the open fireplace to form a compact and relatively efficient heat source. Rumford reduced the hearth recess and the opening to the chimney in order to concentrate the heat source and lessen the heat lost from the room. Gas heaters became available in the mid-nineteenth century, as a result of the increased market created by the piping of gas to more homes. The technology of the gas heater has changed little from its introduction to the present day, the main exception being the development of automatic ignition systems, which use sparks from flint or piezo-electric crystals to light the fire. Like open fires, gas fires needed to stand on the hearth so that the combustion fumes were drawn up the chimney, and they were derivative of the typical fireplace in terms of materials and basic design. The fireclay radiant bars were placed in black cast-iron surrounds, set on cabriole legs, and embellished with modest decoration, for example in the form of raised or recessed scrolls. Gas companies made little effort to promote the sale of gas heating or cooking appliances until the advent of electricity challenged gas lighting, which was responsible for the bulk of gas consumption.

Unlike gas heaters, electric heaters have undergone a complex evolution. An early patent for an electric heating system was granted to its American inventor, Dr. W. Leigh Burton, in 1887. The first British electric heaters were made by Crompton & Company and exhibited at the 1891 Crystal Palace Electric Fair in London. They were based on a type of radiant described in the patent for an electric cooking vessel granted to the British inventor St. George Lane Fox in 1879. The radiants were formed by embedding iron resistance wires in enamel backed by iron panels. The Crompton & Company heaters became commercially available a few years later as wall-mounted appliances or set in wrought-iron screens as floor-standing appliances. This type of radiant was slow to heat and cool and thus inferior to contemporary gas heaters. However,

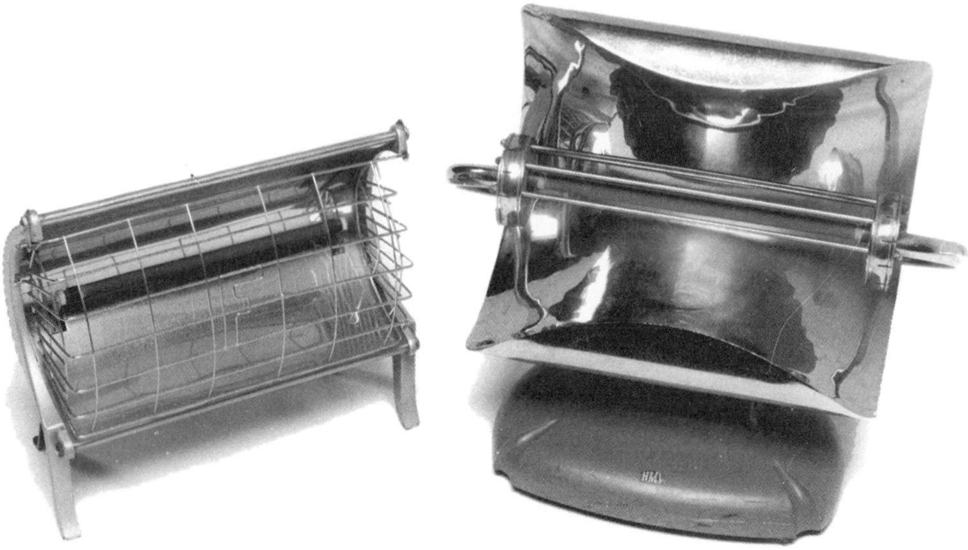

Ferranti single parabolic reflector heater, c. 1930, and HMV Bruton double parabolic reflector heater, 1952 (The Museum of Science and Industry in Manchester)

the great advantage of the electric heater, which was immediately apparent, was that it could be placed anywhere in the room subject only to being within reach of an electric socket. The notion of a centrally placed hearth and chimney was irrelevant to the electric heater, as it did not require venting. In 1896, the British inventor H. J. Dowsing developed the first practical electric radiant heater. Dowsing's radiants were large sausage-shaped carbon filament lamps, each with an output of 250 watts. A reasonable heat source was produced by mounting two, three, or four of the lamps vertically in front of a polished, curved reflector panel of copper or brass. Plain surrounds were the most common, but Dowsing radiators with art nouveau–influenced decorative surrounds were also available. Lightweight and easily portable, they offered modern, stylish, and convenient heating for the relatively few homes that were wired for electricity.

However, Dowsing-type heaters provided only a gentle background heat. Attention therefore returned to resistance wire radiants. The main problem with the iron resistance wires used in the Crompton & Company radiant heaters had been that they rusted and became brittle and wont to break at high temperatures. The solution was provided in 1906, when the American inventor Albert L. Marsh patented a nickel-chrome (or Nichrome) alloy wire, which did not rust or break at high temperatures. In Britain, the first radiant heater using Nichrome wire was produced in 1912 by Charles R. Belling, a former Crompton & Company apprentice who had set up his own factory in Enfield, near London. Belling devised a form of fireclay bar to support the coiled Nichrome wire. Early Belling radiant fires featured surrounds much like those of Victorian gas heaters. Some models had hinged, drop-down trivets for keeping kettles warm. The Bastian heater of 1909 introduced another type of element, which consisted of resistance wire threaded through a quartz glass tube.

In Britain, the prevailing sentimental attachment to the coal fire led electric heater manufacturers to devise simulated coal fires. The Falkirk Iron Company developed an early imitation coal fire in 1913. Seven years later, H. H. Berry introduced the Magicoal heater, which had the added attraction of a flicker effect. The Dowsing radiator was supplanted as the standard type of reflector heater by the parabolic reflector heater, espoused by the British electrical engineer Sebastian Ziani de Ferranti in 1910. A narrow wire element was placed along the focus of the sharply curved, chromed reflector, which was very efficient at distributing heat. Parabolic reflector fires and copper bowl heaters became popular in the 1930s, the most notable model being the HMV Bruton radiant fire heater of 1939, with its chrome double parabolic reflector. In 1926, new improved "Globar" elements, which took the form of silicon carbide rod elements, were launched in the United States. These were essentially more robust versions of the Bastian element.

An alternative to radiant heating was convection heating, whereby elements encased in a perforated cabinet heated the surrounding air, which therefore rose and was circulated around the room. Electric convection heaters first appeared before World War I, but they were bulky and slow to produce a discernible effect. By adding a fan, the resistance wires could carry a higher wattage without overheating, the rate of circulation improved, and the need for a large cabinet was thus reduced. This idea was postulated in 1929 and taken up by Belling in 1937. However, a standard bladed fan was noisy, so the Belling convector was not a great success. Today's small fan heaters were the product of the invention of the tubular tangential fan, a compact and almost silent fan, by the German engineer Bruno Eck in 1953. The idea of convection heating had been tested with little success in gas heaters in the early 1880s, but in the 1950s gas engineers were more successful. They devised a way to draw air from the room over the radiants and recirculate it back to the room without coming into contact with the combustion gases that were passed up the chimney.

By the 1950s, the ideal of the modern scientific home demanded more sophisticated and controllable forms of heating than those provided by stand-alone gas and electric fires. The concept of central heating based on the circulation of hot water through pipes was developed and implemented with great success by the ancient Romans two thousand years earlier. It was revived in eighteenth-century Sweden for heating greenhouses and subsequently adopted in larger public, commercial, and industrial buildings. Its suitability for household use was severely constrained by the size of plant involved. However, by the 1950s, smaller boilers and compact electric pumps were available. Cheap oil made oil-fired central heating an attractive option in the United States, whereas in Britain, coal-fired or gas-fired central heating was more usual.

While electricity could be used as the energy source for conventional central heating, another option was to network electric storage heaters so that they could be centrally controlled. The concept of electric storage heating was based on the desire by electricity utilities to balance the daily generating load by encouraging nighttime use of electricity at cheap rates. By 1900, electrical engineers recognized that all electric radiants retained their heat for some period of time after the current was switched off. The next challenge was to discover which materials held the heat best. Over time, it transpired that a number of materials, including concrete, brick, and soapstone (also known as stearite or magnesium silicate), were suitable, particularly if housed in well-insulated cases. The same effect could be produced by embedding resistance wires in concrete flooring, a form of under-floor heating used in British air raid shelters during World War II. When peak-time demand for electricity rose sharply in the 1950s, the British electricity industry saw storage heating as a way of offsetting the load. In 1962, the industry launched the Unit Plan domestic heating campaign, which encouraged the installation of storage heating by introducing new off-peak tariffs.

Heating is one of the household functions where the evidence suggests that performance has far outweighed design as a factor influencing consumer choice. Manufacturers of electric heaters did use design as a selling point during the promotional heyday of the interwar period, partly to suggest high performance through streamlining and chrome detailing. Lately, they have tended toward discreet and unobtrusive designs that blend in with their surroundings, such as the slim, wall-mounted electric panel heaters of the 1960s. However, there are exceptions to this trend, as designers in the 1990s adopted the radiator as a product that they, and the owners, could use to make an unexpected aesthetic statement through colorful and quirky styling.

See also Air-Conditioning; Belling Company; Cookers; Electricity

References:

Byers, Anthony. *Centenary of Service: A History of Electricity in the Home.* London: Electricity Council, 1981.

Lancaster, Maud. *Electrical Cooking, Heating, Cleaning, etc.: Being a Manual of Electricity in the Service of the Home.* London: Constable, 1914.

Wright, Lawrence. *Home Fires Burning: The History of Domestic Heating and Cooking.* London: Routledge & Kegan Paul, 1964.

Hewlett-Packard

The American high-technology company Hewlett-Packard is a classic example of the Silicon Valley success story. For the first twenty years of its life, the company built its reputation in the specialist electronic fields of sound analysis and microwave instrumentation. Diversification into the fields of medical and computer equipment in the 1960s brought Hewlett-Packard into the major business league. The company entered the Fortune 500 list of U.S. companies in 1962 and rose up the rankings to enter the Fortune top 50 in 1988.

The company was founded in 1939 as a partnership by William R. Hewlett and David Packard, both electrical engineering graduates from Stanford University in California. The two founders had been working part-time on the development of a resistance-capacity audio oscillator, an instrument for testing sound equipment. This became the company's first product, the HP 200A, eight of which were bought by Disney for use in the production of Fantasia. Hewlett-Packard went on to develop other sound analyzers, based, like the audio oscillator, on the principle of negative feedback. In 1942, the company moved from rented premises in Palo Alto to a custom-built factory with offices and laboratory space. During World War II, Hewlett-Packard consolidated its business by gaining orders for electronic instruments for military use, such as microwave signal generators and radar-jamming equipment.

In 1951, four years after becoming incorporated, the company had just over 200 employees and its annual revenue reached $5.5 million. The company continued to specialize in the field of sound analysis. Its high-speed frequency counter (HP 524A) dramatically decreased the time required to measure high frequencies. In 1957, Hewlett-Packard stock was offered to the public for the first time, and the company built a new factory in Stanford Research Park. A year later, Hewlett-Packard made its first acquisition when it took over the F. L. Moseley Company of Pasadena, which produced high-quality graphic recorders. In 1959, the company set up its first overseas operations—a factory in Boeblingen, West Germany, and a European marketing base in Geneva, Switzerland.

The 1960s saw Hewlett-Packard undertake a major diversification program. In 1961, the company moved into the medical equipment field by acquiring the Sanborn Company of Waltham, Massachusetts. The Boeblingen plant also began to manufacture medical equipment, such as its noninvasive fetal heart monitor, introduced in 1967. In 1966, Hewlett-Packard set up a central research facility and produced its first computer, the HP 2116A, which was designed for specialized use as a controller for other Hewlett-Packard instruments. Computer technology soon became an important part of the company's business. In 1968, Hewlett-Packard developed the world's first desktop scientific calculator, the HP 9100A. In 1969, the company produced its first minicomputer, followed in 1972 by its first minicomputer for general business use. In 1974, Hewlett-Packard pioneered the use of dynamic random access memory (DRAM) semiconductors instead of magnetic cores in minicomputers.

Hewlett-Packard was quick to appreciate the potential of the microchip. In 1971, it developed a laser inferometer, a precision measuring instrument that became a standard tool in microchip manufacture. The company's first microchip-based product was the HP-35, the first scientific handheld calculator, launched in 1972. Its first microcomputer, the HP-85, came on the market in 1980. Hewlett-Packard also recognized the growing need for interoperability of computer devices and developed the HP-IB interface bus as a standard interface for peripherals in 1975. This was adopted as an international standard. In 1982, Hewlett-Packard's British subsidiary developed the first electronic mail system for wide-area minicomputer networks.

In 1984, Hewlett-Packard entered the computer printer market with the launch of the HP ThinkJet printer and HP LaserJet printer. Both became phenomenally successful product lines, particularly the LaserJet, which set the standard for laser printing. By 1993, Hewlett-Packard had sold 10 million laser printers. Its first color laser printer was launched in 1994, when it also introduced the OfficeJet combined printer, fax machine, and copier. The company's most expensive research project culminated in the introduction of a series of computers based on Reduced Instruction Set Chip (RISC) architecture in 1986. Hewlett-Packard also continued to develop specialized scientific and medical instruments, such as the atomic emission detector, introduced in 1989, and the HP SONOS 1500 echocardiograph system, a noninvasive ultrasound cardiac analysis instrument introduced in 1991.

In the 1990s, Hewlett-Packard introduced a range of portable computers, beginning with the first of its palmtop computers, the HP 95LX of 1991. Other portable devices included the HP OmniBook 300 portable personal computer (1993) and the HP OmniGo 100 personal organizer (1995). The Pavilion range of personal computers was introduced in 1995 and became the second best-selling range in the U.S. market in 1998. At the higher end of the computer market, Hewlett-Packard acquired the Convex Computer Corporation of Richardson, Texas, in 1995. It took a strong position in the electronic commerce field in 1997 by acquiring VeriFone, a leading supplier of electronic payment systems. By 1998, almost sixty years after its formation, Hewlett-Packard had annual revenues of nearly $50 billion and more than 120,000 employees.

See also Calculators; Computer Printers; Computers

References:

Rogers, Everett M., and Judith K. Larsen. *Silicon Valley Fever.* New York: Basic Books, 1984.

www.hp.com

The Hoover Company

In 1907, William "Boss" Hoover was the owner of a leather-goods factory in New Berlin (later North Canton), Ohio, producing primarily saddles and harnesses. The direction of the business changed dramatically when Hoover acquired the rights to the 1908 patent for an electric suction sweeper, invented by his wife's cousin James Murray Spangler. As an asthmatic, Spangler was troubled by dust when carrying out the sweeping duties that were part of his job as the janitor in a department store. He made a prototype suction cleaner from a broom handle, pillow case, small motor, fan, and brushes. This formed the basis of his successful patent application.

The first commercial version of Spangler's suction sweeper was manufactured by Hoover as the Model 0 in 1908. More lightweight than previous vacuum cleaners, the Hoover suction sweeper was a great success. Hoover set up the

Electric Suction Sweeper Company, later renamed the Hoover Company. Spangler stayed on as partner until he had completed developing the improvements embodied in his 1915 patent, when he assigned the rights to Hoover and retired with a secure income. By 1920, annual sales reached 275,000, so Hoover dispensed with the vestiges of the original business by closing the tannery.

Hoover maintained its place at the forefront of the vacuum-cleaner industry by continuing to improve the product line and expanding its operations. In 1932, Hoover built a British factory at Perivale, on the outskirts of London, that became famous as an art deco landmark. The company moved into the production of other domestic appliances, including washing machines, irons, and toasters. The small appliances division was later sold off to refocus attention on the core floor-care product line. In the 1980s, Hoover's British subsidiary had a disastrous experience when a free-flights promotion boosted sales but at great cost both financially and to its public image.

In 1985, Hoover was acquired by the Chicago Pacific Corporation, which in turn was acquired by the Maytag Corporation in 1989. Today, Hoover has four manufacturing plants in the United States (three in Ohio, one in Texas), two plants in Mexico, and one in Canada. As befits the company whose name was adopted in Britain as a verb for vacuum cleaning, Hoover plays an active role in the industry and helped to develop the industry standard for measuring the effectiveness of dirt extraction from carpets.

See also Vacuum Cleaners

References:

Matranoga, Victoria Kasuba, and Karen Kohn. *America at Home: A Celebration of Twentieth-Century Housewares.* Rosemont, IL: National Housewares Manufacturers Association, 1997.

www.hoover.com

Hot Plates

Electric hot plates were amongst the earliest of the electrical products marketed in the 1890s. The British firm Crompton & Company produced a hot plate in 1891. These simple appliances consisted of a solid circular plate, usually mounted on cast tripod legs. Designed for the table, the legs echoed earlier spirit lamp food warmers.

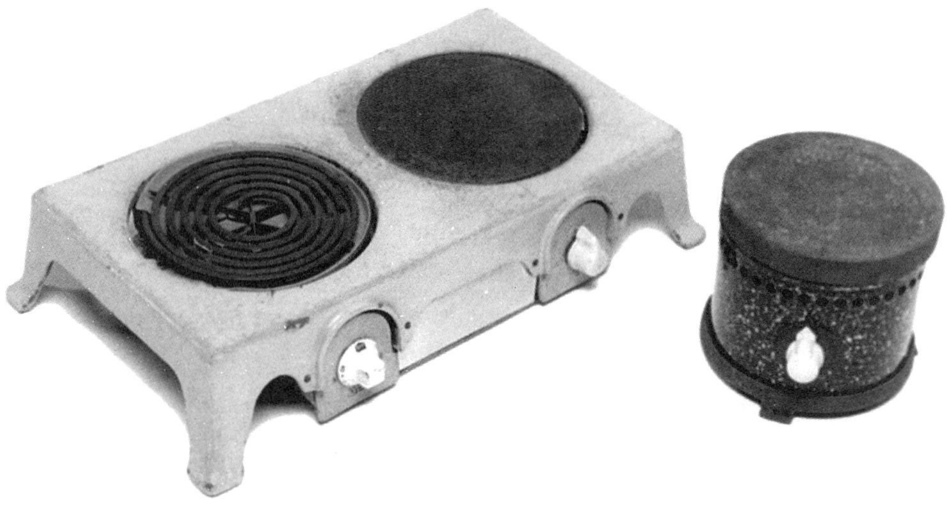

Creda double-burner hot plate, c. 1930, and Belling single hot plate, c. 1920 (The Museum of Science and Industry in Manchester)

Known in the United States as "table stoves," they were produced by Westinghouse, General Electric, and Universal. The introduction of the thermostat in the 1930s saw the hot plate become both a table cooker, like its predecessor the chafing dish, and—turned down low—a food warmer. Double rings were introduced, and some American companies produced a hot plate and coffeepot combination.

Hot plates reflected stylistic changes like other products. 1930s models are usually in chrome, whilst the 1960s saw the introduction of square models with inset ceramic tiles, decorated with colorful designs.

The rise of more convenient "off-stove" appliances like the microwave has led to a decline in the production of hot plates, although they remain useful for small-scale catering or parties.

See also Food Warmers

Houses

The twentieth century was one of continuing urbanization, especially in Europe and the United States. Cities on both sides of the Atlantic entered the twentieth century with a mixture of planned or redeveloped boulevards and neighborhoods, growing suburbs, and decaying slums. The majority of houses were built by private speculators, responding to booms and slumps in demand. The effects of warfare and town planning were to remove both good and bad houses, whilst changing tastes and conservation have helped preserve both exceptional and mediocre ones. The results have been an urban and suburban landscape that has incorporated new buildings such as cinemas and shops whilst retaining large numbers of nineteenth- and early twentieth-century houses and apartments.

The nineteenth century continued a trend that was established in the seventeenth century, an increase in the number of specialized rooms, even in relatively humble houses. In the United Kingdom, the growth of urban suburbs from the 1840s onward, did not lead to a radical change in house design. The majority followed the form of the Georgian terrace, or row house. In the United Kingdom, as opposed to much of Europe, there was also an aversion to flats and apartments, although many houses were sublet and in multiple occupation. By 1911, only about 10 percent of the housing stock was made up of detached or semi-detached houses and only around 3 percent were flats. The vast majority were built in terraces and varied from mean, cheaply built worker's terraces to grand metropolitan rows.

This process can be seen in the typical early twentieth-century suburban villa, which had an entrance porch, hall, drawing room or parlor, dining room, breakfast room, kitchen, scullery, washroom, bedrooms, box-room, bathroom, and toilet, as well as an attic and cellar. There were three distinct zones, one for receiving guests and entertaining, one for domestic work like cooking and laundry, and one for private family spaces.

This house was the result of industrialized production. The speculative builder would have purchased components from specialized suppliers and manufacturers. Bricks, tiles, doors, roof tiles, glass, ironmongery, grates, and fireplaces were all mass-produced, often in other parts of the country. The services such as water and gas were piped in from a more local, but centralized source. The main form of heat was from coal fires and the lighting was probably gas. Both were relatively dirty forms of heating and lighting, and the house required constant work if it was to be maintained to a high standard. Although this standard was by no means universal, it does illustrate that the technical solution to supplying centrally distributed water and power was becoming the norm. There was a similar situation in the United States. The home had become "industrialized" before it received electricity. Such an infrastructure allowed the electrical supply companies to develop.

Cities in the United States already had a considerable stock of apartment housing at

Levittown, New York, 1950s (Urban Archives, Temple University)

the beginning of the twentieth century. U.S. cities had more space than many European cities and the suburbs were developing, thanks to a combination of commuter railroads and industries moving out of town.

The electrification of homes occurred at a time when house design was beginning to change. In England, the influence of the Arts and Crafts Movement and architects like C. F. A. Voysey (1857–1941), although drawing inspiration from historical styles, inspired simpler, detached houses with more open plans. In the United States, Frank Lloyd Wright and others of the "Prairie" school were designing more horizontal houses with open interiors. Wright designed his houses for electricity, which was a cleaner form of light and heat and therefore more suited to an open plan. These architects did not pioneer electricity in the home but were certainly aware that it suited their aesthetic better than gas. Although these architect-designed houses were for the rich, their ideas trickled down to the speculators who appropriated some of the features. In the United Kingdom, this can be seen in the suburban houses of the 1930s. In the United States, debased elements of the Prairie style found their way into post-1945 houses in the form of open plans, interior stonework, and picture windows.

Modernism was to take this concept further. In 1914 Le Corbusier was considering the use of reinforced concrete construction as a way of eliminating internal walls. In 1923 he wrote his famous and often misunderstood phrase, "the house is a machine for living in." This is taken from his 1923 book *Towards a New Architecture* and has to be read in the context of contemporary thinking on domestic science, industrialization, and component production. Le Corbusier was promoting a reassessment of domestic design as a problem that, after centuries of evolution, needed reevaluating in order to provide a solution in a machine age. Le Corbusier's vision of architecture spread far beyond the individual house, and his famous dictum is better understood in full:

> The aeroplane is the product of close selection. The lesson of the aeroplane lies in the logic which governed the statement of the problem and its realization. The problem of the house has not been stated. Nevertheless there do exist standards for the dwelling house. Machinery contains in itself the factor of economy, which makes the selection. The house is a machine for living in. . . . As to beauty, that is always present when you have proportion.

Le Corbusier's three-dimensional proof was the Villa Savoye of 1929–1930, a building that was to influence modern house design for the rest of the century. The open-plan house with large windows was also promoted through the Master's Houses built at the Bauhaus, in Dessau, Germany, in the 1920s. This form resurfaced in the more welcoming climate of California and influenced mass housing through the work of Richard Neutra and Gregory Ain in the 1940s and 1950s.

The trend toward fewer specialized rooms is also linked to social and economic conditions. Houses with bathrooms, electricity, and modern kitchens were more expensive to build. Houses in the United States moved to having fewer rooms downstairs and smaller kitchens. Such houses could be ordered from Sears, Roebuck for self-assembly. British interwar housing was also smaller as there was no longer any need to house servants. Kitchens were smaller because they were designed to take a compact gas or electric stove rather than a large cast-iron range.

The arrival of electricity and domestic appliances followed in the wake of growing suburbanization, increasing domestic informality, and changes in house design. They did not start this process but did become integral to it. Electricity was promoted by builders and domestic scientists as being clean, modern, and efficient. By the end of the 1920s it was the major source of domestic lighting, and once a house had been wired for light it was a ready market for other electrical products. The electric light was, in the words of one General Electric manager, the "entering wedge." This was especially true for small, inexpensive appliances such as the electric iron and the toaster, both of which could run off a light socket.

In the United Kingdom, large municipal suburbs of "council houses" were built in the 1920s and 1930s. Available for rent, many of these were based on the "garden city" principles of the late nineteenth and early twentieth centuries. Speculative suburban houses featured a variety of styles, often with historical references, but all were wired for electricity.

The houses being built in the 1920s and 1930s may not have been the result of the rational reevaluation of Le Corbusier but they were becoming more industrialized. If there were fewer specialized rooms there was an increase in the number of machines for specialized tasks, and kitchens were being planned on "scientific" principles. What was equally important was that, by the end of the 1930s, the modern, electrified suburban house had become the aspiration for thousands of city dwellers on both sides of the Atlantic.

This hegemonic desire for a single-family home and private ownership meant that many women appeared to opt for living in labor-intensive houses that they were expected to maintain to ever-rising standards. In reality, there were few options that were better. In the United States, there had been attempts at providing communal resources in apartment blocks, including central kitchens and laundries. One reason they failed was because of the cost of employing people to do this work. There were also broader social reasons. In the United Kingdom, the use of communal facilities such as laundries and bathhouses was forever associated with poverty and poor housing, which lacked these amenities. In the United States, there was also a deeply rooted desire for independent, private homes. A move from the city to the suburbs was equated with status, a sign of moving up the economic ladder.

The suburbanization of the United States picked up speed after 1945. Home ownership doubled in the postwar period. This process was aided by Federal Housing Administration mortgage insurance and Veterans Administration mortgage guarantees. Between 1947 and 1957, these accounted for over 40 percent of new houses, the majority of which were in the suburbs. Initially rather box-like single-story houses, they developed exteriors based on either ranch or Cape Cod styles, with California modern following later. Internally, they borrowed from Wright and the later modernists, with picture windows, low mantel-less fireplaces, indirect lighting, and the use of wood and stone. The overall look was far more relaxed and informal. They were also made more comfortable with central heating.

New rooms appeared, especially the family room and the living kitchen. The family room started out as a playroom, or rumpus room, often in the basement, and evolved into an informal room with the television at its center. This has led to the informal living room becoming more of a formal reception room, not unlike the Victorian parlor. The "living kitchen" was a move away from the kitchen as a science laboratory and into an extension of the family and dining rooms where family members could also eat and relax.

Postwar housing in Britain took two differing directions. Municipal and speculative suburbs continued to be built, and local authorities experimented with high-rise flats in cities. Although very different, what united them was that once behind the front door the home was a private space.

Cultural and technological conditions therefore led to a form of housing that was lighter and cleaner but demanded a worker, usually a woman, to maintain it. By the 1920s, the appliance manufacturers could view this as an important market. This can be seen from their advertising, which from the 1920s to the 1960s depicted the housewife in the home. Most women bought into the promise of "labor-saving" appliances, as they did save time on some tasks, therefore allowing women to use that saved time to undertake other tasks, such as shopping and child care.

Throughout the twentieth century, exhibitions and trade shows have had a fascination with "houses of the future" that featured kitchens and furniture that popped up at the push of a button, lifts instead of stairs, centralized cleaning and waste disposal systems, convertible rooms, solar power, and the like. It is now possible to design and build small houses that can do all these things, but at a cost. In addition people must want to buy them, and public taste has, for the large part, remained conservative. In the United Kingdom and the United States, municipal tower blocks and housing projects have largely been discredited due to a combination of cheap building, lack of community resources, and poor maintenance. The conservation movements of the 1960s and 1970s have led to many people preferring to live in nineteenth- and early twentieth-century houses. Late twentieth-century speculative housing has, with the exception of central heating and insulation, changed remarkably little from the 1950s and 1960s. Externally, most are an architectural mishmash of historical plundering. Internally, all modern houses are built with a central gas and electricity supply and a fitted kitchen. To function properly as self-contained units, they still require the installation of a host of task-specific appliances and that final component, someone to work them.

See also Consumers; Heating; Kitchens; Lighting; The Modern Movement and Appliance Design

References:

Barrett, Helena, and John Phillips. *Suburban Style: The British Home, 1840–1960.* London: Macdonald Orbis, 1987.

Blake, Peter. *The Master Builders.* New York: Alfred A. Knopf, 1960.

Hine, Thomas. *Populuxe.* New York: Alfred A. Knopf, 1986.

Muthesius, Stefan. *The English Terraced House.* New Haven, CT, and London: Yale University Press, 1982.

Nye, David E. *Electrifying America: Social Meanings of a New Technology.* Cambridge, MA: MIT Press, 1990.

I

IBM (International Business Machines Corporation)

In 1911, three companies merged to form the Computing-Tabulating-Recording Company (CTR), which became International Business Machines (IBM) in 1924 after taking over a company with that name. IBM is now often referred to by its nickname, "Big Blue." Based in Armonk, New York, the company manufactures a range of office equipment. Early on, it became a leading manufacturer of time clocks and electric typewriters and was particularly influential in the development of punched-card tabulation systems in the 1930s. Over the course of World War II, IBM's revenue quadrupled.

After the war, IBM continued to grow and dominated the international market in office equipment. Its postwar growth owed much to the new computer industry. A collaboration with Howard Aiken of Harvard University, which began in 1937, resulted in the IBM Automatic Sequence Controlled Calculator (ASCC) of 1944, a 5-ton machine programmed by punch tape. In spite of this, IBM regarded with initial skepticism the development of the world's first electronic computer, Electronic Numerical Integrator and Calculator (ENIAC), and the subsequent attempts of ENIAC's designers, John Presper Eckert and John William Mauchly, at commercial development. Aiken even went as far as declaring in 1950 that "no more then six computers will ever be sold in the commercial market." IBM reconsidered in 1951, and its first commercial mainframe computer, designed by Aiken, was launched in 1953. Sales of the IBM 701 soon overtook those of Eckert-Mauchly's UNIVAC computer. In 1956, IBM developed the computer language FORTRAN (FORmula TRANslation) for the IBM 704 computer. This became the standard language for scientific and mathematical applications. By 1960, IBM controlled 70 percent of the world computer market.

IBM also became known for a distinctive corporate style in terms of structure and management and a strong corporate identity. It operated differentiated plant production, encouraged internal competition, adopted forceful marketing strategies, and was committed to customer care through after-sales support services. After the appointment of the American designer Eliot Noyes as consultant director of design in 1956, IBM embraced a total design philosophy. The American graphic designer Paul Rand was commissioned to create the IBM trademark in 1956, and the Bauhaus-trained German architect and designer Marcel Breuer was ap-

IBM PC XT computer, c. 1983 (The Museum of Science and Industry in Manchester)

pointed to design IBM buildings. Among the products applauded for their design quality were the 1961 Selectric typewriter, the world's first golf ball model, and the modular System/360 computer, introduced in 1964.

In the 1970s, IBM showed its first signs of frailty. It maintained its seemingly unassailable position in the mainframe computer market, but left competitors such as Digital Equipment Corporation to develop the new market in minicomputers. Nevertheless, in 1978, it was the world's seventh largest corporation with an annual turnover of $21 billion, 275,000 employees worldwide, and an annual research budget of $1 billion. IBM was initially slow to respond to Apple's launch of the world's first personal computer in 1977 because of the extent of its investment in the mainframe market and its service commitments to the big businesses that were its core mainframe clients. When the IBM PC XT personal computer was launched in 1981, it was an immediate commercial success. The IBM PC and the associated Microsoft Disk Operating System (MS-DOS) quickly became the industry standard. Recognizing the importance of software availability, IBM fostered the development of software for its PCs by publishing details of the system architecture. This strategy was successful, but it enabled rival hardware companies to quickly launch cheaper PC clones.

IBM was also affected by the U.S. government's decision to introduce deregulation in the computer and telecommunications industries in the early 1980s. Previously, companies were unable to engage in both industries, and the removal of this restraint simultaneously created new business opportunities and increased competitive threats. An early challenge to IBM arose when American Telephone and Telegraph's Bell Laboratories developed the UNIX operating system, designed as an "open platform" compatible with different computer architectures. With Microsoft, IBM developed a new operating system, OS/2, and introduced its PS/2 line of computers in 1987. However, in 1991, Microsoft dropped OS/2 in favor of developing a network version its Windows platform, leaving IBM to continue with OS/2 alone. IBM's OS/2 strategy has been a notable failure.

IBM's reputation and sheer size were not enough to bolster its business performance indefinitely in the face of increasing pressure from both U.S. and Japanese competitors. In 1992, in spite of a major reorganization and job cuts, IBM experienced record losses and was forced to take the unprecedented step of reducing stock dividends. Decisive action was needed and in 1993 Louis V. Gerstner Jr. moved from the U.S. company Nabisco to become chairman and chief executive officer of IBM. While further job cuts were still necessary, Gerstner focused on other elements of cost-cutting and the redevelopment of IBM's product line. The growing importance of computer networking created a demand for the kind of integrated business solutions in which IBM had always specialized. IBM's market value began to revive. In 1995, IBM acquired the Lotus Development Corporation, which gave it a sizable stake in the business-applications market. The company scored a notable publicity coup in 1997 when its Deep Blue computer defeated the Russian World Chess Champion Gary Kasparov in a six-game match. The chief architect of Deep Blue was Feng-Hsiung Hsu, who joined IBM from Carnegie-Mellon University, Pittsburgh, in 1989. Deep Blue was the first computer to defeat a top-ranked player and achieve grand-master status. While Deep Blue's chess application was important in attracting attention, its fundamental importance was that it demonstrated a degree of processing power, or virtual intelligence, that could be valuable in more serious applications, such as weather forecasting and financial modeling. Today, IBM remains a major manufacturer of computers and office equipment, but is no longer dominant in shaping the market or product development.

See also AT&T; Apple Computer, Inc.; Computer Printers; Computers; Microsoft Corporation; Typewriters

References:

Pugh, Emerson W. *Building IBM: Shaping an Industry and Its Technology.* Cambridge, MA: MIT Press, 1995.

www.ibm.com/ibm/history/

Ice Cream Makers

Ice cream was popularized in the later nineteenth century by Italian immigrants who made and sold ice cream from street carts. The "penny-lick" of ice cream (known as hokeypokey) was a popular treat in industrial cities.

In America the first patent for an ice-cream freezer was taken out in 1848; a smaller domestic version appeared in the 1860s and became especially popular in the 1880s. These models were simple hand-operated machines and remained in production well into the 1930s. The main brands were the White Mountain, Reliance, Crown, Paragon, Glaciator, and the Peerless, the latter made by the Cincinnati Gooch Freezer Company.

They consisted of a coopered wooden bucket that contained an inner pot of either tinned steel or pewter. The inner pot was filled with cream and the space between it and the outer wall filled with a mixture of ice (around 6 lbs) and salt (around 10 oz). The salt lowers the temperature of the ice. Fixed above the inner pot was a crank operated handle that rotated wooden or aluminum paddles, thus mixing the cream as it froze. Water ices as well as cream ices could be made and fresh fruit or juices could be added. Some models had divided inner pots so that two flavors could be made at once. The freezers came in a number of sizes; the White Mountain ranged from one to fifteen quarts. Once frozen, the bucket was sealed and the ice cream left to "ripen."

There were alternatives to these models. The Auto Vacuum Freezer was a japanned steel, hollow-walled canister with a drop-in container for the cream on the top. The ice and salt were fed in first from an opening in the base. It could freeze in thirty minutes, ten minutes longer than a crank-operated machine required for the production of a gallon. A similar model was Schafer's patent ice-cream machine. The Improved American Ice Cream Freezer (made in England!) also used ice and salt, but was fixed on a two-legged stand and operated in the same way as an end-over-end barrel churn. You could even make your own freezer. The British *Book of the Home* (1904) suggested using a small zinc bath filled with ice and salt and the cream in a milk pan, which was turned rapidly within the ice.

The 1950s saw the introduction of electric models. The earliest, such as the Handy Ice Cream Maker of 1953, looked similar to the bucket models, but the paddles were powered by an electric motor. Later 1960s models were designed to be used in a domestic refrigerator's icebox, which saved the labor of first buying or making the ice and breaking it up with an ice pick. They had paddles operated from an electric motor in the center that was powered via a flat flex, which passed through the seal on the door.

Home ice cream makers developed into two types during the 1960s, both using mechanized techniques, but both still following the same principles as the White Mountain. The simpler "pre-freeze" types are essentially plastic bowls with hollow walls that contain a liquid refrigerant. They have to be left overnight in a freezer compartment until this solidifies. When this is done the lid and electrically powered paddle are attached and the mixture poured in ready to be mixed. The more expensive self-refrigerating models have their own small freezing units, which take about ten minutes to become cold enough to begin churning the mixture.

Ice cream makers became popular during the late 1980s and early 1990s thanks to the rise of premium, or "designer," ice creams such as Häagen-Dazs and Ben and Jerry's. Ben and Jerry's was founded in 1978 by Ben Cohen and Jerry Greenfield in Burlington, Vermont. They began with a hand-operated ice cream maker. Haagen-Dazs ice cream was being produced in New York in the 1930s by Reuben Mattus. He decided to call his product Haagen-Dazs so that it would sound Danish, because Danish ice cream had a reputation for quality. The company was acquired by the Pillsbury Corporation in 1983 and began to market itself as a world brand. The appeal of a home ice cream maker was that it could make similar quality ices for much less (once you had purchased the machine).

Modern machines can make ice creams and sorbets. The main manufacturers are Magimix, Philips, Ariete, and Gaggia.

See also Refrigerators

References:

Campbell, Susan. *The Cook's Companion.* London: Chancellor Press, 1980.

Davidson, H. C., ed. *The Book of the Home.* London: Gresham Publishing, 1904.

de Haan, David. *Antique Household Gadgets and Appliances, 1860–1930.* Poole, UK: Blandford Press, 1977.

Ice Crushers

Crushed ice became popular for cooling drinks and cocktails in the 1920s and 1930s. Simple ice crushers were usually hinged presses of cast aluminum. The upper handle had a head of spiked teeth that crushed the ice into the lower pan. Another version, popular in the 1950s, had a hopper above a set of hand-cranked teeth with a plastic container for the crushed ice to fall into below. Capable of producing coarse and fine granules, they can crush a quart of ice in two minutes. These models were available in reds and yellows with chromed lids. They are still on sale as retro kitchenware. Electrically operated models are also available.

See also Refrigerators

Ideal Home Show

The Ideal Home Show was founded in 1908 by Wareham Smith, the advertising manager of the popular English newspaper the *Daily Mail.* The paper had been founded by Alfred Harmsworth in 1896. It had been successfully aimed at the rising lower middle classes of office workers. Harmsworth also targeted women, including a woman's page, and after 1918, employed Mrs. C. S. Peel, an expert on household management and therefore knowledgeable about the new "labor-saving" devices. The *Daily Mail* promoted a national conservatism and individual aspiration.

Wareham Smith was an innovative newspaperman, and the Ideal Home Show grew out of his desire to attract new advertisers. He saw the building trade, with its dependence on architects' recommendations for suppliers and subcontractors as a target area. He wanted them to cut out the architects and advertise in the *Daily Mail.* His plan was to hold a competition for an ideal home that would be built and then surrounded with all the materials and appliances that would make it so. It was the first of its kind, although it naturally owed a debt to the great European and American trade fairs of the nineteenth century. Harmsworth was uncertain of its value, but hoped it would attract more readers. Undeterred, Smith organized the 1908 show, a form of three-dimensional domestic manual with lifesize houses and promotional stores as well as "Babyland," a whole section devoted to childcare, at Olympia, in Hammersmith, which was to be its home until the 1980s. The Ideal Home Show was a great success, attracting over 150,000 people in 14 days. It was repeated in 1910 and 1912, returning as a regular event in 1920.

The show was financially driven and as a result included all popular tastes. It did not seek to teach the public lessons in taste; rather it saw its visitors as consumers and was happy to include mock Tudor houses along with streets built in the modernist style. The 1913 exhibition included a room designed by the Omega workshops, the group founded by the artist, critic, and writer Roger Fry. The Ideal Home Show was as happy to look back to the past with Tudor villages in 1910 as it was to look to the future in 1928. The House of the Future (projected to the year 2000) would have automated labor and color television and be served by atomic energy. The Village of Tomorrow (1934) featured a range of modern houses, including one by Wells Coates.

The concept of labor-saving and the promise of the new electrical appliances was promoted by the *Daily Mail* to its largely servantless women readers. The labor-saving section was introduced in the exhibition in 1913. The concept of the gadget was celebrated by the artist William Heath Robinson in his house, The Gadgets, a life-size re-creation of his delightfully absurd domestic solutions to the problems of the owners, Mr. & Mrs. Glowmutton! New appliances and gadgets remain one of the most popular sections of the exhibition.

The Ideal Home Show returned after World War II in 1947, one year after the more didactic Britain Can Make It exhibition at the Victoria and Albert Museum. For both events it was more a case of "Britain can't have it," with continued rationing and an acute housing shortage. The show became more aspirational than ever, especially as new designs and fitted kitchens from Europe and America were displayed. For many people in the 1950s, Do-It-Yourself was a way to make your own ideal home, and the show was a place to come and pick up ideas. Attendance reached an all-time high of 1,329,644 in 1957.

The show continued to shuffle its themes throughout the late twentieth century. The theme of the future reappeared in 1956 and 1960, with Kenwood sponsoring a kitchen of the future. Traditional (the Georgian House of 1960) and contemporary houses (the McLean split level house of 1963 and Nigel Coates house of 1997) were built and displayed, increasingly with gardens, reflecting another growing leisure activity. Other areas that have been promoted are sustainable building, energy conservation, home security, and the safety of appliances.

The Ideal Home Show has been imitated around the world, most notably the French Salon des Arts Menagers, which ran between 1923 and 1976. The English original is now an institution, one that has influenced and inspired hundreds of thousands of people throughout the twentieth century. Never designed to offer only the best, it remains a snapshot of culture and consumption, making material that circular relationship between consumer and manufacturer.

See also Consumers; Do-It-Yourself; The Modern Movement and Appliance Design

References:

Ryan, Deborah S. *The Ideal Home through the Twentieth Century.* London: Hazar Publishing, 1997.

IKEA

The Swedish IKEA Company was founded in 1943 by Ingvar Kamprad. It is one of the largest furniture manufacturers and retailers in the world, with stores in twenty-six countries and an annual gross sales of £3.4 billion (1996). The son of a farmer from Småland, Kamprad was born in 1926. He started by selling matches and then progressed to mail-order goods. In 1943, the seventeen-year-old had a business significant enough to be incorporated. He chose the name IKEA, the first two letters being his initials and the last two being those of his birthplace, the farm of Elmtaryd in the village of Agunnaryd, near Almhult.

Kamprad moved into designing and selling furniture after visiting a trade fair in Milan. His decision to sell at factory prices by mail order meant that costs had to be kept low. He realized that, by designing to suit production techniques, he could use materials more economically than his competitors. He was therefore able to undercut their prices, sometimes by 50 percent. This led to resistance from the established furniture trade, and some suppliers tried to boycott IKEA, but Kamprad outmaneuvered them by sourcing materials from Denmark and Poland.

The first IKEA showroom opened in 1953, followed by a store in 1958: it utilized the ideas of self-service that were emerging in Europe at the time, the first time this approach had been applied to selling furniture. IKEA customers were expected to select their goods in the warehouse, take them to the checkout, carry them home, and then assemble them. From 1956, many of the items were designed to "flat-pack" (a legacy from mail order), so this was not too difficult. Kamprad chose cheap out-of-town sites with ample car parking. From the start the stores had nurseries, play areas, and restaurants.

The IKEA philosophy is based around building relationships, hard work, humility, and thrift. "The IKEA Way" is a compulsory training course for all employees, who are known as co-workers. Most of this is based on Kamprad's own personal philosophy, "to create a better everyday life for the majority of people" through a company where it is "better to sell 600 chairs at a low price than 50 at a higher one." This has led to claims that IKEA has been a huge force for the democratization of culture. Paradoxically, Kamprad flirted with extreme right wing politics in the late 1940s and early 1950s. He later regretted his actions and made a very public apology when the story was revealed by a journalist in 1994. IKEA by then had become a major industrial force; by 1987 it accounted for over 40 percent of Swedish furniture production.

IKEA has become such a success due to the vision of its founder and its ability to source manufacturers throughout the world, especially in China and the former Soviet countries where labor costs are low. It opened stores across Europe in the 1970s and the United States and the United Kingdom in the 1980s. It employs around 30,000 people, carries almost no factories, and operates on a minimal infrastructure with a just-in-time delivery structure. It retains its very low prices through economies of scale and a continuing investment in research and design. The company is still run by Kamprad, who placed most of it in trust in the Netherlands in 1994. Its annual catalogue is printed in seventeen languages in over twenty-eight countries and has a print run of over 50 million.

IKEA does not sell electrical domestic appliances, but its "Marketplace" within each store has a wide range of kitchen equipment, cookware, storage, and lighting. The stores also sell a line of self-assembly fitted kitchens.

IKEA has done much to democratize design. It has not innovated in the same way as either the English Arts and Crafts Movement or the internationalist Bauhaus, but neither of these were able to tackle the critical problem of competitive pricing, and IKEA has. Terence Conran's Habitat (which IKEA now owns) also achieved this, but lacked the structures to go global. IKEA is now the world's largest furniture chain store, with over 80 outlets throughout the world (including recently opened stores in Russia and China). IKEA has also remained resolutely Swedish in its outlook. It refuses to give products different names in different countries, and the very Swedishness of these names is now a distinctive selling point. It is currently addressing environmentally sustainable production methods.

Twentieth-century Scandinavian designers developed humane forms of modernism, as exemplified by the Finnish Alvar Aalto and the Danish Arne Jacobsen. IKEA has taken this tradition and industrialized and marketed it in such a way that it can claim to offer Kamprad's ideal of "beauty for everyone."

See also Habitat; Kitchens; The Modern Movement and Appliance Design; Designers (p. 337)

References:

www.ikea.com

Industrial Design and Designers

The debate over what constitutes "good" design in everyday domestic products is one that has been raging since the mid-nineteenth century. All products, whether hand or machine made, require a quality design for them to succeed, either functionally or aesthetically. The industrialization of many of the older craft skills separated the design from the production process. If the original design was weak, then so would be the thousands of multiples that flowed from it. In the United Kingdom, men like Henry Cole realized this problem required a successful marriage of art and industry. The 1851 Great Exhibition was a watershed for many like Henry Cole. Although an organizer of the event, he was critical of the standard of design of many of the products on display. The exhibition led to a polarization of opinions, with the likes of John Ruskin and William Morris, who proposed a return to craft production, set against the manufacturers of industrial England.

It is fair to say that the Arts and Crafts Movement had little real impact on the design of early electrical appliances. Most of the early appliances showed their laboratory or industrial origins. Small production runs in small workshops usually dictated their look. Those appliances that were intended for the living room did have some form of decoration. There are examples such as Dowsing lamp fires, which featured electric bulbs as heating elements, set in beaten copper frames with enameled roundels. Some electric light fittings also reflected Art Nouveau shapes and decorative motifs.

The first company to take industrial design seriously was AEG (Allgemeine Elektrizitats Gesellschaft), with the work of Peter Behrens. Behrens understood what was later to be known as the "machine aesthetic" and was aware of the processes of industrialized production and the benefits of integrated components. The Bauhaus and the Modern Movement investigated how art and industry could work together but had little tangible success. With these two exceptions it is fair to say that for the first thirty years of the twentieth century there was little real effort put into the design of domestic appliances. Those that were electric versions of earlier designs such as kettles took traditional forms, while even the genuinely new objects (like mixers, vacuum cleaners, and washing machines) would have looked like industrial machines.

The real advances in industrial design took place in the United States during the 1920s and 1930s. They were triggered by social and economic trends. The period from 1919 to 1929 saw industrial production double and purchasing power rise by one-fifth. The centers of population were increasingly urban. Americans were buying more technological goods, especially cars. The availability of credit increased, as did the volume of advertising. Magazine advertising rose from \$25 million in 1915 to \$150 million in 1929. By 1925, the Metropolitan Museum in New York had, under its curator Richard F. Bach, established a collection of over a thousand examples of what it called "industrial art"—examples of what it considered to be the best in modern design. Bach had caught the mood of the industry, as large companies like General Electric were establishing design departments. It was also looking to the new breed of commercial free-lance designers to help reshape products such as refrigerators. General Electric was not alone, companies like Sears also called in commercial designers to restyle their appliances. The example of Ford having to replace the Model T due to increasing competition had shown that the look of a product was important. Manufacturers invested in design primarily to maintain or increase their market share. The Depression reinforced this further.

During the 1930s the most influential designers were Henry Dreyfuss, Norman Bel Geddes, Raymond Loewy, and Walter Dorwin Teague. Their work varies in style; Loewy and Bel Geddes embraced streamlining products while Dreyfuss was more restrained. What unites them is that they gave the products a more unified and confident look. In many cases they supplied stylish envelopes for the technology, often hiding it in stove-enameled pressed steel. Their work was a mixture of Bauhaus-influenced European purity, the U.S. automobile industry, and a response to corporate desires for novelty. It must be remembered that they were operating during the Depression and their work was judged on whether it sold or not. They were also subject to, and complicit in, the development of planned obsolescence in order to stimulate demand. Manufacturers wanted to stimulate demand through yearly model changes, like the automobile industry. By the eve of World War II, the American appliance industry had reached a stage that would be picked up in the postwar period, when products would sell as much on their looks and identity value as their performance.

Europe at this time was nowhere near as sophisticated, although the influence of streamlining did cross the Atlantic. The British and U.S. plastics industries both experimented with modernist and art deco inspired designs. Ekco used architects such as Wells Coates and Serge Chermayeff, and His Master's Voice, a UK company, commissioned the modernist architect Christian Barman to design electric heaters and irons, but these were isolated examples.

The immediate postwar period saw the reemerging industries of Italy and Germany taking a lead in the design of domestic products. Braun made use of the restrained work of Dieter Rams, Fritz Eichler, and Hans Gugelot, which combined modernist purity with genuine innovation, exemplified in the Kitchen Machine of 1957. The Italian domestic appliance industries began to flourish in the 1950s, thanks to American financial aid and a rising home market. Many companies invested in employing consultant designers who responded to new opportunities and materials with style. The Italian plastics industry pioneered the use of colorful quality plastics. The Milanese Kartell Company is an outstanding example: founded in 1949 it has used a variety of designers, including Gino Colombini, Marco Zanuso, and Philippe Starck. The Compasso d'Oro awards were established in 1954 to encourage higher standards in design and production. It was the brainchild of Aldo Borletti of the Milan department store La Rinascente and was important as it encouraged designers to take a fresh look at everyday domestic products. This environment has allowed the likes of Kartell and Alessi to flourish.

The United States in the 1950s continued the trends set in the late 1930s. Domestic appliances aped some of the styling of cars and also became more colorful. Some in-house design teams did manage to create notable products, the Russell Hobbs electric kettle being a prime example. The 1960s and 1970s witnessed a move to standardization and stagnation in design terms. The look of products was now seen as important as the spending power of individuals increased. Nowhere was this more so than in the world of fashion and graphic design, which were heavily influenced by pop culture. In Britain, the Council of Industrial Design maintained an adherence to modernist principles and criticized these changes, but their influence was unstoppable. Italian product design also continued to innovate throughout the 1960s. As a result, many appliances received minor fashion changes such as the introduction of fashionable colors, inappropriate floral transfers, or geometric patterns. The 1970s witnessed a revival of rustic and floral patterns.

The 1980s saw a reassessment of the design of many domestic appliances based on the influence of postmodernist designers such as Ettore Sottsass and his Memphis group (whose playful merging of styles was influential). Color was introduced on smaller appliances such as toasters and kettles and a more lighthearted attitude was beginning to influence the larger manufacturers, especially Sony, Sharp, and Panasonic. Much of the three dimensional design in what has been termed the designer decade was often a cheerful plundering and blending of previous styles. The 1980s also saw a revival of interest in what had become known as design classics like Dualit toasters and KitchenAid mixers. Linked to this has been the migration of professional equipment into the domestic kitchen. This is a move generated by fashion and TV celebrity chefs. *Consumer Reports* "Best Buys for 2000" proclaimed, "Commercial is Cool." This has led companies to produce pro-style products such as stainless steel stoves for the top end of their lines.

Britain in the 1980s and 1990s also saw the few companies remaining in British ownership failing to respond to homegrown design talent. The Trevor Bayliss–designed Baygen radio and the Dyson vacuum cleaner both struggled to find financial backing in the short-term/high-return environment of British capitalism. A more recent example is Jonathan Ive, designer of the Apple iMac, who now works in San Francisco.

The design process is now a central part of the circular nature of consumption and production. Current styles are either retro chrome-plated models or contemporary curvy types whose switches and dials continue to owe much to automobile design. Companies employ their own in-house design teams and also resort to "big-names." The industrial designer is expected to produce products that perform, look good, and sell. As technology gets more and more reliable the identity value of the product, its look and its meaning, is now even more important than ever.

See also AEG; Braun; Consumers; Design Council; Ekco; General Electric; Heating; The Modern Movement and Appliance Design; Obsolescence; Plastics; Refrigerators; Sears, Roebuck and Company; Toasters; Vacuum Cleaners

References:

Bayley, Stephen. *In Good Shape: Style in Industrial Products, 1900–1960*. London: Design Council, 1979.

Bayley, Stephen, ed. *The Conran Dictionary of Design*. London: Octopus Conran, 1985.

Meikle, Jeffrey L. *Twentieth Century Limited: Industrial Design in America, 1925–1939*. Philadelphia: Temple University Press, 1979.

Sparke, Penny. *A Century of Design: Design Pioneers of the Twentieth Century*. London: Mitchell Beazley, 1998.

Sparke, Penny. *Electrical Appliances*. London & Sydney: Unwin Hyman, 1987.

Internet

See Modems

Irons

Like many other appliances, the electric iron was invented in the nineteenth century and developed in the twentieth. The range of irons available in 1880 would have been familiar in 1800. Flatirons or sadirons, heated on a stove or open fire, were the most common, along with charcoal and heated-slug irons.

The first electric iron appeared in France in 1880, using an electric arc to create the heat. It needed constant adjustment and had a tendency to shed hot pieces of carbon on to the clothes! A safer alternative was to use an electrical element to heat the soleplate. The English Crompton Company produced such models in 1891, made of cast iron with a turned wooden handle. The socket was either at the front or the back of the handle and ran from a connector for an electric light. A U.S. model was patented by Dyer and Seeley in 1882. This featured a stand in which the iron was placed to heat up, but was both expensive and unreliable. An AEG (Allgemeine Elektrizitats Gesellschaft) catalog of 1896 featured eight electric models. An alternative was the

Maid using electric iron, 1908 (Corbis)

gas-heated iron, which could be connected to a gas light supply. Such models looked similar to earlier charcoal irons, both having a chimney to allow the fumes to escape.

The basic pattern had been established by 1900 with the traditional shield shape heated by an element. Such models were sold by Crompton and GEC in the United Kingdom and the Hotpoint Company and the Pacific Electric Heating Company in the United States. They had no heat control and weighed around 9 lb (4 kg). Hotpoint was so named because its iron gave off as much heat at the tip as in the middle. Most of these models were aimed at commercial laundries in cities that had an ample electricity supply. An early design problem was that the socket for the plug was placed at the back of the iron, which made it difficult to stand on end like the traditional flatiron. This was solved by a direct internal connection, which placed the cord up higher, allowing the user to stand the iron on the broad end.

The American Beauty produced by the American Heating Company of Detroit in 1912 was directly aimed at the domestic market. It looked different from commercial models, having a chrome-plated body, a style that was to last nearly sixty years. Such electric irons became popular in both Europe and the United States as more and more households installed electric lighting in the 1920s and 1930s. As with other appliances, the real breakthrough came with variable heat control in 1926. This was a U.S. invention, produced by the Liberty Gauge and Instrument Company of Cleveland and based on the experiments a fourteen-year-old boy carried out in 1912. The potential of the Liberty iron was seen by Proctor and Schwartz (later Proctor-Silex), which bought the company. Thermostatically controlled irons were a considerable improvement because they did not burn clothes so easily and could iron at lower temperatures. The steam iron was also introduced in 1926 by the Eldec Company of New York. The Steam-O-Matic with its over-the-top water container was introduced in 1938. Electric irons were a real help to housework. Ironing was still a chore, but the need to constantly heat a set of irons had gone—a great relief, especially in summer. Compared to larger luxuries like refrigerators and washing machines, they were relatively inexpensive, which made them popular. In 1929, a survey of 100 Ford Motor Company employees revealed that 98 of them had electric irons at home. In the UK only 25 percent of households had an electric iron in 1935, by 1939 this figure had risen to 75 percent.

British and U.S. irons of the 1930s and 1940s were mainly chrome plated with black Bakelite handles. The British HMV Company employed the architect and designer Christian Barman to design a streamlined iron finished in yellow enamel. In 1946 the Thorn Electrical Company produced the Mary Ann, which revived the double pointed soleplate of the Mrs. Potts Enterprise iron of the 1870s in an electric form. Like its predecessor, the shape did not catch on.

An alternative to the hand-held iron was the flat presser. These developed in the United States in the 1930s and were domestic versions of the large electric ironers used by commercial laundries and dry cleaners. Another alternative was the rotary ironer, in which the fabric was passed, under pressure, over a padded heated cylinder. The Horton Manufacturing Company of Indiana introduced their domestic model in the 1920s. It featured a gas-heated ironing surface and an electrically operated roller. Horton moved to an all-electrical model, the Beauty Aid, in the 1930s. Both appliances were bulky and expensive and therefore failed to find a large market. Rotary ironers were produced in the United Kingdom in the 1960s by Kenwood and Morphy Richards, but again failed to catch on due to size and relatively high price. Smaller steam pressers are currently available for the domestic market but remain expensive at around $200, compared to the good quality hand irons that cost between $10 and $50.

In 1950, Universal produced the Stroke-Sav'r, which used vaporised steam. Hoover introduced a steam iron with automatic temperature control in 1952. It also featured a cast aluminum soleplate and a steam cut-off when the iron was standing upright. Most irons during this decade were steam-or-dry models that used distilled water, although the Presto Company of Wisconsin had introduced the first steam iron capable of using tap water in 1949. Steam irons were able to use lighter materials without a loss in performance and were therefore easier to use. They were not universally reliable, however; Consumer Reports found that eight out of the ten steam irons it tested in 1951 had hazards such as faulty thermostats and melting soleplates.

The British Pifco Company produced the Chereton, an early cordless iron, in 1950. Travel steam irons arrived in 1953. General Electric produced a steam and spray iron in 1957 that shot a fine mist of water onto the clothes ahead of the iron. Teflon-coated soleplates appeared in the 1960s. By the early 1980s, the whole body was plastic, while the soleplate remained of metal, usually cast aluminum. Plastics allowed a wider range of colors to be used, either for the body or for the trim. The 1980s also saw further refinements such as steam irons that could use tap water and cordless models. The cordless iron sat in a holder, which acted as a recharger. The 1990s saw irons being re-styled with post-modern influences. The German Rowenta Company's Surfline irons of 1994 featured a transparent turquoise blue plastic casing. Contemporary irons now come in a range of blues, yellows, and greens.

Travel irons have developed to use non-electric power. A Japanese version uses "heat-pack" technology, in which two liquids combine to crystallize and produce heat. This miniature iron stays hot long enough to iron a set of clothes.

Developments in minimum-iron artificial fibers and "non-iron" cottons have led to claims that the iron's days are numbered. Nevertheless we still endure "the ironing" as we continue to prefer natural fabrics and to believe they look best when ironed.

See also Betty Crocker; General Electric; Hamilton Beach; Pifco; Proctor-Silex; Sunbeam; Teflon

References:

Goldberg, Michael J. *Groovy Kitchen Designs for Collectors, 1935–1965*. Atglen, PA: Schiffer Publishing, 1996.

Heskett, John. *Industrial Design*. New York: Oxford University Press, 1980.

Sparke, Penny. *Electrical Appliances*. London and Sydney: Unwin Hyman, 1987.

J

Jacuzzi

Named after its inventor, Candido Jacuzzi, an American engineer based in California, the Jacuzzi is a whirlpool bath. The whirlpool action is created by a pump. When it was introduced in 1956, the Jacuzzi was just a small portable pump, designed to stand in the bath tub. It was conceived as a therapeutic device, inspired by the fact that Jacuzzi's son suffered from rheumatism. The massaging effect of the pumped water acted as a relaxant and soothed pain. The second phase in the development of the Jacuzzi came in 1968, when Roy Jacuzzi invented the integrated whirlpool bath. The concealed pump fed a series of jets in the sides of the bath. In this form, the Jacuzzi was promoted as a more pleasurable form of bathing. In 1970, the Jacuzzi evolved into the spa bath, a tub large enough for several people and equipped with built-in heating and filtration. It became associated with a wealthy Hollywood lifestyle, an image that has persisted.

See also Showers
References:
www.jacuzzi.com/history/

Skiers take a break from the slopes to enjoy a Jacuzzi at the Squaw Valley Lodge in California, c. 1994 (Morton Beebe, S.F. / Corbis)

Jar Openers

Tin cans and can openers were in use throughout much of the late nineteenth century. The commercial bottling of preserves such as jams and marmalade was established by the beginning of the twentieth century, with most of the preserves being sold in stoneware pots sealed with paper lids. These were superseded by glass jars with screw-on metal lids. These are still in use and, although they give an excellent seal, the lids can become stuck.

The most common solution is still to heat the lid under hot water and try again with a tea towel, but a variety of small appliances have been developed to assist the process. One is a rubber cone with a ribbed inside to grip as it fits over the jar. Metal versions include pincers and an adjustable steel hoop that fits around the lid. Rubber straps are also available. Another variant is a metal wedge-shape opener that can be mounted on a wall or under a shelf. The lid is placed into the flanged, serrated edges, which give a greater grip.

K

Kenwood

A young entrepreneur called Kenneth Wood established the Kenwood Company in 1947. Wood had been in the Merchant Navy during World War II and had been impressed by the American appliances he had seen. He began in a garage in Woking, England. The first Kenwood product was the A 100 electric toaster. A year later the company produced the first all-British food mixer.

Its most famous product, the Kenwood A 700 Chef food mixer, was devised by Wood and his colleague Roger Laurence and introduced in 1948. Following the lead of U.S. models such as the Sunbeam Mixmaster, it had an impressive array of attachments for juicing, shredding, mincing, and peeling. It had a metal body and was a scaled-down version of a commercial model that Kenwood marketed to caterers and hotels. The Chef had a plastic nose cap and control switches in a range of "gay" colors to match the look of the owner's kitchen. It's advertising was aimed at women with the slogan "Kenwood. Your Servant, Madam." Although expensive for the time, it was well made, and by 1956 the company had a turnover of £1.5 million.

By 1960 the Chef was beginning to look old fashioned, especially when compared to the Braun Kitchen Machine of 1957. This led to a redesign by the British designer Kenneth Grange in 1960. He produced the new look in just four days. It incorporated some of the characteristics of the Braun with simpler, cleaner lines and the incorporation of the power head and the base into a single form. Unlike the Braun, the case was still of white stove-enameled metal, with the separate elements and the control knob in blue plastic.

The Chef was a great success and was followed by an equally popular smaller version, the Chefette, which could double as a stand and hand-held mixer.

The company became a part of the Thorn Group in the 1960s and reemerged as Kenwood Appliances in 1989. It is still based in the south of England at Havant, Hampshire, and some manufacturing carries on in the United Kingdom, although the bulk of production is now in China. Kenwood produces a range of small appliances, including mixers, toasters, kettles, blenders, food processors, rice steamers, mini ovens, fryers, and can openers. It remains a brand leader in food-preparation products. Kenwood also has an Italian presence with the Ariete Company, which produces coffeemakers, ironing systems, and steam cleaners.

See also Food Mixers; Food Processors; Industrial Design and Designers; Toasters

References:
Forty, Adrian. *Objects of Desire: Design and Society, 1750–1980.* London: Thames and Hudson, 1986.
Katz, Sylvia, and Jeremy Myerson. *Kitchenware.* London: Conran Octopus, 1990.

Kettles

The boiling of water has long been a basic domestic requirement. The shape of the kettle, with a round-shouldered body, serpentine spout, and horizontal handle fixed on curving brackets was developed in the seventeenth century. They were made of raised copper or brass sheet. Cast-iron models followed in the nineteenth century. Smaller, more refined models were introduced as the habit of tea and coffee drinking grew in the eighteenth and nineteenth centuries. Boiling water was poured into the kettle, which was brought to the table and kept warm by a spirit-burner. Stylish examples were produced by designers like the English W. A. S. Benson and the Dutch Jan Eisenloffel. The kettle was a basic piece of kitchen and table equipment and therefore one of the earliest to be converted to electricity, despite the fears of introducing electricity to water.

The earliest examples were small electric hot plates for traditional kettles. The first true electric kettle was made in Chicago in 1894. British models appeared soon afterwards. Crompton & Company of Chelmsford advertised their 1897 model, claiming that it could boil three pints of water in 15 to 20 minutes. This illustrates the problem that electricity had to overcome. It was slow and expensive; a gas ring could do the same job in less than half the time. The problem for early electric kettles was that to avoid the hazard of water and electricity meeting, the heating element was separated from the water at the base of the kettle. As a result, much of the heat was lost, leading to the slow and costly boiling time.

The manufacturers were certainly aware of this problem. Some models had domed, not flat, bases that gave a greater surface area

Landers, Frary & Clark electric kettle, c. 1930 (The Museum of Science and Industry in Manchester)

of metal for the heating element to act upon. The range of elegant kettles designed between 1908–1909 for AEG (Allgemeine Elektrizitats Gesellschaft) by Peter Behrens featured the first immersed element. In 1922, the Englishman Leslie Large patented an element that could be used in water. It featured wire wound around flat mica inserted into a flat copper tube. Large worked for Bulpitt & Sons of Birmingham, UK, and this kettle was marketed as a part of their Swan brand (a name derived from Swansea Tin Plate). The only drawback was that in hard water areas the elements became "furred-up" with lime-scale deposits. This problem was alleviated with a wire ball kept inside the kettle.

The remaining problem was how to ensure that the kettle did not boil dry. A variety of safety devices were introduced, including whistles, buzzers, and flashing lights. In 1931 Bulpitt & Sons were the first to introduce an electric safety plug that self ejected when overloaded. The first "automatic" kettle that switched itself off was successfully developed by Bill Russell and Peter Hobbs in 1955. A small hole near the top let the steam from the boiling water into a switch unit. The heat bent a bimetallic strip, which switched off the electricity. The reliability and high boiling speed of the resulting K series made it a popular choice for over thirty years.

The 1950s also saw stainless steel become the predominant material for electric kettles, taking over from nickel-plated steel and copper. Stove-top kettles continued to be produced; a popular American model was the West Bend Trig. This was a squat aluminum kettle whose name derived from the trigger that released the spout cap. It was later available in colored anodized aluminum. Manufacturers also began to experiment with shapes. Although not as radical as the German Wilhelm Wagenfeld's geometric stove-top kettle of 1930, some kettles began to move away from the traditional round shoulders and serpentine spout. The Eralite Chadeau of 1946 was a more squat electric model. The British HMV Company produced the striking Bentink model in the late 1930s, designed by Christian Barman. Ten years later came the Maddox, a hybrid between a kettle and coffee percolator; it anticipated later kettles that were shaped like jugs. The 1960s saw more streamlined models with broad spouts for filling and pouring.

Plastics began to be used more and more in the 1970s. Bakelite had long been used for handles, due to its good insulating qualities, but it could not withstand the rapid heat changes that took place in the body of the kettle. The 1973 Hoover Rainbow Autoboil illustrated the potential of acetal. Russell Hobbs produced the Futura in 1973, the first electric kettle to have its body molded in acetal. It was designed by Julius Thalman, an in-house designer. The company had a strong design team that also produced the restrained and highly functional stainless steel K series of kettles. The 1980s saw the rise of the jug kettle. The first was the Redring Autoboil of 1979. Usually made of polypropylene, these kettles featured water level gauges and allowed small amounts of water to be boiled efficiently. Ceramic jug kettles had been produced in the 1930s, and Braun had experimented with a metal model in the 1960s. Energy-saving features were highlighted along with safety. In 1986, Tefal introduced the cordless Freeline kettle, which sat on a static base. By the late 1990s, cordless models accounted for half the British sales. Philips introduced the Filterline in 1992, which featured a removable mesh filter to trap lime scale that dissolved in the water.

Paradoxically the 1980s also saw the rise of the kettle as a style object. Artist- and architect-designed stove-top kettles were produced, mainly by Alessi. Richard Sapper's Bollitore kettle of 1983 became an international (and expensive) cult object, despite having a handle that overheats. Its spout features a brass note whistle with two pitch pipes in E and B, operated by the rising steam. Sapper was inspired by the sound of an Amtrak train. The postmodernist architect Michael Graves produced a more whimsical kettle with a plastic bird whistle in 1985. The Japanese K600 revived the bright enamels of the English New Maid of the 1950s. These products have led to the large manufacturers like Philips and Tefal producing mass-produced models clearly inspired by the designer models. The Danish company Bodum has produced the Ibis range in a variety of colors.

Russell Hobbs continued to innovate in the 1990s with their Millennium model, featuring a flat heating element designed to cut down on lime-scale deposits.

Most electric kettles are now inexpensive throwaway items, yet they remain an essential piece of equipment for both the home and the office.

See also AEG; Alessi; Philips Electronics; West Bend

References:

de Haan, David. *Antique Household Gadgets and Appliances, 1860–1930.* Poole, UK: Blandford Press, 1977.

Woodham, John. *Kettle: An Appreciation.* London: Arum Press, 1997.

KitchenAid

The KitchenAid range was produced and developed by the Hobart Company of Troy, Ohio, which specialized in heavy-duty commercial appliances. It was one of the largest manufacturers of food equipment in the United States at the beginning of the twentieth century. Its first electric mixer was designed in 1908 by Herbert Johnson, an engineer and, later, company president. It was an 80-quart mixer for bread dough. The company turned to war production in 1917.

The company returned to peacetime production in 1919, and one of its subsidiaries, the Troy Manufacturing Company, produced the first KitchenAid product in that year—the H5, a commercial electric dough beater designed for the domestic market. This was a "stand" mixer that utilized a planetary action: the mixer rotated the beater in one direction while moving the bowl in the opposite one. This action was patented in 1920.

The name KitchenAid was the result of what could be called an early focus-group. While the H5 was being tested, the wives of the Troy Company management discussed what to call it. One woman commented, "I don't care what you call it, but I know it's the best kitchen aid I have ever had." The name stuck, and H5s rolled off the assembly lines at the rate of four a day. They were robust and well engineered, a fact reflected in the $189.50 price tag.

KitchenAid concentrated on the domestic market, advertising in women's magazines in the 1920s. By the late 1920s, it has ceased production of the bulky H5 and produced smaller, lighter, and more affordable machines. Like other appliance manufacturers, it embraced industrial design and styling during the 1930s. KitchenAid approached the consultant designer Egmont Arens (1888–1966) to restyle three of their mixers. Arens was an advocate of streamlining, and the result was the familiar rounded motor case with the protruding attachment nozzle. It scarcely had a hard angle in it. Arens also produced a less successful, but equally curvy, design for the Hobart commercial meat-slicer in 1935.

The company expanded its range in the 1930s and 1940s. It introduced a coffee grinder in 1937 and moved into the larger appliance market in 1949 with the first automatic dishwasher. It also introduced a range of colored models in the 1950s, rejoicing in names such as Petal Pink, Sunny Yellow, Island Green, Satin Chrome, and Antique Copper. The Whirlpool Corporation purchased KitchenAid in 1986. It produces a range of washers, dryers, dishwashers, stoves, and small appliances including food processors, blenders, and toasters. Its products maintain a stylish industrial feel and are marketed as the corporation's upscale brand. Arens's mixer design is still in production and has been hailed by many design writers as an American classic, up there with the Coca Cola bottle and the Zippo cigarette lighter.

See also Coffee Grinders; Food Mixers; Industrial Design and Designers

References:
www.kitchenaid.com

Kitchens

The concept of the fitted kitchen—the designed kitchen with clearly defined working areas—began with the observations of two American women. Catherine Beecher, the principal of the Hartford Female Seminary, published *A Treatise on Domestic Economy for the Use of Young Ladies at Home and at School* in 1841. She advocated the need for a "light, neat, and agreeable" working space. She expanded these ideas and suggested changes in kitchen layout based on ship's galleys in *The American Woman's Home* of 1869. Christine Frederick later applied industrial time and motion studies to suggest the most efficient way of arranging the components of a kitchen in her *Scientific Management in the Home* of 1915. She borrowed from the contemporary studies of factory processes and favored a sequential organization of work surfaces and the routing of tasks. The preface to the book was by Frank Gilbreth, an advocate of scientific management. He wrote that "Mrs. Frederick has seen the necessity for making the home a laboratory."

This scientific approach was favored at the Bauhaus and given form in the Haus am Horn

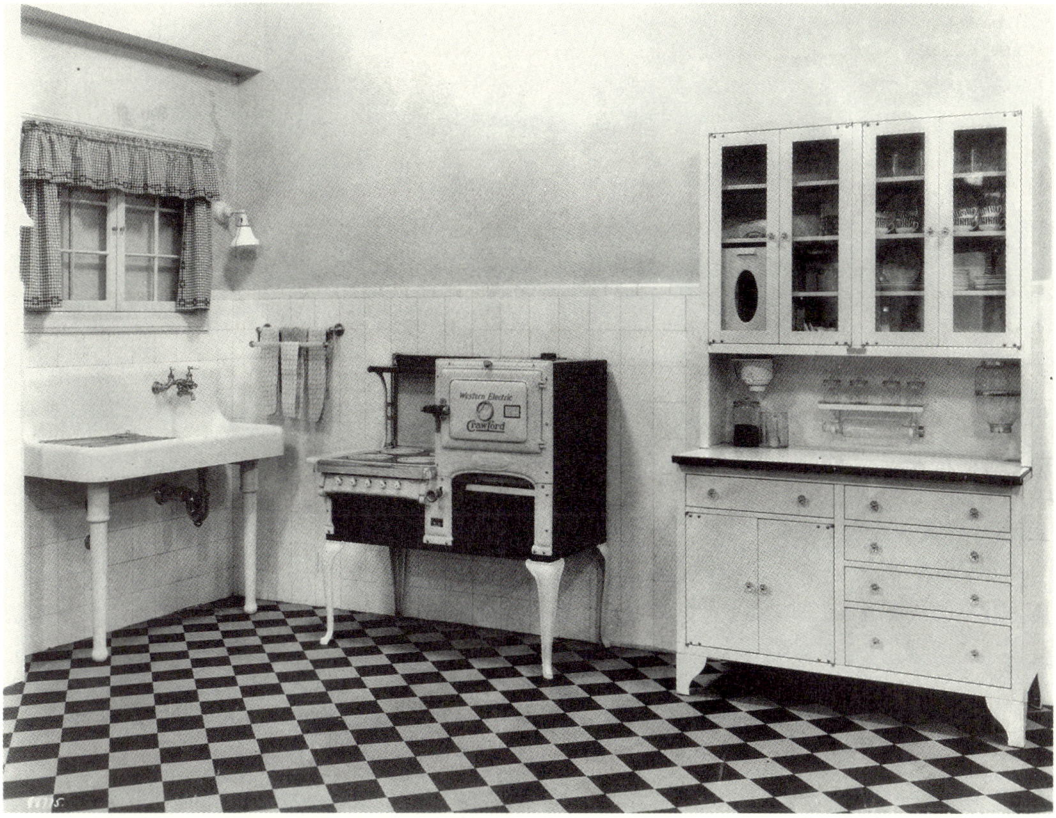

A model kitchen, 1924 (Library of Congress)

of 1923. This exhibition house was built as a mass production prototype using modern materials. The kitchen was a galley type designed by Marcel Breuer. Other early examples of designed kitchens were in houses designed by Modern Movement architects such as Le Corbusier and Mies van der Rohe in the 1920s.

Frederick's ideas were adopted on a large scale in the 1920s housing projects in Frankfurt, Germany. The "Frankfurt Kitchen" was designed as a total unit with built-in storage space and continuous working surfaces. The project involved a group of architects led by city architect Ernst May. One of them, Ferdinand Kramer, recalled that Christine Frederick's book was "our Bible." They took her principles of standardization and applied them to the heights of both work tops and large appliances like stoves. The storage of cookware and small appliances was in cupboards above and below the work surfaces. The results were truly "fitted" kitchens.

Frankfurt was exceptional. The 1926 British Ideal Home Exhibition displayed an "American" kitchen as the most modern contrast to a display of old kitchens from around the world. This featured compact, multipurpose cabinets and small electrical appliances but still used separate appliances and furniture. By 1935 "top-of-the-range" British houses featured kitchens with built-in cupboards and drawers of either wood or stainless steel. These modular kitchens were also appearing in Europe. The Dutch designer Piet Zwart had designed a line of cupboards and cabinets that could fit under a running work-surface. He developed this idea further for the Bruynzeel Company, producing the highly advanced Bruynzeel Kitchen of 1938. It was also ahead of its time in that it featured modular ovens and

A 1980s-style kitchen with a large island section in the middle (Michael Dunne; Elizabeth Whiting &Associate/Corbis)

refrigerators and a line of matching fixtures and fittings.

The first mass-produced British fitted kitchens came with the combined kitchen/bathroom unit built into prefabricated emergency housing after World War II. George Fejer, a Hungarian architect, proposed a scheme with a separate stove-top and a built-in oven at elbow height. This innovative solution was rejected on economic grounds. The British Iron and Steel Federation's Kitchen of Tomorrow in the 1949 Ideal Home Show featured a similar arrangement.

As prosperity and consumption increased, the demand for fitted kitchens increased both within Europe and the United States. The British public looked enviously at the dream kitchens of Hollywood films with bar units, fitted cupboards, and pedal opening refrigerators. The United States gave Europe two examples, the gleaming laboratory kitchen and the live-in kitchen, which became more and more popular after 1945. The live-in kitchen was decorated in a more traditional manner and was often the center of the home, where meals were eaten.

In the United Kingdom, companies such as Hygena and Wrighton became market leaders in the production of fitted kitchens. Both companies employed aspirational advertising in magazines. Hygena employed George Fejer who designed their successful System 70. The companies continued to use soft wood frames and plywood but increasingly moved to chipboard and "wipe-clean" plastic laminates. West German companies such as Poggenpohl expanded into the top end of the British market in the 1970s. The self-assembly market also grew with the development of "flat-pack" units that could be combined to make a unified whole. The 1960s had seen designers such as the Italian Joe Colombo produce futuristic kitchens, but modernity was no longer the only option, and the 1970s saw the introduction of "farmhouse" styles that gave the illusion of rustic style while maintaining the benefits of modern construction techniques and electrical appliances.

The 1980s and 1990s have witnessed the continuing popularity of the fitted kitchen, with fitted stove-tops, ovens, and microwaves. There has also been a reaction to the total "off-the-peg" look, with more options for customizing. The late 1990s also saw a less fitted look becoming popular with the return of some freestanding units and pro-style stainless steel stoves and cupboards. A leader in this field is the German Bulthaup Company, which introduced units that could be used separately or as an integrated line in the early 1980s. They also reintroduced the butcher's block table that was placed in the center of the kitchen. Bulthaup followed this with the System 25 of 1992, with units in wood, aluminum, and stainless steel.

The choice of finishes and materials is now huge. Consumers can buy from large stores and smaller design-and-build kitchen specialists. Wood remains a popular material, along with MDF (medium density fiberboard), which can be either painted or laminated.

The elements of continuous working services, fitted drawers, and cupboards are taken for granted. The style of the kitchen is now most important, especially if it also doubles as a dining or family room. The kitchen reflects the personality of the householder as much as living rooms do.

See also Consumers; Do-It-Yourself; Habitat; Ideal Home Show; IKEA

References:

Elena, Pierre. "The Rise of the Fitted Kitchen." *Social History Curators Group Journal* 18 (1990–1991).

Heskett, John. *Industrial Design.* New York: Oxford University Press, 1980.

McDermott, Catherine. *The Design Museum Book of Twentieth-Century Design.* London: Carlton Books, 1998.

Ryan, Deborah S. *The Ideal Home through the Twentieth Century.* London: Hazar Publishing, 1997.

Kmart

The American retail company Kmart was founded in 1899 by Sebastian Spering Kresge. Today, Kmart has more than 2,000 stores and about 265,000 employees. Based in Troy, Michigan, it ranks second to Wal-Mart among American retailers.

Kresge seized on the idea of the five-and-dime variety store that had been pioneered by Frank Winfield Woolworth in 1879. His first variety store opened in downtown Detroit. By 1912, Kresge had 85 stores and annual sales had reached $10 million. The company was incorporated in Delaware in 1912 and in Michigan in 1916. Its formula for success was based on convenient locations and open displays. In the mid-1920s, a new chain of Kresge stores selling goods priced at $1 or less was established. These were familiarly known as the "green-front stores," as distinct from the original five-and-dime stores, which had red fronts. In 1929, Kresge opened its first stores outside the United States, in Canada.

One of the key factors in the company's continued growth and success has been its responsiveness to changing retail patterns. In the mid-1930s, Kresge opened a store in America's first suburban shopping center, the Country Club Plaza in Kansas City, Missouri. In 1962, nine years after the first discount store, the Ann and Hope Mill Outlet in Rhode Island, had opened, Kresge launched its first Kmart discount department store in Garden City, Michigan. Seventeen more Kmart stores opened in the first year and sales reached almost $500 million. By 1966, sales from the 162 Kmart stores and 753 Kresge stores hit the $1 billion mark. The success of the Kmart stores was such that by 1977 they generated 95 percent of the company's sales, prompting a change of name to the Kmart Corporation.

In the 1980s, Kmart consolidated its position through a series of corporate acquisitions. These included the Walden Book Company, the largest American retail book chain; Builders Square, Inc., a home improvement chain; Pay Less Drug Stores Northwest; and PACE Membership Warehouse. The company's focus on discount retailing meant that the Kresge stores no longer fit the core business profile and were sold off in 1987. The process of acquisition continued in the early part of the 1990s, alongside an expansion of its international retailing, but soon after the acquisition of the book chain Borders in 1992, the process went into reverse. Divestiture was necessary to pay for the program of modernizing the Kmart stores that began in 1990. In 1991, the first Super Kmart Center opened in Medina, Ohio. The Super Kmart chain, which expanded to 102 stores, stocked groceries and fresh food as well as traditional product lines and operated 24 hours a day, 7 days a week. Between 1993 and 1997, Kmart sold off most of its newer businesses and most of its overseas stores, and restructured itself to halt declining profits.

In the 1990s, Kmart began to shift from an earlier focus on lower-income customers and targeted its stores at middle-income families with children. This has been reflected in a changed store layout, whereby brand-name Pantry items are placed at the front of the store to facilitate "high-frequency" product sales. Tested successfully in 1996, the new "Big Kmart" format was extended to 1,245 stores by the end of 1998, with the remainder scheduled for similar remodeling. Other initiatives to foster customer loyalty included the negotiation of exclusive Kmart ranges of leading brands, such as Martha Stewart and White-Westinghouse, and the introduction of a private credit card in 1996, which attracted over a million users. Kmart has offered secure on-line shopping since 1998 and in-store electronic shopping. Central to these recent changes has been the policy of putting the customer first, which has also been demonstrated through liberal practices in relation to refunds and exchanges and a guarantee to match competitors' prices.

See also Sears, Roebuck and Company; Wal-Mart

References:
www.kmart.com/corp/

L

Landers, Frary and Clark

The company Landers, Frary and Clark (LFC) was founded in 1862 when the Landers and Smith Manufacturing Company took over the Frary and Clark Company of Connecticut. LFC grew with the rising star of American industrialization and expansion, producing goods as varied as skates, mousetraps, meat hooks, and window fittings. Not surprisingly, LFC took the brand name Universal in the 1890s. It began to introduce a number of groundbreaking products such as the Universal food chopper (1897), the Universal breadmaker (1895) and the Universal coffee percolator (1905).

LFC continued to expand its line into the 1950s. By then it faced increasing competition and was bought by General Electric in 1965.

See also Coffeemakers; Kettles

Lavatories

See Water Closets

Lawn Mowers

Most town houses in British cities did not have gardens. The larger Georgian and Regency terraces had squares or gardens reserved for the occupants. It was the later development of suburbs that led to single or terraced houses with self-contained gardens. The rising middle classes provided a market for advice, and the writer and publisher John Loudon published his *The Gardeners' Magazine* in 1826, followed by *The Suburban Gardener and Villa Companion* in 1838. The twentieth-century expansion of suburbia, with houses designed for single-family occupation, led to gardening becoming a popular occupation. It was regarded as being a "Christian" pastime and also suitable for women. New homes seldom had established gardens, and it was usually left to the owners to develop them.

Larger gardens could accommodate sections for fruit and vegetables but most twentieth-century suburban gardens were a mix of grass lawns, shrubs, borders, and flower beds. As a result, the lawn mower was to become an important piece of domestic equipment, essential for keeping the homeowner's garden and social standing in good order.

In the agricultural and country-house world, large expanses of grass had been cut by scythe. The first lawn mower, "machinery for the purpose of cropping or shearing the vegetable surface of lawns," was invented by Edwin Budding of Stroud, Gloucestershire, United Kingdom, in 1830. Budding was a tex-

Man mowing lawn with his Toro lawn mower, 1999 (Courtesy Toro)

tile worker and may have been inspired by the process of napping cloth. His design incorporated the main features that are still in use today, namely helical cutting blades, a small wooden roller, and a larger metal roller. This powered the cutting blades, which were geared to it. The original machines were manufactured locally by the Phoenix Iron Works of Thrupp, near Stroud. They were large and clumsy machines costing 10 guineas (£10 10s.) An early customer was Regent's Park Zoo, London. A smaller mower was available, being described by Budding as "an amusing, useful and healthy exercise" for country gentleman. From 1832 other companies, including Ransomes of Ipswich, began to manufacture Budding machines under license.

The later nineteenth century saw a greater variety of smaller machines. The rising number of small suburban gardens and the popularity of lawn tennis and croquet stimulated their production in the 1870s. The American New Excelsior of 1874 was a lightweight machine compared to the heavier English models. By the 1890s the standard form of the suburban lawn mower could be seen in the New Automaton by Ransomes, Sims and Jefferies (an amalgamation of companies) that featured a green painted cast-iron frame, red painted blades, and a detachable grass box. Other companies such as Qualcast were to produce such muscle-powered models for much of the twentieth century.

Automation had begun at the very end of the nineteenth century. The Benz Company of Stuttgart, Germany, and the Coldwell Lawn Mower Company of Newbury, New York, produced experimental gas-driven models in 1897. Grimsley and Son of Leicester, United Kingdom, also produced an experimental model with a six-horsepower engine in 1899. The first commercial model, by Ransomes, Sims and Jefferies, appeared in 1902 and also used a six-horsepower engine.

By 1930 a variety of models were available to suit a variety of pockets. Simple models such as the Shanks Britisher sold at around £2. Standard models like the New Automaton or the Shanks Caledonia were between £5 and £8. The blades were still powered through the back roller but were becoming chain driven. At the top end of the market were gasoline-engine-powered models by Dennis and Atco that could cost as much as £80.

Electric models had always suffered a mobility problem because of the need to have a power cable. Nevertheless lightweight models became more popular in the 1960s. Again they used helical cutting blades. The development of the Hovercraft in the 1950s inspired the Swedish inventor Karl Dahlman to develop a lawnmower with a rotary blade that could hover on a cushion of air. The British version, the Flymo, went on sale in 1963. The company became part of the Electrolux Group in 1969. Black & Decker also produced a line of hover and rotary mowers. Although they were light and easy to use, their main disadvantage was the lack of a grass box. Nevertheless hover mowers became very popular for small gardens, being easy to use and less temperamental than some gasoline-powered mowers. Rotary blades are also used on gasoline machines.

See also Black & Decker; Electrolux; Houses

References:

Barrett, Helena, and John Phillips. *Suburban Style: The British Home, 1840–1960.* London: Macdonald Orbis, 1987.

Robertson, Patrick. *The Shell Book of Firsts.* London: Ebury Press and Michael Joseph Ltd., 1983.

Lighting

Lighting was the first aspect of home life to benefit from the application of electricity. However, while the electric light bulb was widely available by 1900, only a tiny percentage of homes had been wired for electricity. Electric lighting did not become commonplace in the home until the 1920s and 1930s.

In 1900, most homes were lit by gas, kerosene (also known as paraffin oil), or candles. Until the late 1880s, the main advantage of gas lighting was the convenience of the fuel being piped into the home. Gas burners were more efficient in producing heat than light, and the flame exuded sooty smoke, which left deposits. Gaslights also consumed oxygen and produced water vapor, making rooms stuffy and damp. The first petroleum deposits were discovered in Ontario, Canada, in 1857, but the first commercial oil well was drilled in Titusville, western Pennsylvania, in 1859. Kerosene, a readily flammable light oil, was extracted by distilling the crude petroleum and became popular as a lighting and heating fuel by the mid-1860s. It was a far superior fuel to the animal and plant oils that had previously been used for lighting, not only in terms of combustion, but also because it was stable, had little odor, and produced little smoke. It was also comparatively cheap.

The Incandescent Filament Lamp

The first type of electric lighting was the carbon arc lamp, demonstrated by the English chemist Humphrey Davy in 1808. It produced a very intense light, suitable for street lighting and other public lighting. It was unsuitable for domestic use, however, because it required a high voltage. Another English chemist, Joseph Wilson Swan, carried out the first experiments that led to the development of the incandescent filament lamp or electric light bulb. The basic principle of the filament lamp is that when electric current is passed through a conductive filament (a fine thread), the filament becomes white-hot and glows brightly. Swan began in the mid-1840s by creating filaments by baking strips of paper at high temperatures to produce carbon fiber. He placed the carbon filament in a glass bottle and connected each end of the filament to wires attached to a battery. The next step, which proved problematic at first, was to evacuate the bottle and seal it with a cork stopper. If air remained in the bottle, the carbon would burn away quickly. The invention of the mercury pump in 1865 made it possible to create an effective vacuum. However,

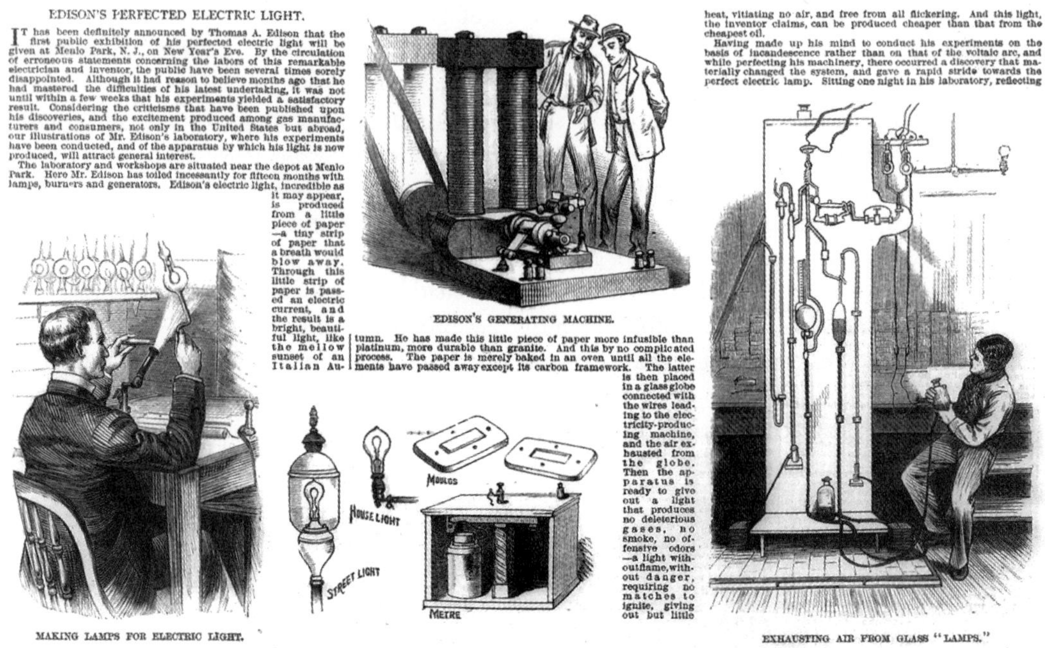
EDISON'S PERFECTED ELECTRIC LIGHT.

IT has been definitely announced by Thomas A. Edison that the first public exhibition of his perfected electric light will be given at Menlo Park, N. J., on New Year's Eve. By the circulation of erroneous statements concerning the labors of this remarkable electrician and inventor, the public have been several times sorely disappointed. Although it had reason to believe months ago that he had mastered the difficulties of his latest undertaking, it was not until within a few weeks that his experiments yielded a satisfactory result. Considering the criticisms that have been published upon his discoveries, and the excitement produced among gas manufacturers and consumers, not only in the United States but abroad, our illustrations of Mr. Edison's laboratory, where his experiments have been conducted, and of the apparatus by which his light is now produced, will attract general interest.

The laboratory and workshops are situated near the depot at Menlo Park. Here Mr. Edison has toiled incessantly for fifteen months with lamps, burners and generators. Edison's electric light, incredible as it may appear, is produced from a little piece of paper—a tiny strip of paper that a breath would blow away. Through this little strip of paper is passed an electric current, and the result is a bright, beautiful light, like the mellow sunset of an Italian Autumn. He has made this little piece of paper more infusible than platinum, more durable than granite. And this by no complicated process. The paper is merely baked in an oven until all the elements have passed away except its carbon framework. The latter is then placed in a glass globe connected with the wires leading to the electricity-producing machine, and the air exhausted from the globe. Then the apparatus is ready to give out a light that produces no deleterious gases, no smoke, no offensive odors—a light without flame, without danger, requiring no matches to ignite, giving out but little heat, vitiating no air, and free from all flickering. And this light, the inventor claims, can be produced cheaper than that from the cheapest oil.

Having made up his mind to conduct his experiments on the basis of incandescence rather than on that of the voltaic arc, and while perfecting his machinery, there occurred a discovery that materially changed the system, and gave a rapid stride towards the perfect electric lamp. Sitting one night in his laboratory, reflecting

"*New Jersey—the wizard of electricity—Thomas A. Edison's system of electric illumination." Wood engraving in Frank Leslie's* Illustrated Newsletter, *January 10, 1880 (Library of Congress)*

Swan still struggled to find a filament that was suitably durable. In 1878, he demonstrated a carbon filament lamp that lasted for a few hours.

Meanwhile, the American inventor Thomas Alva Edison had also begun work on an incandescent lamp. He formed the Edison Electric Light Company in 1878, before he had produced a viable lamp. At his laboratory in Menlo Park, New Jersey, Edison experimented with various types of filament. In October 1879, he succeeded in keeping a lamp with a filament of carbonized cotton sewing thread alight for 45 hours. However, the production of the lamps was a laborious process because the fragility of the carbonized thread demanded careful handling. Edison then developed a more robust bamboo fiber filament that had a life of up to 1,000 hours. Across the Atlantic, Swan was using a nitro-cellulose carbon filament, created by partially dissolving cotton thread in sulfuric acid. Swan patented his incandescent lamp in 1880. Edison, whose patent was approved in 1879, challenged Swan's patent rights, but the British courts upheld Swan's patent in 1882. A year later, Edison and Swan pooled their British patents and formed the Edison and Swan United Electric Light Company, which used Ediswan as its brand name.

In the 1880s and 1890s, electric lighting was expensive and few homes were wired for electricity, so it became a source of prestige and a symbol of ostentation. The first Swan lamps sold in Britain in 1879 cost 25 shillings (£1.25) each, which represented about two weeks' wages for an unskilled worker. In 1880, the wealthy English engineer Sir William Armstrong built the world's first hydroelectric plant at his home, Cragside, in Northumberland. The plant supplied electricity to light his picture gallery, which was fitted with Swan's incandescent lamps. In the United States, amongst the first customers of Edison's Pearl Street power station in New York were the banker and philanthropist John Pierpont Morgan Jr. and the financier and railway entrepreneur William Henry Vanderbilt. Electric lighting was valued not just for its practical benefits, but also for its aesthetic qualities. In 1882, Edward H. Johnson, the vice-president of the New York Edison Company, showed off a Christmas tree adorned with twinkling lights at his home. Sets of Christmas tree lights were a commercial product by 1900. By 1881, the price of a light bulb had fallen to 5 shillings in Britain. In the United States, the price was about $1, equivalent to about half the average daily wage.

The Incandescent Gas Mantle

The electric light bulb was still merely a novelty in 1885, when an incandescent form of gas lighting emerged. Baron Karl Auer von Welsbach, an Austrian chemist, had developed a particular interest in rare earths. He discovered that cotton fabric impregnated with cerium and thorium nitrates became incandescent when heated by a gas flame. The incandescence was produced by the oxidation of the cerium and thorium. This discovery was marketed in the form of the Welsbach gas mantle, a sheath of stiffened, impregnated fabric that fitted over a standard gas burner. As the gas mantle and carbon filament light bulb were roughly equivalent in their lighting value, cost was a significant factor. In Britain, where the gas industry was run by local public utilities, gas was much cheaper than in the United States, so the growth of the electricity industry was slower.

New Filaments

In order for electric lighting to become more widespread, it had to become more cost-effective for the consumer. This hinged on the installation costs, the unit price of electricity, and the cost of light bulbs. Light-bulb manufacturers formed close relationships with electricity utility companies, particularly in the United States, as their success was interdependent. Licensing and cross-licensing arrangements led to the adoption of brand names, signifying that the product was of a standard specification, even if made by different companies. In the United States, two companies, General Electric (successor to the Edison General Electric Company) and Westinghouse, held the bulk of electrical engineering patents. In 1895, they agreed to pool their patents. To identify their light bulbs, they adopted the trademark, Mazda, which means "wise spirit" in Sanskrit and is the good god of the Zoroastrian religion. In Britain, from 1911, the Mazda trademark and related patent rights were assigned to the British Thomson-Houston Company, part owned by General Electric. While the relative price of light bulbs remained high, there was a niche in the market for small companies that specialized in refitting dead light bulbs with new filaments. One such American company was the Bay State Lamp Company, formed in 1901, that later became the Hygrade Sylvania Corporation.

As more companies entered the light-bulb manufacturing business, the pace of technical development quickened. The carbon filament was an inefficient illuminant, producing only 1.5 to 3.5 lumens per watt, and the light quality declined steadily as carbon vapor, released by the hot filament, condensed on the glass envelope, leaving a black deposit. By 1900, carbon filament lamps typically had a life of 400 to 500 hours. A metallic alternative was not immediately obvious, as only metals with a very high melting point were suitable. Davy had found that platinum was suitable, but it was very expensive. In 1897, the German chemist Walther Hermann Nernst developed a new type of electric lamp with a zirconium oxide filament that produced 6 lumens per watt. However, the Nernst lamp was less compact and more expensive than a light bulb. In 1898, Baron Karl Auer von Welsbach made another contribution to lighting technology by developing the osmium filament, the first commercial metal filament. It was used in light bulbs produced by the German company Auer Gesellschaft. In 1905, the German company Siemens and Halske introduced another new filament, the tantalum filament.

The metal that proved to be most successful was tungsten, the metal with the highest melting point (3410°C). The tungsten fila-

Father reading to his son by electric light (Warren Morgan/Corbis)

ment required less than half of the electricity consumed by the carbon filament to produce the same amount of light. It first became available in light bulbs sold under Auer Gesellschaft's new Osram trademark in 1906. The name, Osram, was a construct of "osmium" and "wolfram," the German name for tungsten. In Britain, the General Electric Company produced light bulbs under the Osram trademark from 1909.

In the United States, an employee of General Electric improved tungsten filament technology in 1908. William David Coolidge developed a method of processing tungsten using a powder technique. The tungsten was ground to a powder that was compressed and fused electrically. The fused tungsten rod was then heated and hammered repeatedly until it was sufficiently ductile to be drawn into fine wires. With an efficiency of 8 lumens per watt, the tungsten filament lamp reestablished the superiority of the incandescent filament lamp over the incandescent gas mantle.

In 1912, another General Electric employee, Irving Langmuir, demonstrated that the efficiency of the tungsten filament lamp could be raised to 12 lumens per watt if the bulb was filled with an inert gas, such as nitrogen or argon. The presence of the inert gas allowed the filament to get hotter, thus releasing more light, and extended the life of the filament by reducing evaporation. Langmuir was also able to fit in a longer filament by coiling it. Langmuir's gas-filled, coiled tungsten filament lamps were introduced commercially in 1913. By 1920, it was standard for light bulbs to have a life expectancy of 1,000 hours. The last major advance in filament lamp technology was the coiled-coil filament lamp, which used a very tightly coiled filament rather than a straight wire. Introduced in 1934, it was less prone to evaporation, lost less energy through convection and was 20 percent more efficient than the standard coil filament. However, less than 10 percent of the energy released by even the most efficient incandescent lamp was in the form of light.

Other technological developments were directed more toward reducing production costs. Before World War I, light bulbs were hand-blown. These are recognizable by the pip in the roof of the bulb. Soda-lime glass was used for the body of the light bulb, but potash lead glass was used for the base, which had to be fused to "pinch" the filament supports. The pinch needed to be more chemically durable because there was a tendency for short circuits to occur, which weakened the less-resistant soda-lime glass. Initially, platinum was used for filament supports because its thermal expansion coefficient was similar to that of glass, but after 1900 cheaper substitutes were developed, particularly molybdenum and an iron-nickel alloy. Bayonet caps and screw caps, which fitted securely into lamp holders, were introduced in the 1880s. When high-wattage lamps became possible, borosilicate glass was used to give increased resistance to temperature changes.

Automatic bottle-blowing machines were devised before 1900, but this technology was not immediately adopted for light bulb manufacture because demand did not justify it. As more homes became wired, demand grew and light bulbs began to be mass-produced, which brought prices down and thus stimulated more demand. The earliest automatic light bulb machines, such as the Westlake, used the same suction technique as the first bottle machines. By 1921, when about a third of American homes were wired for electricity, the Corning Glass Works in the United States had developed a "ribbon" machine that was capable of producing 20,000 to 30,000 light bulbs an hour. The ribbon machine fed a continuous stream of glass over moulds. By 1950, an automatic glass-blowing machine served by one operator could produce 100,000 light bulbs an hour.

Fluorescent Lamps

The electric lamp manufacturers encouraged the acceptance of electric lighting by stressing its cost efficiency and enhanced lighting quality. This approach also accorded with the objectives of the electricity utilities, as more customers meant greater use of electricity. As people became used to the brighter light produced by the electric lamp, manufacturers introduced higher wattage light bulbs. Electricity showrooms would mount lighting schemes aimed at persuading customers to buy more light fittings in order to obtain different lighting effects.

This strategy of promoting forms of lighting that would increase electricity consumption was challenged by the emergence of fluorescent lighting in the late 1930s. The fluorescent lamp was an evolution of the gas discharge technology pioneered by a number of scientists, including Heinrich Geissler in Germany and William Crookes in Britain, from the mid-nineteenth century. Switching on electric current causes the cathode of the gas-filled discharge tube to release a stream of electrons, which ionize the gas atoms, producing both visible and invisible light. The gas ions become luminous and take on a color characteristic to the particular gas. The invisible ultraviolet light that is created can be rendered visible by coating the tube with phosphors. Like gases, particular phosphors have their own color properties. Edison experimented with gas discharge technology in 1896, but abandoned his experiments at the research stage. In 1901, the American electrical engineer Peter Cooper-Hewitt developed a mercury discharge lamp that gave off greenish-blue light. Georges Claude, a French industrial chemist, demonstrated the first neon discharge tube in 1910. With its bright red light, the neon tube was adopted for illuminated signs. In 1930, the sodium vapor lamp, which gave off yellow light, came into use for street lighting. One of the problems with discharge lamps was that the vaporization process was corrosive. This was solved when manufacturers discovered that glass with a higher alumina and lime content was more resistant.

General Electric demonstrated the first modern fluorescent lamp in 1935 at an industry convention. It was a tube, with electrodes at each end that emitted bright green light. General Electric originally conceived that the fluorescent lamp would be used for colored lighting, not general lighting. In 1938, it introduced a line of "lumiline" tubes in a range of colors. However, experiments revealed that careful blending and coating of phosphors could produce white light. As the fluorescent tube was four to five times more efficient than the incandescent lamp, the electricity utilities were concerned that widespread adoption of the daylight fluorescent tube would severely dent demand for electricity. They encouraged manufacturers to target the business market and stressed the disadvantages for domestic consumers—the high cost of installation, the higher price of the tubes, and the harsher quality of the light. They downplayed the fact that the high initial costs, a consequence of the need for the light fitting to incorporate a starter that boosted the voltage, were offset by the long life of the tubes (up to fifteen times longer than incandescent tubes) and lower running costs. This strategy was largely successful. While fluorescent lighting was supplying about 80 percent of the world's artificial light by 1980, it was far less widespread in the home than the office, with most home use being confined to the kitchen and bathroom, and as built-in lighting in cupboards and cabinets.

Energy Efficiency

The energy efficiency of the fluorescent lamp only became a selling point when environmentally aware consumers began to demand products that were less wasteful of electricity and resources. Even when the halogen lamp was introduced in the early 1960s, it was promoted more on the basis of its added brightness than its durability. Halogen-filled light bulbs give off twice as much light as standard light bulbs and last for two to three times as many hours. In the 1970s when environmentalism became prominent, lamp manufacturers began to investigate energy efficiency. The leader in this field was the Dutch company Philips, which began producing light bulbs in 1891. Philips recognized that the energy efficiency of the fluorescent tube needed to be combined with the convenience and flexibility of the incandescent lamp. The solution was to create a narrow fluorescent tube that could be looped compactly and that was designed to fit into a standard lamp holder. The Philips long-life SL18W light bulb was introduced in 1980. Although its price was about five times that of a tungsten filament light bulb, it lasted five times longer—5,000 hours—and used about a fifth as much electricity to produce the same amount of light. Optical-fiber technology offers great potential for lighting that is both energy-efficient and creative, but although it has been used widely in display lighting, in the home its use has been confined to novelty table lamps.

Lighting Design

Before the tungsten-filament light bulb was introduced, incandescent lamps—whether gas or electric—only produced low levels of light. A 60-watt carbon filament lamp was the equivalent of 20 candles, whereas a 60-watt gas-filled, tungsten filament lamp was equivalent to 120 candles. While illuminance (amount of light) was low, the main concern was to maximize light, not to shade it. Glass light shades helped to diffuse light by reflecting it more widely and, in the case of gaslights, provided a shield against smoke. From the 1890s, designers and manufacturers of "art glass" began to produce very decorative electric lights, both in terms of color and shape, which drew on art nouveau influences. The leaders in this field were the American Louis Comfort Tiffany, and the Frenchman René Lalique. Art Nouveau influences were also visible in the Mazda advertisements produced by the American illustrator Maxfield Parrish. However, mass-produced light shades were more modestly designed. They tended to be in pastel shades with cut designs, such as floral motifs, or molded fluting or ribbing.

The American architect Frank Lloyd Wright was the first architect to fully embrace the notion of lighting schemes as a means of enhancing the home. Wright liked to provide a holistic service, embracing furniture, fittings, and all elements of interior design, in order to ensure the integrity of his vision. The Robie House in Chicago, completed in 1908, was the first house known to feature indirect lighting and to use rheostats so that lights could be dimmed. From the 1920s, as electricity became more widespread, manufacturers and designers paid more attention to differentiating lighting in terms of function. The ceiling light was the provider of ambient lighting, so the main concern was the effective dispersal of light. The main function of the table lamp or desk lamp was to provide focused lighting for tasks such as reading, writing and sewing. Integrated lighting schemes also accommodated other concepts, such as feature lighting (lighting as a visual feature in its own right or to enhance an object or structure) and mood lighting, which could be provided through wall lights and floor lamps.

Task lighting provoked a number of innovative designs in the 1920s and 1930s. The German Bauhaus school, with its emphasis on rational, functional design, produced a number of influential lamp designs. The designers in the Bauhaus Metal Workshop worked closely with two German lighting manufacturers, Körting & Mathieson and Schwinter & Gräff, which then mass-produced the successful designs. Marianne Brandt designed the Kandem table lamp for Körting & Mathieson as a bedside lamp. It had a small footprint to suit the bedside table or cupboard, a push-button switch and a hinged head, allowing the shade to be angled to vary the direction of light. The principle of adjustability was taken to greater extremes in the Anglepoise desk lamp, designed in 1932 by the British engineer George Carwardine and manufactured by Herbert Terry & Sons. The Anglepoise lamp had an articulated stem comprising three levers kept in constant tension by springs so that the head could be angled in any direction. A more aesthetically refined version, the Luxo L-1, was designed by the Norwegian engineer Jacob Jacobsen in 1937. The flexibility and functionality of the Anglepoise lamp was only bettered when the U.S. company Black & Decker introduced its SnakeLight in 1994. The SnakeLight is a battery-operated flashlight with a high-intensity xenon bulb and a bendable stem that can form a coiled base or be draped on a support. It is therefore portable as well as omnidirectional.

After World War II, lighting design became more extravagant and specialized. While most named designs only reached an elite market, many of the ideas filtered down to the mass market. Two examples of mass-market impact are found in the work of Isamu Noguchi and Richard Sapper. The Japanese American sculptor Isamu Noguchi began to create paper light sculptures in the 1940s and from the 1950s applied the same basic ideas in a series of lamp designs. His Akari lampshades used traditional Japanese materials, mulberry bark paper and bamboo, and were produced in a variety of organic shapes. As they were collapsible, they could be sold as flat packs, which was an added convenience for the purchaser. Noguchi's basic ideas were widely copied and popularized from the 1960s by chain stores such as Habitat. From the 1970s, the two main mass-market trends in lighting design have been high tech and retro. One of the most influential early high-tech designs was the Tizio table lamp designed by the German designer Richard Sapper in 1972. This was a strikingly modern rendering of the articulated desk lamp, made from matte black plastic and aluminum. High-tech lighting design has brought styles of lighting previously restricted to the office or public space, such as spotlight tracks and recessed ceiling lights, into the home. Retro lighting has resurrected a variety of period designs, from Art Nouveau fluted shades to 1960s lava lamps.

See also Electricity; General Electric; General Electric Company; Habitat; National Electric Light Association; Philips Electronics

References:

Bijker, Wiebe E. "The Social Construction of Fluorescent Lighting, Or How an Artifact Was Invented in Its Diffusion Stage." In *Shaping Technology/Building Society,* edited by Wiebe E. Bijker and John Law. Cambridge, MA, and London: MIT Press, 1992.

Byers, Anthony. *Centenary of Service: A History of Electricity in the Home.* London: Electricity Council, 1981.

Nye, David E. *Electrifying America: Social Meanings of a New Technology.* Cambridge, MA: MIT Press, 1990.

O'Dea, William. *The Social History of Lighting.* London: Routledge & Kegan Paul, 1958.

www.ge.com
www.gec.com
www.osram.com
www.philips.com
www.siemens.com
www.siemens.de
www.sylvania.com

Loudspeakers

A loudspeaker consists of a drive unit for converting electrical energy to sound waves and a diaphragm that transmits the sound. In the late nineteenth century, simple acoustic horns were fitted to mechanical Edison-type cylinder phonographs and Berliner-type disc gramophones to amplify the recorded sound. In the twentieth century, loudspeakers were developed for use with a range of audio equipment.

In the 1920s, the introduction of radio broadcasting created a new market for loudspeakers. During the same period, the manufacture of gramophones driven by an electric motor rather than mechanical clockwork and with an electric pick-up fitted to the stylus required the use of loudspeakers in place of simple horns. Early radios did not have integral speakers, so loudspeakers were marketed separately. Most horn loudspeakers, including the British Thomson-Houston C2 type, of which over a million had been sold by 1930, had an electromagnet as the simple drive unit and a small metal diaphragm placed in the narrow opening of the horn. Electrical energy causes the "voice coil" of the electromagnet to oscillate, causing the diaphragm to vibrate. The natural resonance of the horn, which could be wood, metal, papier–mâché, or ebonite, amplified the sound. The demand for loudspeakers gave impetus to improvements in design. One example was the change from the straight-necked or curved-neck design to the swan's neck design. More significantly, in 1922, the U.S. company Sterling Telephone & Electric brought out the Magnavox R2B model with a moving coil drive unit.

In 1924, Sterling Telephone & Electric Company launched a truly innovative loudspeaker, the Primax, which was the first commercial hornless loudspeaker. Based on a design patented in France by Lumière in 1909, it consisted of a moving iron drive unit behind a circular pleated paper diaphragm. The moving iron provided the amplification, with improved bass range, and the sound was dispersed through the diaphragm omnidirectionally. However, it was expensive and was soon superseded by another new design.

Around the same time, the cone type of loudspeaker, introduced in Britain by the British Electrical Manufacturing Company as the Beco model, has proved to be a durable design. The Beco featured a moving iron drive unit and a conical, aluminum foil diaphragm. While the cone diaphragm became standard, although more commonly made of paper or linen, the moving iron drive unit was short-lived. In 1924, Chester W. Rice and Edward W. Kellogg designed a prototype cone loudspeaker with a moving coil drive unit for General Electric in the United States. This design became known as the dynamic loudspeaker owing to its superior sound reproduction and, with modifications, was standard by the mid-1930s.

Another advantage of the dynamic loudspeaker was that it could be housed in a cabinet. From 1925 to the mid-1930s, dynamic loudspeakers were produced in decorative wood or Bakelite cabinets. The decorative effect was mainly achieved through the design of the speaker grille rather than through the shape of the cabinet. Perhaps the most striking and immediately recognizable grille design was the rising sun motif adopted by the British company W. G. Pye & Co. From 1930, the radio with integral loudspeaker became standard and the market for separate loudspeakers slumped.

Loudspeakers reemerged as objects in their own right with the popularization of hi-fi (high fidelity) stereophonic sound systems in the 1960s. While the cone design remains, speakers no longer consist of a single cone. For high-quality sound reproduction, there are separate low-frequency (bass or woofer) and high-frequency (treble or tweeter) cones, and, in some cases, midrange cones too. The main factor affecting the quality and price of a loudspeaker is the sensitivity of the woofer to deep bass tones. Modern speaker cabinets are typically restrained in design, the emphasis being on technical performance.

While dynamic loudspeakers remain the most common type, more advanced loudspeaker designs are available. For example, the Bose Corporation was founded in the United States in 1964 with the objective of creating loudspeakers that gave a more natural sound. One of Bose's innovative designs is the Acoustic Waveguide speaker, introduced in 1984. The Waveguide system is based on a long mathematically defined tube, folded to fit within a small speaker cabinet. The speaker cone is placed within the Waveguide so that the acoustic waves produce sound with rich bass tones. A more recent development in loudspeaker technology has been the development of the digital loudspeaker, pioneered by two British companies—New Transducers Limited (NXT), which launched the core technology in 1996, and 1 . . . Limited, which holds a manufacturing license. The logic behind this development is that conventional loudspeakers diminish the performance of digital sound technologies, such as compact disc players or digital audiotape players. The digital sound recording is near perfect, but the conventional loudspeaker is always subject to a degree of distortion. The digital loudspeaker is a thin panel that vibrates in response to digital pulses that can be transmitted as electromagnetic waves or infrared signals without power amplification. Highly portable, unobtrusive, and energy-saving, digital loudspeakers represent a significant advance. However, most consumers currently use music systems that include analog components, such as cassette recorders, so digital loudspeakers are not viable option.

See also Compact Disc Players; Radio; Record Players; Tape Recorders

References:

Hill, Jonathan. *Radio! Radio!* Bampton, UK: Sunrise Press, 1986.

M

Marconi Company

The British company Marconi was founded in 1897 by Guglielmo Marconi to develop and market wireless communications systems. While the theoretical groundwork for wireless communications was laid by scientists such as the Germans Heinrich Hertz and Karl Ferdinand Braun, it was Marconi who translated the theory into a practical application. He is therefore widely acknowledged as the inventor of wireless communication. After World War II, the company was swept up in the series of mergers and acquisitions that transformed the British electrical industry. By the end of the twentieth century, the original Marconi company had lost the Marconi name, which was instead adopted by the parent General Electric Company.

Guglielmo Marconi was born in 1874 in Bologna, Italy. As a young man, he developed an interest in physics and received private tuition, but was rejected by Bologna University, where he had wished to study. In 1894, he began to repeat and refine recent experiments relating to electromagnetic waves at the family home. Marconi saw the potential for a system of wireless telegraphy for communicating with ships. He felt that his ideas were more likely to be appreciated in England, the center of the international shipping trade. In 1896, he traveled to London, where his Irish mother had useful contacts, and filed his first patent a few months after his arrival. Marconi received funding from the General Post Office to conduct further experiments. He formed the Wireless Telegraph and Signal Company in 1897 and set up the world's first radio factory in Chelmsford in 1899.

In 1900, the company was renamed Marconi's Wireless Telegraph Company and Marconi received a patent for a system for making simultaneous broadcasts on different frequencies. His next step was to expand internationally by setting up new companies, including the Marconi Wireless Telegraph Company of America. The first transatlantic wireless transmission took place in 1901 from Poldhu Cove, Cornwall, to Signal Hill, Newfoundland. In 1909, Marconi and Braun were jointly awarded the Nobel Prize for physics "in recognition of their contribution to the development of wireless telegraphy." Marconi returned to Italy after the outbreak of World War I and thereafter spent much of his time there. After the United States entered the war in 1917, the U.S. government took control of all wireless stations in order to control strategic communications. This paved the way for the U.S. wireless industry

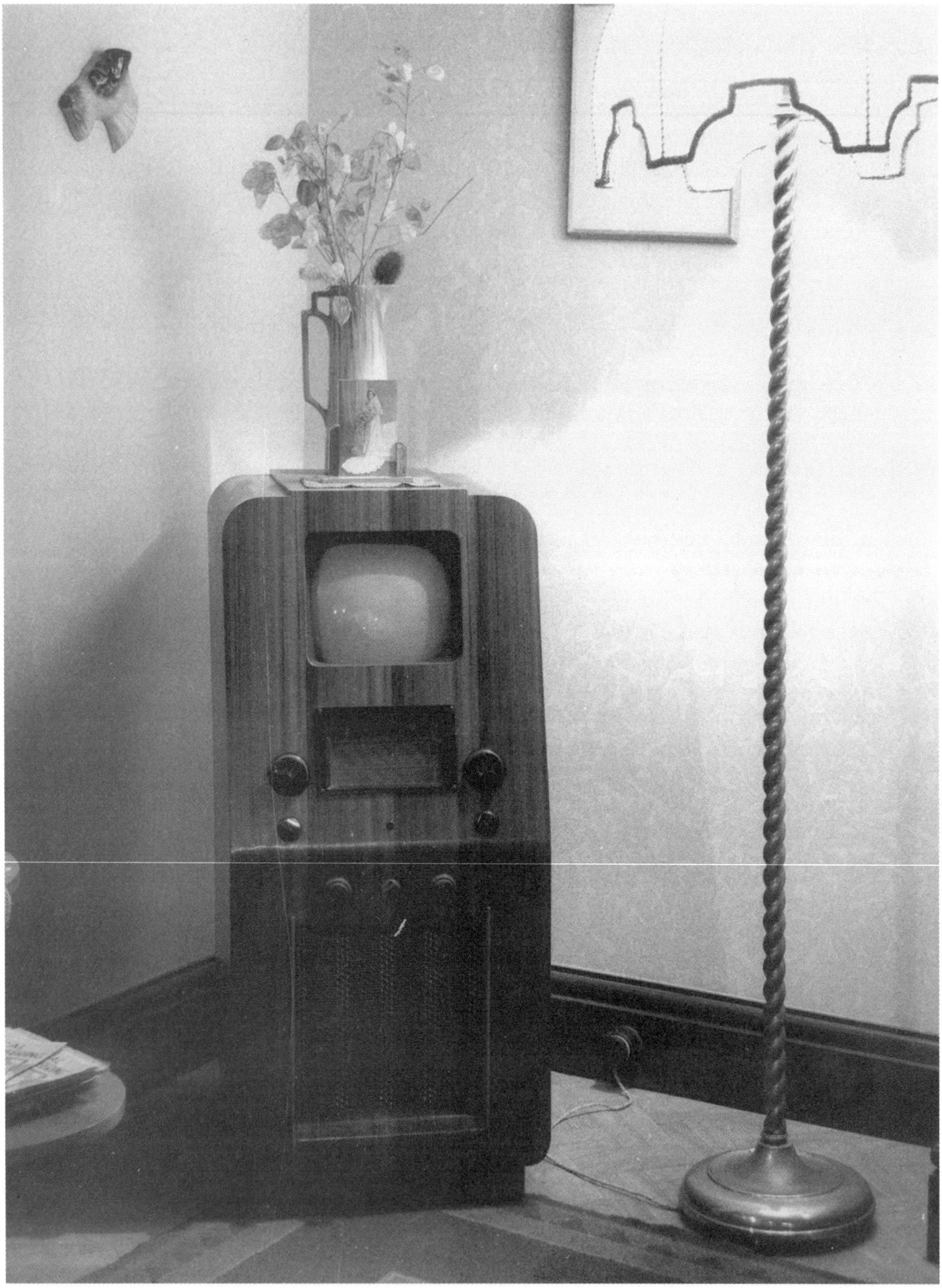

Marconi model 709 combined television and radio console, 1938 (The Museum of Science and Industry in Manchester)

to be brought under private U.S. control after the war, when the government brokered the acquisition of Marconi's U.S. subsidiary by a consortium of U.S. companies led by General Electric and including Westinghouse and American Telephone & Telegraph. The wireless company was renamed the Radio Corporation of America (RCA) in 1919.

In Britain, the Marconi company built a high power transmitter at Chelmsford in order to experiment with public radio broadcasting. The broadcasts began in 1920 and included a concert given by the Australian soprano Nellie Melba in July. However, they were suspended in November because they were interfering with telegram services. In order to satisfy popular demand while the concept of a single licensed broadcasting company was investigated, Marconi's Wireless Telegraph Company was granted two broadcasting licenses in 1922. Later that year, 300 manufacturers and dealers, including Marconi, formed the British Broadcasting Company (BBC). The service was funded by a new license fee payable by all owners of radio receivers. In 1927, the BBC became a public corporation by royal charter, as it remains today.

Marconi himself continued to work on improving radio communications and became particularly active in the field of short-wave transmission. In 1924, the company introduced the Beam System of directional short-wave transmission. During the 1930s, the Marconi company moved into two new areas of wireless technology—television and radar. A team of researchers at Electrical & Musical Industries (EMI) began working on developing a system of television broadcasting in 1931 and contracted Marconi to provide transmitters for their experiments. The system that they finally settled on was an electronic system, more akin to those being developed in the United States than to the rival British system developed by John Logie Baird. In 1934, Marconi and EMI formed a consortium in readiness to compete for public television broadcasting contracts. In February 1937, following four months of trial broadcasts by Marconi-EMI and Baird Television, the BBC chose the Marconi-EMI 405-line system. Marconi began to manufacture both television transmission equipment and television receivers. In the same year, Marconi obtained the contracts for Britain's first air-defense radar network, a chain of coastal stations completed in 1938. That was also the year of the death of Guglielmo Marconi, marked by two minutes' silence on radio transmitters across the world.

In 1939, the company set up the Marconi Research Laboratories. The value of radar systems became clear early in World War II. The 1940 Battle of Britain was proof that radar dispensed with the need for airborne patrols to detect incoming enemy planes. From then on, defense technology became an increasingly important area of business for Marconi. In 1946, Marconi was taken over by English Electric, a company formed in 1918 by the merger of a number of manufacturers of electrical engineering factory equipment. The acquisition of Marconi was the latest move in a program of diversification that English Electric had begun in the late 1930s, when it began producing diesel-electric locomotives and aircraft. Marconi retained a separate corporate identity and in 1963, reflecting the changed emphasis of its business, shortened its name to the Marconi Company. In 1964, it moved to new premises in Chelmsford. The British electrical industry experienced a major upheaval in the late 1960s, when the General Electric Company (GEC) first acquired Associated Electrical Industries (AEI) in 1967 and then merged with English Electric in 1968.

Marconi continued to operate as a self-contained business within GEC. It was renamed GEC-Marconi Limited in 1987. In 1996, under a new chief executive, GEC embarked on yet another major restructuring. GEC-Marconi was renamed Marconi Electronic Systems in 1998 and, in 1999, was divested and merged with British Aerospace to form British Aerospace Systems. With 100,000 employees and a combined turnover of £12 billion in 1998, British Aerospace Systems is the largest defense and aerospace company in Europe and the third largest in the world. The Marconi name lives on with the renaming of GEC as the Marconi Corporation in 1999.

See also General Electric Company; Radio; RCA; Television

References:

Hill, Jonathan. *Radio! Radio!* Bampton, UK: Sunrise Press, 1986.

www.marconi.com/about_marconi/heritage/

Matsushita

The Japanese company Matsushita is one of world's leading consumer electronics companies, although it is known to consumers through its major brand names, which include National, Panasonic, Technics, and JVC. The company's founder, Konosuke Matsushita, set up the Matsushita Electric Housewares Manufacturing Works in 1918. Its first product was an electric plug. The company built its first office and factory in 1922. The product line expanded slowly in the 1920s with the introduction of bicycle lamps in 1923 and electric irons in 1927. The company was renamed the Matsushita Electric Manufacturing Works in 1929. It began to produce radio and dry cell batteries in 1931. In response to the growth of production, a divisional system was introduced in 1933 and a new office and plant opened in Kadoma. In the following year, Matsushita created a Training Institute for staff. The company was incorporated as the Matsushita Electric Industrial Company in 1935 and also launched a new trading company. New manufacturing subsidiaries followed—the National Electric Lamp Company in 1936 and the Matsushita Electric Motor Company in 1938.

World War II halted further expansion. After the war, Matsushita resumed research and development into television, which had begun in 1935, and began to take an interest in the new transistor technology following Konosuke Matsushita's visit to the United States in 1951. It began to produce larger domestic appliances, beginning with washing machines. In 1952, Matsushita acquired the Nakagawa Electric Company, which became the Matsushita Refrigeration Company after the company's first refrigerators went on the market in 1953. The signing of a technical cooperation agreement with the Dutch company Philips was another sign of Matsushita's aspirations to be at the forefront of electrical and electronic developments, supported by the creation of its Central Research Laboratories in 1953. The launch of its first black-and-white television in 1952 marked the beginning of its electronic product line, reflected organizationally by the formation of the Matsushita Electronics Corporation.

Matsushita took its first steps to carve out an international market for its products when it opened an office in New York in 1953 and acquired a 50 percent stake in the Japanese Victor Company (JVC) in 1954. JVC became the American branch of Matsushita. It had been founded in 1927, when it began to manufacture phonographs for the Japanese market. JVC introduced Japan's first television in 1939 and thereafter consolidated its position in the audio and video product markets. In the late 1950s and 1960s, Matsushita continued to grow by adding products to its existing lines and branching out into new areas of electrical and electronics manufacturing. New divisions included the Osaka Precision Machinery Company (later Matsushita Seiko), while new products included tape recorders, color televisions, microwave ovens, and videotape recorders.

Matsushita established itself as a major player in the world consumer electronics market in the 1970s when it was able to impose the VHS (Video Home System) videocassette format as the world consumer standard. The VHS format was developed by Matsushita and JVC in competition with the Philips N-1500 format and Sony's Betamax format. While VHS was the last of the three formats to hit the market, in 1976, it achieved almost total market dominance within ten years. Matsushita's winning strategy was to gain a price advantage by tailoring performance to the desired price; Sony had attempted to optimize performance and price the product accordingly.

Matsushita's pattern of growth continued to follow its established path. By 1990, it owned almost ninety companies in Japan alone and had worldwide sales of $45 billion. At this time, Matsushita was investing heavily in a new field, the movie industry. In 1989, it spent $100 million on a new film studio, Largo, in Hollywood and in 1990 it bought the American entertainment corporation MCA, for $6.5 billion. MCA's diversified entertainment interests included MCA Records, Universal movie and television production, book and music publishing, and theme parks. While some of these interests coincided with or complemented Matsushita's core businesses, others did not mesh so well. It was therefore not entirely surprising that in 1995 Matsushita sold 80 percent of MCA (renamed Universal Studios) to the Seagram Company Ltd. for $5.7 billion. In the late 1990s, Matsushita was quick to launch new digital electronic appliances, including its first consumer digital camcorder in 1995 and its first DVD (digital versatile disc) player in 1996. It ended the century ranked twenty-sixth in the global Fortune 500 list.

See also Camcorders; Television; Videotape Recorders

References:
www.matsushita.com
www.mei.co.jp/corp/hist/hist-e.html
www.panasonic.com

Maytag Corporation

Founded in Newton, Iowa, by F. L. Maytag in 1893, the Maytag Company began as a manufacturer of farm tools. The company's sales pattern reflected the seasonal nature of farming, so Maytag looked for a new product line that would be less susceptible to fluctuations in demand. In 1907, Maytag produced its first washing machine, a manual model. Maytag soon began to develop a line of washing machines, launching its first electric washing machine in 1911 and, in 1915, a washer powered by gasoline for homes without electricity. By the mid-1920s, Maytag had a 20 percent share of the U.S. washing machine market.

After World War II, Maytag diversified into the production of other white goods (stoves, washing machines, and refrigerators). Although Maytag continued to have a healthy share of the washing-machine market, its strategy seems to have been based on gaining a reputation for reliability rather than innovation. Its first automatic washing machines hit the market in 1948, ten years after these machines first appeared. Five years later, it added automatic clothes dryers to its product line. Maytag was also slow to move into the dishwasher market, introducing its first countertop dishwashers in 1966. Maytag did achieve an industry first in 1985, with the introduction of a stacking washer and dryer.

In the 1980s, Maytag embarked on a program of diversifying its product base through acquisition. The first step was the purchase of Jenn-Air, manufacturer of a line of large kitchen appliances, in 1982. The evolution from the Maytag Company to the Maytag Corporation took place in 1986, following the acquisition of Magic Chef, with its product line of smaller kitchen appliances. In 1987, Maytag acquired the Hoover Company, thus moving into the floor-care appliance market. Maytag also strengthened its commercial appliance business, which had begun with the manufacture of launderette-scale washers in the late 1950s. In 1997, it acquired the Blodgett Corporation, a manufacturer of ovens for the catering industry. Maytag ended the twentieth century as North America's third largest manufacturer of home appliances.

See also Hoover Company; Washing Machines

References:
www.maytagcorp.com

McGraw Electric

McGraw Electric was founded in Sioux City, Iowa, by the 17-year-old Max McGraw in

1900. It grew as the Interstate Supply Company and the Interstate Manufacturing Company. During the 1920s it acquired the Clark Water Heater Company and a fuse manufacturer, the Bussmann Company. In 1929 it bought the Waters Genter Company, which had been producing the Toastmaster automatic toaster since 1926. In the 1930s the company consolidated its position as a manufacturer of domestic appliances and added improvements to the Toastmaster line. It converted to ammunition production during World War II.

In 1947 it produced the popular and long-lived 1B 14 toaster, which remained in production until 1961. The late 1940s and 1950s saw further expansion and acquisitions of other appliance manufacturers, including the appliance division of General Mills in 1954. The company changed its name to the McGraw-Edison Company with the acquisition of Thomas A. Edison Industries in 1957.

The company produced another long-lived product, the best-selling electric waffle iron, the W252, in 1966. It was not discontinued until 1997. In 1978 the company introduced the B700, which became the best-selling toaster in the United States, achieving sales of over 16 million in ten years. It also joined forces with Prestige, a subsidiary of the English Ekco Company, in 1965.

In 1980 a group of executives formed Toastmaster Incorporated following a management buy-out from McGraw-Edison. Toastmaster Incorporated was then bought by Magic Chef Inc. and then by Maytag. Another management buy-out from Maytag in 1987 led to Toastmaster's becoming a publicly held corporation in 1992. It introduced the Bagel Perfect in 1996 and a toaster designed by the German Porsche Design group in 1998.

Toastmaster was acquired by Salton Inc. in 1999. The brand name is strong and continues to be used for lines of breadmakers, toasters, waffle irons, coffeemakers, blenders, and irons.

See also Ekco; Maytag Corporation; Toasters; Waffle Irons

References: Norcross, Eric. "Toastmaster—A History with Dates," *www.toaster.org*. 1999.

Microsoft Corporation

Bill Gates (William Henry Gates III) and Paul Allen formed the partnership that later became Microsoft in 1975 in Albuquerque, New Mexico. As high school students in Seattle in 1967, they were brought together by their mutual interest in computer programming. In 1975, their hobby became a commercial venture when they produced a version of the BASIC computer programming language for the Micro Instrumentation and Telemetry Systems' Altair 8800, the first microcomputer. Gates and Allen registered the Microsoft trade name in November 1976 to market the new language, MS-BASIC. In March 1977, the partnership was formalized and in 1979 Microsoft moved its offices to Bellevue, a suburb of Seattle. The partnership evolved into a privately held corporation in June 1981.

The next big step was the launch of the Microsoft Disk Operating System (MS-DOS) on the new IBM personal computer in August 1981. The success of the IBM PC led to its adoption as the industry standard for personal computer architecture. As other manufacturers flocked to produce PC clones, Microsoft was able to convince them to adopt MS-DOS as the standard operating system (OS), guaranteeing its rapid expansion. It also began a relationship with Apple, the U.S. company that had launched the first commercial personal computer in 1977. Apple developed its own operating systems but made arrangements with other companies to procure applications software. The agreement with Apple brought Microsoft into the business-applications side of the software market. Its first application, the Multiplan spreadsheet, was developed in 1982. The first version of Word, later to become the world's premier word-processing package, followed in 1983, together with the IBM-compatible Microsoft Mouse.

In the course of creating applications for the embryonic Apple Mac, Microsoft gained familiarity with the Mac's graphical user interface (GUI) and began working on its own GUI operating system, Windows, in 1983. However, this led to a dispute with Apple over alleged infringement of Apple's design rights. Microsoft secured a favorable settlement that, as subsequent legal action confirmed, restricted the design of only Windows 1.0, not later releases. Windows 1.0 became available in late 1985. Early in 1986, Microsoft moved to its current site in another Seattle suburb, Redmond. Microsoft introduced its first application for Windows, the Excel spreadsheet (first developed for the Mac), in 1987. It also entered into an agreement with IBM to jointly develop OS/2, an operating system intended for networked business computers. By the end of 1988, Microsoft overtook Lotus as the world's leading software producer.

When Windows version 3.0 appeared in 1990, Microsoft had an operating system of comparable user-friendliness to the Mac OS. Annual sales revenue reached $1 billion. Convinced that its main market lay with the evolution of Windows, Microsoft withdrew from OS/2 development and began developing a Windows-based networking OS, Windows NT, which was launched in 1993. While Microsoft continued to develop new applications, such as the Access database introduced in 1992, it also began to branch out into electronic publication. It acquired a stake in the British publishing company Dorling Kindersley in 1991 and brought out Encarta, the first multimedia encyclopedia, on CD-ROM (compact disc with read-only memory) in 1993. Microsoft then moved into the U.S. broadcasting arena by forming a joint venture with NBC in late 1995 to create complementary MSNBC news and information channels on cable television and the Internet. In 1997, this strategy was furthered by the acquisition of WebTV Networks, a television-based Internet-access service, and investment in the cable television operator Comcast.

Microsoft's ownership of the key personal computer operating systems has made it a formidable competitor in the applications software market. Windows has moved on through version 95 to 98 to 2000. Microsoft can develop applications such as Word and Access in tandem with the evolution of Windows. This competitive advantage has created intense concern about monopolistic practices. Producing Mac versions of its applications has been a way for Microsoft to stifle criticism. In 1997, Microsoft and Apple announced a product development agreement, which also involved Microsoft investing $150 million in Apple. However, this did not deflect concern about Microsoft's actions to promote its Internet Explorer web browser. The rival product, Netscape Navigator, was launched earlier and established itself as the leading web browser. By bundling Internet Explorer with Windows 95, Microsoft was deemed by the U.S. Department of Justice to have abused fair trade practices and was required in 1997 to cease that practice. In 1998, the Department of Justice and twenty states charged Microsoft with breaching antitrust laws. The resulting judgment in 1999 found against Microsoft.

See also Apple Computer, Inc.; Computers; IBM

References:
www.microsoft.com/MSCorp/Museum/

Microwave Ovens

The microwave oven can trace its origins back to the development of radar technology during World War II. The electromagnetic energy (microwaves) produced by a high-powered magnetron was found to heat liquid. Percy Le Baron Spencer, a physicist working for the U.S. Raytheon Company, developed the idea further in the 1940s. He allegedly found that his sandwiches, which had been sitting by the radar equipment, had become warm. To test this further he placed some maize in a paper bag and placed it next to the magnetron. He soon had some popcorn!

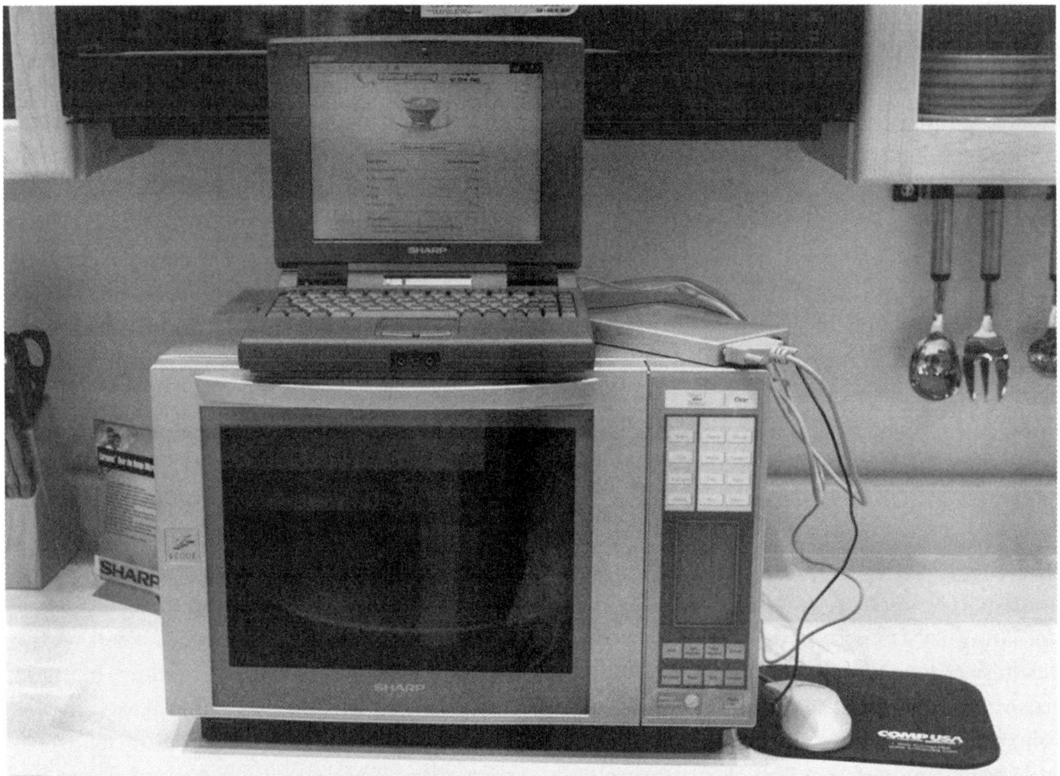

An Internet-capable convection microwave oven made by Sharp Electronics Corporation is displayed at the International Housewares Show in Chicago, 2000. (Reuters Newmedia Inc. / Corbis)

Raytheon patented a refined version in 1945. Its first cooker, the Radarange, was marketed to military and hospital kitchens. They were bulky and expensive. The first British microwave was the Artic Cooker by Lewis Point Holdings, introduced in 1947. Raytheon began to produce domestic models in 1967.

The revolutionary aspect of the microwave is that it created a new way of cooking food. All traditional methods required that heat be created and transferred to the food. The microwaves heated the food only. Created by the magnetron, they were moved around by a fan, penetrating the food and agitating the molecules within it, especially water, fats, and sugars. Not only were they clean, but also they were fast. Dishes such as baked potatoes and steamed puddings could be done in a fraction of the time.

Like other new products, there were some popular misconceptions regarding safety. There was some confusion as to whether the ovens were radioactive or used X-rays. The public also needed to know that although the microwaves could penetrate glass, paper, ceramics, and polyethylene plastics, they were blocked by metals.

Early American domestic machines were still bulky. Their popularity within Europe and the United States was stimulated by the rising Japanese companies of the 1970s. Hitachi, Panasonic, Sanyo, and Sharp produced smaller machines with an increasing range of features. Sunbeam entered the microwave market in 1978. The rise of the microwave coincided with the accelerating speed of life within the capitalist West. The microwave was the perfect cooker for the convenience food manufacturers thanks to its ability to defrost and cook frozen food in minutes. The power of the ovens has increased; modern compact models deliver 600 to 700 watts, while larger ones deliver between 800 and 1,100 watts. The higher the wattage, the faster the food cooks.

Despite their cooking power traditional food aesthetics did not work in their favor. Microwave-cooked meats just did not look right; they were cooked but appeared pale and uninviting. The companies responded with combination microwaves and grills, which can be used simultaneously. Also there was the fear that the food was not cooked all the way through. Developments to ensure this does not happen have led to the introduction of rotating trays and more recently "showerwave" systems to ensure a more even distribution of microwaves. The rise of digital technology has also given the ovens even more sophisticated controls and features.

By the mid-1980s, the microwave had become an accepted part of the kitchen, used to partner, but not replace the traditional stove. The modern, split-level, divided-function kitchen now features separate stove-tops, ovens, and microwaves, all built in to an integrated design. Stand-alone microwaves remain popular and, like most appliances now come in a range of colors.

In the United States the most popular brand is Sharp, followed by General Electric, Kenmore, and Panasonic.

In 1998, the Knowledge Lab, an international technology research consortium attached to NCR, the automated teller machine (ATM) manufacturer, began investigating combining a microwave oven with an on-line Internet banking service. This initially startling combination is the result of "relationship technology," based on market research that shows that many people sit in their kitchens, watch television, and do their households accounts while cooking food. The microwave could become a multifunctional appliance for cooking, Internet banking, and shopping as well as a television.

See also Kitchens; Relationship Technology; Tupperware

References:

Byers, Anthony. *Centenary of Service: A History of Electricity in the Home.* London: Electricity Council, 1981.

MiniDisc Recorders

The MiniDisc is the first recordable audio disc format, introduced by the Japanese company Sony in 1992. A 5 cm (2 in) square cartridge holds a magneto-optical disc that provides up to 74 minutes of playing time. Sony had interpreted the decline in sales of blank audiocassettes since 1989 as a sign that the time might be ripe for a new, more-compact recording format. However, while MiniDisc recording has several advantages over analog or digital-tape recording—instant track access, ability to edit recordings, no loss of quality after rerecording—it has yet to establish itself as a mainstream format. This is probably a result of the introduction of a number of new audio formats in rapid succession, leaving consumers uncertain as to the best choice for them.

See also Compact Disc Players; Tape Recorders

Mobile Phones

The mobile phone is a two-way radio that can send and receive messages within a specific frequency band. The difference between a mobile phone and any other two-way radio is that the mobile phone is served by a network of special transmitters and is identified by a number, allocated by the service provider, that is also recognized by land-line telephones. In its modern cellular form, the mobile phone is perhaps the classic lifestyle product of the late twentieth century. Yet, even as recently as 1990, the current status of the mobile phone as a commonplace personal accessory was barely conceivable.

For the first twenty years of its life, before the concept of radio broadcasting evolved, the radio was used primarily as a means of one-to-one communication by code or sound. The radio set was known as the wireless set to distinguish it from the wired telegraph and telephone. During World War I, when radio became an important method of military communication, portable radio transmitters and receivers were developed for use in the field. However, portability was

A man on the beach using a cellular phone, Delaware, in the late 1990s (Richard T. Nowitz/Corbis)

limited as the power was supplied by wet batteries (or accumulators) that were bulky and heavy.

Mobile two-way radio, installed in motorcars and powered by the car engine, came into being in the 1920s. The police department in Detroit, Michigan, adopted the system in 1921, and it spread to other police forces in the United States. In the early 1940s, more frequency bands (channels) were assigned for police radio, and other types of business users began to operate private mobile radio systems. The next step was to introduce a mobile system for public subscribers, akin to the wired telephone system. In 1945, the U.S. Federal Communications Commission (FCC) licensed the first public mobile telephone system in St. Louis, Missouri. Four years later, the FCC assigned five radio channels to provide interconnection, through an intermediary operator, between mobile phones and the wired telephone network. However, it was in Sweden that mobile direct dialing made its first appearance in 1956, when Ericsson set up an automatic mobile phone system in Stockholm. It was another eight years before American Telephone and Telegraph (AT&T) introduced a similar system in the United States.

These early mobile phone services suffered from serious practical and technical disadvantages. Ericsson's first automatic mobile phone cost as much as a car, and even by the mid-1970s, its mobile phones weighed a hefty 9 kg (20 lb). Each local network was served by just one high-power transmitter, so each channel could only carry one phone call. In New York, 700 subscribers shared 12 channels, bringing inevitable frustration. AT&T's research and development wing, Bell Laboratories, had conceptualized a cellular solution to the capacity problem in 1947, but at that time the technology was not capable of delivering the solution. The basic premise of cellular telephony is that by dividing the service area into small cells, each served by a low-power transmitter (radio base station), the maximum number of concurrent calls becomes the number of channels multiplied by the number of cells. The range of the transmitter is tailored to the size of the cell, so there is only marginal overlap and no cross-cell interference. As mobile users move between cells, their calls are passed from one transmitter to the next. To achieve this seamlessly required sophisticated electronic switching and frequency matching, which only became commercially feasible when microprocessors were developed in the 1970s.

The cellular concept was revived in the early 1970s when the FCC invited proposals for a new mobile phone service to operate in an ultrahigh frequency band around 800 MHz. AT&T was the only company to submit a proposal. Progress toward implementation was complicated by the instigation of antitrust proceedings against AT&T in 1974. In order to proceed, AT&T agreed to drop its own manufacturing subsidiary, Western Electric, from the proposal and to work with other telephone equipment manufacturers. In 1977, AT&T was granted a trial license and began its first public trial, in Chicago, in 1978. The U.S. communications and electronics company Motorola was one of the equipment manufacturers working on cellular phone equipment. Motorola was a leading manufacturer of car radios and had developed the first hand-held two-way radio during World War II. It entered the microprocessor business in 1974, and in 1975 was granted its first patent for cellular phone equipment. In 1981, in partnership with American Radio Telephone, Motorola conducted cellular phone trials in the Washington–Baltimore area. The system developed by AT&T and Motorola became known as the Advanced Mobile Phone Service (AMPS).

While the FCC was deliberating over full commercial cellular phone standards and services in the United States, other countries were already going ahead with commercial implementation. The Japanese state telephone service, Nippon Telephone and Telegraph (NTT), introduced a cellular phone service in Tokyo in 1979. In 1981, the Scandinavian countries (Sweden, Norway, Denmark, and Finland) agreed on a common cross-border cellular system, the Nordic Mobile Telephone system (NMT) developed by Ericsson. By 1984, NMT had more than 100,000 subscribers, making it the world's largest cellular phone service. The NMT standard was also adopted by a number of other European countries.

In the United States, the FCC finally decided on a structure and strategy for AMPS in 1982. In each local area, there would be two licensed operators—the existing local telephone company and a mobile phone carrier. In order to manage the application process, the FCC decided to issue licenses in batches, beginning with the thirty largest cities. The first commercial cellular phone system began operating in Chicago in 1983. In Britain, the government decided in 1982 to grant two nationwide licenses for cellular phone services. One license was awarded to the Cellnet consortium led by British Telecom (BT), the state telephone operator, and the other went to the Vodafone consortium led by Racal Electronics. The government opted for the U.S. AMPS standard, renamed TACS (Total Access Communications System), and cellular phone services began in 1985.

Cellular phones were completely in tune with the competitive business environment of the 1980s. The mobile phone became the definitive lifestyle accessory for the aspiring yuppie (young upwardly mobile professional). At this stage, mobile phones were still too expensive to be seen as a tool for social communication. Moreover, while mobile phone handsets had shrunk to hand-held size by the mid-1980s, they were far from pocket-size. In 1986, Ericsson launched its first truly lightweight mobile phone, the Hotline, which weighed 700 g (1.5 lb). Another problem with the original analog mobile phones was that phone calls could be tapped by the use of radio scanners. This problem received high-profile exposure when transcripts of mobile phone conversations involving members of the British Royal family were published in tabloid newspapers.

The rapid growth in demand for mobile phone services meant that the cellular network was becoming overcrowded by the late 1980s. The solution was to switch from analog sound transmission to digital transmission. The analog transmission system used by the first phase of cellular phone networks is based on frequency modulation (FM) of the carrier radio waves. The maximum number of channels within the cellular frequency band is determined by the width of channel required to carry an FM signal. Digital transmission can pack more channels into the same width frequency band, because the sound signal is compressed by sampling, allowing narrower channels. It uses a form of modulation called pulse-amplitude modulation, which involves the amplitude of the sound wave input being sampled at frequent intervals to produce a binary code output. The binary code serves as instructions to switch the transmitted radio waves on and off, producing radio signal pulses of varying length and depth.

Three standards for digital sampling and transmission for cellular phone services emerged between 1987 and 1993. These standards are mutually incompatible. The first standard, the Groupe Spécial Mobile (GSM), was developed in Europe between 1982 and 1987 with the backing of the Conference of European Posts and Telecommunications. The first digital GSM systems were introduced in Germany, Scandinavia, and Britain in 1991. The GSM acronym has since been reinterpreted as the Global System for Mobile Communications. The first digital standard to be authorized in the United States was CDMA (Code Division Multiple Access) in 1993. TDMA (Time Division Multiple Access) followed in 1994, and GSM was endorsed in 1995. GSM has two clear advantages over the other two digital standards: it is the only system that operates internationally, and it is also the only system that provides other digital services, including e-mail (text messaging), fax, Internet access, and computer-compatibility.

The latest advance in mobile telephony services is the improved international service available through the Globalstar network of low earth-orbiting satellites. Globalstar is a multinational venture, founded in 1991. The Globalstar service uses the CDMA transmission standard, but is also compatible with GSM and AMPS mobile phones, and landline telephones. From 2000, Globalstar will provide not only voice transmission but also full data communication services.

Although the increasing sophistication of mobile phone services has meant that mobile phones have to accommodate more functions, the phone handsets have become smaller, lighter, and more streamlined. The miniaturization of the mobile phone is the outcome of advances in microelectronics and in rechargeable battery technology. Lightweight batteries, typically composed of nickel-cadmium or lithium compounds, can provide at least two hours of mobile phone talk time and can be recharged in as little as one hour. Mobile phones may now weigh little more than 130 g (4 oz). The Finnish company Nokia has been one of the leaders in producing compact and stylish mobile phones.

During the 1990s, increasing competition in the deregulated telecommunications industry made mobile telephones more affordable, as companies began to give consumers a wider choice of payment formats. However, in recent years, concerns have grown about the side effects of mobile phones in terms of health. The main issue is radiation leakage from mobile phone handsets. While there are regulatory standards on emissions, tests have demonstrated that some of the mobile phones on the market do not meet these standards. Standards also vary from country to country, with, for example, Britain applying a more stringent standard than the United States. The issue is not just the level of radiation emitted, but where it is emitted. It has been shown that it is not atypical for the earpiece to emit four times as much radiation as the antenna. Researchers have argued that headaches reported by regular mobile phone users are just one symptom of the effects of excessive radiation absorption on the human body. One immediate solution being adopted by an increasing number of mobile phone users is to attach wired microphone and earpiece sets, thus keeping the phone handset away from the head.

See also AT&T; Motorola, Inc.; Radio; Telephones; Vodafone

References:

Forester, Tom. *High-Tech Society: The Story of the Information Technology Revolution.* Cambridge, MA: MIT Press, 1987.

Goodman, Robert. *How Electronic Things Work . . . and What to Do When They Don't.* New York: McGraw-Hill, 1999.

http://inventors.about.com

www.wave-guide.org

Modems

A modem is a device that enables computers to send and receive information by phone line. It is an abbreviation for modulator-demodulator, which describes the processes of converting (or modulating) the digital signals produced by the computer into the analog signals required for phone transmission and vice versa (demodulating the analog signals) for reception. The main factor in the performance of a modem is the speed at which it can transmit and receive data, usually expressed in bits per second. Invented in 1958, the modem did not achieve mass-market status until the 1990s when the Internet became a major medium of communication. Demand for improved performance saw typical modem speed increase from 33 kilobits per second in 1997 to 56 kilobits per second in 1998. A fax-modem enables communication with fax machines.

A modem allows users to send and receive electronic mail (e-mail) and access the World Wide Web. The Internet began as a network called ARPANET (Advanced Research Projects Agency Network), set up by the U.S. Department of Defense in 1969 for use by researchers. The first two access points (nodes) on ARPANET were UCLA and Stanford University. By 1972, when electronic mail was introduced, there were 23 nodes. The next stage came in 1986 when the U.S. government funded the National Science Foundation to set up a network (NSFNet) for all academic users. The expansion of NSFNet beyond the academic community created the Internet. The World Wide Web was conceived in 1989 by Tim Berners-Lee, a British software engineer working at the CERN (Conseil Europèen pour la Recherche Nucléaire, or European Organization for Nuclear Research), the particle physics laboratory in Geneva, Switzerland. By the end of 1990, his prototype World Wide Web was running at CERN, and it was made available on the Internet in 1991. Berners-Lee continued to develop the basic components of Web technology—HTTP (hypertext transfer protocol), URLs (universal resource locators) and HTML (hypertext markup language)—with the help of feedback from Web users. In 1994, a structure for managing the development of the Web was created when CERN and the Massachusetts Institute of Technology (MIT) agreed to set up W3. the World Wide Web Consortium. Tim Berners-Lee is now based at MIT and is director of W3.

The Internet consists of a hierarchy of access points, the top level of which is a backbone of linked supercomputers in the United States. Public access is managed by commercial Internet service providers (ISPs), such as America Online and CompuServe. In a short time, the Web has become an incredibly diverse resource—a combination of virtual library, virtual retail outlet, virtual meeting place, and more. The number of server computers hosting Internet sites grew from 1,000 in 1984 to 6 million in 1996. Predictably, given its origins and the practice of not charging for local phone calls, subscription to the Internet initially grew fastest in the United States, with an estimated 35 percent of homes connected by the end of 1998. Internet access in Britain mushroomed during 1999 as a result of the sudden proliferation of service providers offering Internet access with no subscription charge. It was estimated that half of British homes had access to the Internet by the end of 1999.

See also Computers; Fax Machines; Telephones

References:

Gardner, Robert, and Dennis Shortelle. *From Talking Drums to the Internet: An Encyclopedia of Communications Technology.* Santa Barbara, CA: ABC-CLIO, 1997.

www.w3.org

The Modern Movement and Appliance Design

The "Modern Movement" never really existed as an association of artists and designers, yet it is a useful term to describe those who saw design and social purpose as a holistic process. Its early exponents came, some-

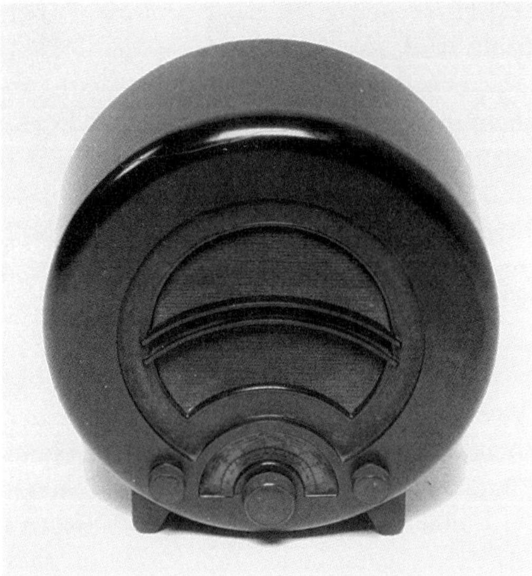

Ekco AD36 radio, 1935, designed by Wells Coates (The Museum of Science and Industry in Manchester)

what surprisingly, from the Arts and Crafts Movement of the late nineteenth century (which rejected machine production), but were also influenced by Russian constructivism, expressionist painting, the Dutch de Stijl group, the Deutsche Werkbund, and Peter Behrens.

The Bauhaus, the art school founded by Walter Gropius at Weimar, Germany, in 1919 has been widely perceived as one of the key pioneers of the Modern Movement in art, architecture, and design. Although its methods were founded on the Arts and Crafts Movement, it did make a significant contribution to the development of modernism and the International Style through the work and the theory of its teachers such as Laszlo Moholy-Nagy, Marianne Brandt, Marcel Breuer, Mies van der Rohe, and Josef Albers. The school moved to Dessau in 1925 and into its innovative new building in 1926 but was eventually closed down by the Nazis in 1933. Many of the Bauhaus teachers fled to America and were to have a significant effect on postwar architecture.

One of its founding principles was that artists should be trained to work with industry, allowing Germany to have a skilled labor force capable of producing sophisticated, high-quality goods for export.

The Bauhaus was always short of money and in 1922 Gropius began to change direction, investigating the possibility of founding a limited company to sell Bauhaus products. The school was moving toward a new direction, contemporary design for industrial production, a vision that was by no means shared by a number of the teachers. Gropius appointed the Hungarian Moholy-Nagy to lead this effort. Students visited factories and businesses and familiarized themselves with simple tools.

The results were an influential range of handmade products that looked as though they had been mass-produced. Some, such as a table lamp of 1923–1924, Brandt's tea sets, and Albers's bowls, were indeed capable of being machine made. During the life of the Bauhaus, however, few were mass-produced and they remained expensive luxury items. The main exception was a small 1928 bedside lamp with a flexible stem, designed by Brandt. The Leipzig firm of Körting & Mathieson produced 50,000 of them between 1928 and 1932. Despite this, the Bauhaus workshops tended to investigate traditional objects and did little work with the emerging electrical appliance industries; indeed, Gropius proscribed the use of plastics as he felt they were imitative and derivative materials.

It was only after the Bauhaus had closed that the influence of the Modern Movement was felt on appliance design. The British designer and architect Wells Coates and the Russian-born Serge Chermayeff designed truly modern Bakelite radio cases for the Ekco Company in the 1930s. Gordon Russell designed a wide range of modern wooden radio cases for the Murphy Radio Company during the same period, eventually setting up a new factory in London in 1935. A Council for Art and Industry was established by the Board of Trade in 1934 and became the Council of Industrial Design in 1944. Its role was to encourage manufacturers to make more of professional designers.

Meanwhile, the Swiss born architect Le Corbusier was also designing and advocating the future in terms of the machine aesthetic, a world where modern technology and beauty could create a socially beneficial environment.

In 1930s United States, the marriage between art and industry developed through the work of modern designers such as Henry Dreyfuss, Norman Bel Geddes, Raymond Loewy, Walter Dorwin Teague, and Harold van Doren. Although their work on domestic appliances did not strictly apply the "form follows function" maxim of the Bauhaus, they were influenced by it, and their designs and restylings produced products with cleaner, simpler lines in tune with the machine age. The purer work of Eliot Noyes at International Business Machines was closer to the Bauhaus spirit.

The legacy of the Bauhaus lived on within Germany with the establishment of the Hochschule fur Gestaltung Ulm, the high school for design, in Ulm, Germany, in 1951. Hans Gugelot became head of the product-design department in 1954 and established links with Braun. Braun employed the best exponent of what is now called neofunctionalist design. Dieter Rams was born in 1932; he trained as a carpenter, architect, and interior designer and joined the Braun design team in 1955. He was director of the design department from 1960 to 1980. His simple, understated designs, which enveloped the likes of radios, heaters, and food mixers, remain almost timeless and were an inspiration to the product-design courses established in Europe throughout the 1960s and 1970s.

The success of Terence Conran's Habitat stores in the 1960s and 1970s also promoted the concept of well-designed, affordable products, another unrealized Bauhaus aim. It even sold Marcel Breuer–designed tubular steel furniture.

The 1980s witnessed an increasingly sophisticated market, which began to care about how things looked as much as how well they worked and directed the attention of all the large appliance companies. The Italian company Alessi has continued the Bauhaus tradition of actively working with architects and artists on product design. Ironically, some of the products that began life in the Bauhaus workshops are now being machine produced by Alessi as expensive limited editions. Equally amusing is the de Stijl vacuum cleaner produced by Dyson in the late 1990s, an homage to one of the movements that influenced the Bauhaus pioneers some eighty years earlier.

See also AEG; Alessi; Braun; Consumers; Dyson Appliances; Ekco; Habitat; IBM; Obsolescence; Radio; Refrigerators; Vacuum Cleaners; Washing Machines

References:

Bayley, Stephen, ed. *The Conran Dictionary of Design.* London: Octopus Conran, 1985.

Droste, Magdalena. *Bauhaus, 1919–1933.* Cologne, Germany: Benedikt Taschen, 1990.

Meikle, Jeffrey L. *Twentieth Century Limited: Industrial Design in America, 1925–1939.* Philadelphia: Temple University Press, 1979.

Sparke, Penny. *A Century of Design: Design Pioneers of the Twentieth Century.* London: Mitchell Beazley, 1998.

Morphy Richards

The British company Morphy Richards was founded in 1936 as a manufacturer of electric irons. It soon became a successful manufacturer of a range of small electrical appliances, such as kettles, toasters, and hair dryers. Its success has been based on the functional design, affordability, and reliability of its products. While not notable as a world innovator, the company has been responsible for introducing new technical advances to the British market. Examples of this include steam irons and pop-up toasters. The Morphy Richards streamlined toaster of 1956 was one of the most popular British models, coming in either a chrome or a yellow enameled body. Its PA75 model, also available in chrome or yellow enamel, was equally successful.

Morphy Richards expanded into consumer electronics in 1982, launching into the portable radio and audio market. It became part of the Glen Dimplex Group in 1985. In the 1990s, sales benefited from the introduction of matching lines of small appliances in deep colors. Today, Morphy Richards is the leading British manufacturer of small electrical appliances and has market shares of about 33 percent for kettles and 25 percent for toasters. The audio division has grown to include telecommunications products.

See also Blenders/Juicers; Frying Pans; Glen Dimplex; Hair Dryers; Irons; Kettles; Sandwich Toasters; Toasters; Trouser Presses.

References:

www.morphyrichards.co.uk

Motorola, Inc.

The communications and semiconductor company Motorola is one of the United States' top fifty companies. It began life in 1928 when Paul V. Galvin and his brother, Joseph E. Galvin, bought the bankrupt Stewart Storage Battery Company in Chicago and set up the Galvin Manufacturing Corporation. Two years later, the company produced an affordable car radio, which it marketed under the new brand name Motorola. The company quickly became established as the market leader and initially focused on developing products that met the needs of its core client group, the police. In 1937, it began to produce radios for home use and moved to a new plant and offices. By 1940, annual sales were close to $10 million and the company employed nearly 1,000 employees. In 1941, the first complete Motorola two-way police radio system was installed in Bowling Green, Kentucky. During World War II, the Galvin Manufacturing Corporation was engaged almost exclusively in the production of radios for military use, the only exception being the conversion of existing stocks of car radios for home use. The Motorola Handie-Talkie, produced for the U.S. Army Signal Corps, was the first hand-held two-way radio.

In 1947, the company changed its name to Motorola to capitalize on the high profile of the brand name. The first Motorola television, the Golden View VT71 model, was launched in 1948 and was the first television to retail for under $200. Sales of 100,000 televisions in the first year of production placed Motorola fourth in the U.S. television market. Motorola introduced a color television in 1954, but the combination of the high price and technical problems with color transmissions meant that it was a commercial failure and was withdrawn in 1956. Meanwhile, the growing demand for car radios was reflected in the negotiation of contracts with Ford, Chrysler, and General Motors for the pre-installation of car radios and the opening of a new car radio plant in Quincy, Illinois, in 1948. Motorola took advantage of the increased frequency allocation for car-radio communications by introducing the Dispatcher two-way FM system in 1949.

Recognizing the potential of the new transistor technology, Motorola set up a research and development unit in Phoenix, Arizona. In 1952, the company's researchers produced a 3-amp power transistor. Its first commercial transistorized product was a car radio containing germanium transistors, introduced in 1956. Motorola not only produced transistors for its own products but also became a commercial supplier of semiconductors to other companies through its Semiconductor Products Division. In 1959, Motorola moved to new corporate headquarters in Franklin Park, Illinois, and introduced its first all-transistor radio, the pocket-size X11 model.

During the 1960s, Motorola continued to grow through consolidation of its traditional product lines, diversification into related new product areas, and overseas expansion. From 1962, it supplied transponders for the U.S. space exploration program. The company reentered the color television market in 1964 through a joint venture with National Video, which resulted in the first rectangular color picture tube. In 1967, Motorola produced the first U.S. solid-state color television, the Quasar model. In partnership with Ford and the Radio Corporation of America (RCA), it developed the first 8-track tape recorders for car use. In 1969, Motorola moved into integrated circuit technology, which brought further reductions in the size and weight of its radios and opened up new markets for its semiconductors. From 1972, it began producing components for quartz watches for companies such as Timex and Bulova.

Two emergent technologies, microprocessors and cellular phone systems, shaped Motorola's business development in the 1970s and 1980s. In 1974, the same year that the first Motorola microprocessor, the 6800, appeared, Motorola sold its television manufacturing business and the Quincy plant to the Japanese electronics company Matsushita. Two years later, the company headquarters were relocated to Schaumburg, Illinois. Motorola acquired a foothold in data communications through acquiring the Codex Corporation in 1977 and Universal Data Systems in 1978. These interests were consolidated by the acquisition of Four-Phase Systems in 1984. Its first 16-bit microprocessor, the 68000, came on the market in 1979. Not surprisingly, given Motorola's long relationship with the car industry, car-engine control systems provided the first major sales success for its microprocessors. The 32-bit MC68020 microprocessor containing 200,000 transistors, launched in 1984, reached a wider market, being taken up by more than 125 companies within 2 years. The introduction of the 88000 family of RISC microprocessors in 1988 and the third-generation 32-bit microprocessor, the 68040, in 1990 consolidated Motorola's position in the microprocessor market.

Motorola was granted its first patent for cellular phone equipment in 1975. The Federal Communications Commission granted the company a development license for this technology in 1977. In partnership with American Radio Telephone, Motorola carried out cellular phone trials in the Washington-Baltimore area in 1981. Commercial implementation began in 1983. By 1985, Motorola had overseas cellular phone contracts in Hong Kong, China, Japan, Britain, and Scandinavia. Its Micro-TAC mobile phone of 1989 was the smallest and lightest available. Confirmation of the reorientation of Motorola's business away from its traditional radio business came in 1987, when Motorola produced its last car radio and sold off some of its car product lines.

By 1990, Motorola's annual sales exceeded $10 billion, and it had more than 100,000 employees. The company benefited from the rapid growth of the markets for both mobile phone systems and microprocessors in the 1990s. By the end of the century, Motorola was ranked thirty-seventh in the Fortune 500 list of U.S. companies and hundredth in the global Fortune 500 list.

See also Computers; Mobile Phones; Radio; Television

References:

www.motorola.com/General/Timeline/

Motors, Electric

One of the principal trends underlying the development of many domestic appliances throughout the twentieth century has been the application of instant energy, which has improved their performance. For many appliances, this energy source has been electricity. The first essential factor in this development was the slow but steady spread of an efficient and cost-effective electricity supply in Europe, Canada, and the United States beginning in the 1880s. The second key factor was the development of a small and efficient electric motor.

The first electric motor was built by the Englishman Michael Faraday in 1831 and consisted of a copper disk rotating between the poles of a powerful magnet. Faraday did not exploit his discovery industrially. Further work was carried out by Faraday's fellow countryman James Clerk Maxwell, but the first real commercial model was developed by the Croatian-born, U.S.-based Nicola Tesla in 1889. Working with the Westinghouse Company his 1/6 horsepower motor was used to drive a three-bladed domestic fan. Although it had no speed or direction controls it entered a market eager to alleviate hot and humid summers, but for many Americans the lack of electrical supply remained a problem.

At the beginning of the twentieth century most small electric motors were used in commercial and laboratory environments. They were sold to the domestic market as individual power sources capable of being attached to appliances such as wringers and sewing machines, with the power being transferred by belting. An American variant was the "kitchen power table" that had a 1/4 horsepower motor fitted below it. It was connected to a power head to which the appliances were connected by pulleys or horizontal shafts. An equally remarkable British example was an electric posser, which resembled an agitator washing machine. The zinc washtub sat upon a base plate and the copper posser was attached to a cast-iron arm. Both are powered by an ungrounded electric motor, which lifted the posser up and down while partially rotating the base plate. It also featured an electrically powered wringer.

Electric motors became smaller and could be factory-fitted as a part of the appliance. They were not initially integrated into the design and, as with the case of early electric whisks and food mixers, had an industrial appearance. The motor was hidden within a pressed metal envelope during the 1920s.

The impact of the small electric motor cannot be underestimated. Writing in 1948, Siegfried Giedion stated, "It meant to the mechanization of the household what the invention of the wheel meant to moving loads. It set everything rolling."

See also Electricity; Food Mixers; Hamilton Beach; Westinghouse

References:

Giedion, Siegfried. *Mechanization Takes Command.* New York: Oxford University Press, 1948.

N

National Electric Light Association (NELA)

The National Electric Light Association (NELA) was the body that provided the fledgling U.S. electricity industry with promotional focus and coordination during the late nineteenth century and early twentieth century. As the interface between the industry and consumers, the NELA was dependent on its image and reputation.

The NELA was formed in 1885, and the first of its twice-yearly meetings was held at the Grand Pacific Hotel in Chicago. It drew its members from both the electricity utilities and electrical engineering companies. Its commercial orientation meant that members tended to be sales, marketing, and advertising staff rather than technical staff. Even before World War I, when only about a sixth of American homes were wired for electricity, the NELA was active in promoting domestic use of electricity. Recognizing that women were the key decision-makers in the home, the NELA set up its own women's committee. It also developed close relationships with other women's organizations and invited their representatives to address NELA conventions. In the 1920s, it collaborated with the Federation of Women's Clubs to survey the use of electricity in 4 million homes. The NELA also urged electrical goods retailers to employ women, especially trained home economists, as sales staff and to encourage them to play an ambassadorial role by lecturing to women's organizations and demonstrating at trade fairs. Looking to creating awareness amongst the next generation of consumers, the NELA also involved children in its market research, for example by sponsoring a competition that involved children surveying lighting in their own homes and those of neighbors. Survey findings and detailed product research informed the direction and content of publicity campaigns, such as the 1931 refrigerator campaign. That campaign, like others, was conducted through a barrage of advertising in local and national papers.

Criticism of the NELA hinged on the contention that it was not representative of the interests of the electricity industry as a whole. In particular, the NELA was accused of being antagonistic toward municipal utilities. Almost inevitably, given their manufacturing dominance and private utility investments, General Electric and Westinghouse exerted a strong influence on the NELA. Biases were most apparent in the NELA's attitude toward rural electrification, which was dismissed as unprofitable. In 1928, the Senate instructed the Federal Trade Commission to conduct an investigation into the public relations practices of the NELA. The hearings lasted for six years and exposed a hidden pro–private utilities agenda, propounded at an advertising cost of about $1 million a year. This damaged the image of the NELA and helped to make public opinion favorable to the subsequent New Deal legislation on electricity provision. The 1929 Wall Street Crash and the ensuing economic deprivation led the government to take a more active role in regulating the electricity industry. After the election of President Franklin Delano Roosevelt in 1933, the National Power Policy Committee closely scrutinized electricity pricing policies, particularly in relation to rural electrification. The NELA's resistance to government regulation was in marked contrast to the response of its younger sibling, the National Electrical Manufacturers Association (NEMA), founded in 1926. NEMA cooperated with the New Deal by setting up its own National Recovery Administration code to monitor production and prices. Having alienated the public, the NELA lost its much of its market influence.

See also British Electrical Development Association; Electricity

References:

Nye, David E. *Electrifying America: Social Meanings of a New Technology.* Cambridge, MA: MIT Press, 1990.

Nintendo

Founded in the late nineteenth century as a manufacturer of playing cards, the Japanese games company Nintendo was a world leader in the electronic games market 100 years later. It has sold more than 1 billion video games worldwide and 40 percent of U.S. households have a Nintendo game system.

The company began life in 1889 when Fusajiro Yamauchi began making Hanafuda playing cards. In 1933, it became a partnership, Yamauchi Nintendo & Company. Another change of name to the Nintendo Playing Card Company came in 1951. The introduction of playing cards featuring Walt Disney characters in 1959 was a big success. In 1963, the company started manufacturing games, a change reflected in the shortening of its name to Nintendo Company. By 1969, it was ready to expand production of games and built a new factory in Uji City, Kyoto. Nintendo's Beam Gun series of games, launched in 1970, were the first electronic games produced in Japan. In 1974, the company developed an image projection system using 16 mm film projectors for the amusement-arcade market and began exporting to the United States and Europe. A collaboration with another Japanese company, Mitsubishi Electric, resulted in the introduction in 1976 of an electronic video recording system for arcade games. In 1977, Nintendo entered the home videogames market. Having adopted microprocessor technology for its new home video games, in 1978, Nintendo introduced a new series of microprocessor-based arcade video games.

In 1980, reflecting the growth of the company's international sales, it set up a U.S. subsidiary, Nintendo of America. Nintendo launched a new arcade game, Donkey Kong, in 1981, which became the most popular game on the market. The company's growing success prompted the expansion of its facilities at home and abroad. In 1982, Nintendo of America was reincorporated and moved to Seattle. In Japan, a new Nintendo plant in Uji City opened in 1983. In the same year, a new subsidiary, Nintendo Entertainment Centres, was formed in Canada, based in Vancouver. The Family Computer games console featured customized data and picture processing units. While Nintendo continued to produce arcade games, the launch of the Nintendo Entertainment System (NES) in 1985 marked the arrival of a new generation of home video games that would dramatically enlarge consumer sales. NES introduced the character-based game Super Mario Brothers, which proved hugely influential. In 1987, the NES became the number-one-selling toy in the United States, and a million units of a new character game, the Legend of Zelda, were sold. Nintendo brought out

65 new NES games, including games intended to appeal to adults, in 1988.

Nintendo's next innovation was a games console that made video-game playing a mobile activity. The Game Boy, introduced in 1989, was the first hand-held video game system and was designed to be used with interchangeable game packs. A still smaller pocket version appeared in 1996, while, in 1998, a color version and add-on camera and printer extended the line. The home games console was upgraded in 1990 with the launch of the Super Famicom 16-bit system in Japan. For the international market, Nintendo developed the 16-bit Super NES and the new Super Mario World game in 1991. It continued to enhance the core technology, developing the improved Super FX Chip in 1993 and the Advanced Computer Modeling (ACM) system of graphics in 1994. The game Donkey Kong Country was the first with ACM graphics, followed by Donkey Kong Land for the Game Boy in 1995, the year that total cumulative sales of Nintendo games packs reached 1 billion units. In 1996, Nintendo produced the Nintendo 64, the world's first commercially successful 64-bit home video game system. This created an unprecedented sales frenzy, with half a million sold in Japan on the launch day and the first U.S. shipment of 350,000 units sold in 3 days.

In 1998, Nintendo introduced its most successful Game Boy concept to date—Pokémon. The concept took the essence of what had made Virtual Pets an instant hit and put it into a game format. Instead of caring for and training passive pets, Pokémon was based on the training of pocket monsters to compete with each other. For Nintendo, with its origins as a playing-card manufacturer, Pokémon also presented an opportunity to fuse past and present games formats. Packs of Pokémon character cards complemented the video game. The production of limited numbers of some characters fueled a frantic collecting craze amongst children and concern amongst parents about unhealthy competitiveness. New Pokémon games—Pokémon Snap and Pokémon Pinball—came on the market in 1999, and the franchise expanded with the release of a Pokémon feature movie and television series. However, at the end of 1999, Nintendo was lagging behind Sega and Sony in the drive to enhance the appeal of home games consoles by incorporating DVD (digital versatile disc) and Internet technologies.

See also Electronic Games; Virtual Pets

References:

Herz, J. C. *Joystick Nation*. London: Abacus, 1997.
www.nintendo.com/corp/history.html

O

Obsolescence

"Built-in obsolescence" was a term coined by Vance Packard in 1957 in his critique of American marketing, *The Hidden Persuaders*. Despite his slightly oversimplistic depiction of the American consumer as an unthinking vessel for the campaigns of the advertising industry, he did draw attention to a phenomenon, which had been developing since the 1930s, that of manufacturers limiting the life span of their products, especially those directed at the home.

The roots of this lay in the Depression of the 1930s, when U.S. manufacturers faced a bleak future. Between 1929 and 1932, net income from manufacturing fell by more than two-thirds. Recovery began in 1933, but not until 1937 did manufacturing return to pre-crash levels. Corporate America had vast reserves of plants and people—and a belief that underconsumption was a key cause of the crash. It is therefore unsurprising that much thought went into projects to stimulate consumption.

Henry Ford held this view, as did Charles F. Abbott, of the American Institute of Steel. He advocated "more creative wasting in order to gain more progress . . . remaking our whole world through progressive obsolescence." The advertiser Earnest Elmo Calkins looked to "artificial obsolescence" through the redesign of products and recognized the need for restyling as "something entirely apart from any mechanical improvement, to make them markedly new, and encourage new buying, exactly as the fashion designers make skirts longer so you can no longer be happy with your short ones." This concept was developed by Calkins's admen Egmont Arens and Roy Sheldon in their book *Consumer Engineering: A New Technique for Prosperity* (1932). They promoted market research into what consumers wanted and industrial designers to translate those dreams into realities.

Arens and Sheldon leaned toward obsolescence by technological improvement, but by the postwar boom of the late 1940s and 1950s, U.S. advertisers knew that their arsenal included not only guilt at not having the best for one's family but also fashion, hope, aspiration, and status. Now also termed "psychological obsolescence" it was, and continues to play, a big part in corporate strategies worldwide. In 1950s United States (and later in Europe), most people already owned usable stoves, TVs, cars, washing machines, and so on. The manufacturers' production levels could not rely on "natural wastage" as products broke down, but had to be stimulated by the creation of "psychological obsolescence."

By the time Packard published *The Hidden Persuaders* in 1957, the United States was about to experience another slump. This time it was less severe and receded due to a combination of government and consumer spending. In 1960, Packard published *The Waste Makers,* which charted planned obsolescence where technology and fashion met. Products were designed to wear out after a few years as well as being subjected to periodic restyling. This pattern is now accepted by both manufacturers and consumers.

Planned obsolescence is not only economically important but it has assisted the progress of scientific marketing and the use of industrial designers within the domestic appliance industry. The process, linked to lifestyle marketing and retailing, was both refined and expanded during the 1980s and 1990s. Production is now recognized as being as much a cultural process as it is a technological one.

See also Consumers; Industrial Design and Designers; Sunbeam

References:

Meikle, Jeffrey L. *Twentieth Century Limited: Industrial Design in America, 1925–1939.* Philadelphia: Temple University Press, 1979.

Packard, Vance. *The Hidden Persuaders.* Harmondsworth, UK: Penguin Books, 1957.

Packard, Vance. *The Waste Makers.* Harmondsworth, UK: Penguin Books, 1961.

Omelette Makers

Electric omelette makers employ similar technology to that used in toasted sandwich makers but with deeper nonstick hotplates. They usually make two omelettes from a three-egg mixture. The main manufacturer is Tefal.

See also Sandwich Toasters

P

Philips Electronics

Founded in 1891, the Dutch company Philips is ranked eighth amongst the world's electronics corporations. With factories in more than sixty countries, it is the world leader in a number of product areas, including lighting—its original product field—and electric shavers. Philips has been particularly innovative in the field of audio and video technology, where it is ranked second in Europe and third in the world. It is also a major producer of electronic components and supplies components used in 60 percent of the world's telephones.

Gerard Philips set up the company at Eindhoven in the Netherlands to manufacture incandescent lamps and other electrical products. Philips soon established itself as one of the major lighting manufacturers in Europe. In 1914, the company established a research laboratory as a means of sustaining its development through new products. In particular, it moved into the electronics field by working on X-ray tubes, radio valves, and cathode ray tubes. Philips began to experiment with television in 1925 and introduced its first radios in 1927, followed by gramophones in 1929. Its first television sets appeared in 1937, while in 1939 came the launch of one of Philips's most successful product lines, the electric shaver. By 1995, sales of Philishave electric shavers had reached 300 million units.

While the research laboratory was responsible for evolving the underlying product technology, a design department was created to fulfill the complementary role of developing corporate image and product design. In 1925, Philips appointed the Dutch architect Louis Kalff to provide artistic direction for its advertising section. Kalff soon became involved in product design and influenced the use of new materials such as Bakelite. For example, while the first Philips radio, the 2514 model, was an austere steel cabinet covered with black leatherette, the accompanying loudspeaker was a distinctive circular design in mottled Bakelite. Kalff also produced the stars and waves design for the company logo.

In the 1950s and 1960s, Philips was at the forefront of developments in transistors, integrated circuits, and television camera and receiver tubes. It also moved into new areas of sound and video recording technology. In 1963, Philips introduced the compact audiocassette and cassette recorder, a format that achieved a much greater level of home ownership than its predecessor, the reel-to-reel tape recorder. Next, Philips turned its attention to video technology, simultaneously developing a cassette format for videotape and

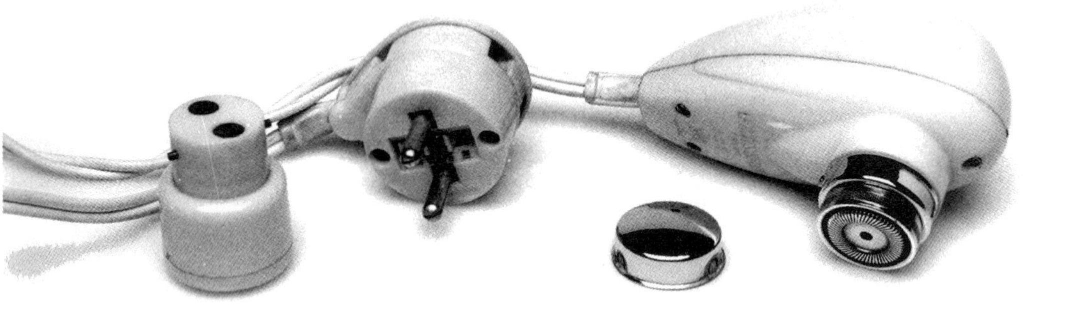

Philishave model 7735 electric shaver, c. 1950 (The Museum of Science and Industry in Manchester)

a disc-based video system. While the Japanese company Sony was leading way in videotape developments, Philips was the leader in the videodisc format. In 1971, Philips introduced its N-1500 videocassette recorder. Affordability gave Philips a healthy share of the early videocassette market in Europe, but the emergence of the VHS format in 1976 marked the beginning of the decline of the Philips system. Philips demonstrated its LaserVision videodisc system in 1972, several years before its commercial launch. Another format battle ensued and none of the rival videodisc systems made much impact on the household market.

Philips had more success with the development of the audio compact disc. Heeding the lessons of the video format rivalries, Philips and Sony formed a joint venture in 1979. This resulted in their joint launch of the compact disc player in 1982 and their joint standard for the computer CD-ROM (compact disk with read-only memory) in 1985.

Meanwhile, Philips consolidated its more established lighting and television interests by continuing to innovate and by making strategic acquisitions. In the lighting field, from the 1970s Philips has developed a range of energy-saving lamps. It strengthened its global position by acquiring a number of U.S. businesses in the 1970s and 1980s, including Magnavox (1971), the television operations of GTE Sylvania (1981), and the light-bulb division of Westinghouse (1983).

Philips has also continued to demonstrate an appreciation of the value of good product design. The management of the design process was refined by the creation of its Industrial Design Bureau in 1960. The initiation of a collaboration with the Italian company Alessi in the mid-1990s was a departure from previous design practice, but echoed the spirit of the earlier technical partnership with Sony. Alessi is noted for the highly aesthetic qualities of its products. The collaboration produced the Philips-Alessi line of small appliances, including a toaster, kettle, and coffeemaker. These products were characterized by their rounded forms and modish colors.

See also Alessi; Compact Disc Players; Lighting; Razors; Videodisc Players; Videotape Recorders

References:
www.philips.com

Pifco

The Pifco Company was founded 1900 as the Provincial Incandescent Fittings Company in Manchester, England, by Joseph Webber. Its early products were gas mantles and oil heaters. It soon expanded, purchasing a U.S. gas lighting company in 1908.

After World War I, the company began to concentrate on electrical products. An early postwar product was Christmas fairy lights. The line grew to include electric blankets, radio valve testers, hair dryers, shavers, trouser presses, fans, and knife sharpeners. Expansion continued with a move to larger premises and new factories in Salford, Birmingham, and Monton, a suburb on the outskirts of Manchester. In 1950 Pifco introduced the Chereton, an early cordless electric iron.

Joseph Webber died in 1955 and was succeeded as chairman by his son Alfred. Pifco became a public company in 1957. It moved to new headquarters in a former cotton mill at Failsworth near Manchester in 1970. The company has continued to develop through the acquisition of others such as Carmen personal care products (mainly known for electric hair rollers), Salton appliances, and Tower cookware. It took over Russell Hobbs in 1991, Mountain Breeze air conditioners in September 1996, Hi-Tech Industries in 1999, and Haden kettles in 2000.

The Webber family is still actively involved in the management of the company. Alfred Webber died in 1983 and was succeeded by Michael Webber.

See also Hair Dryers; Irons

Piped Water

During the twentieth century, the provision of a constant supply of pure water piped into the home became a service that was taken for granted in developed countries like Britain and the United States. However, that was not the case at the beginning of the century when piped water was still a novelty for most householders.

In Europe, water engineering reached an early peak at the height of the Roman Empire. The Romans excelled at civil engineering and water engineering, technologies that they introduced to the peoples that they conquered, along with a more urban way of life. In medieval Europe, people obtained fresh water from local streams or rivers, wells, and springs. As towns grew, extra supplies were laid on by piping water to public conduits. This level of provision became inadequate in both quantity and quality terms when rapid industrialization and urbanization began in the late eighteenth century. Demand for water escalated, for both industrial and private consumption, while, concurrently, natural sources became increasingly contaminated by unregulated industrial and human sewage.

For wealthy households, the solution was to pay water companies for a piped supply, but there was no concerted effort to improve general water provision until the aftermath of the European cholera epidemic of 1831–1832. The medical profession was developing a better understanding of the origin of certain diseases and recognized that unclean water and poor sanitation encouraged the spread of disease. The British social reformer Edwin Chadwick compiled an influential report that led to the introduction of the 1848 Public Health Act, which set up a General Board of Health. In the United States, Lemuel Shattuck waged a similar campaign in Concord, Massachusetts, in 1850. Another American public health reformer, William Jewell of Philadelphia, was the driving force behind the creation of the National Sanitary Convention in 1857.

Widespread improvements were still needed. By the mid-nineteenth century, even in large British towns and cities, only wealthy and middle-class homes had piped water, for which they paid higher rates than the majority of households that received water from a communal standpipe. Moreover, whether you had a direct household supply or relied on communal supplies, water was only made available at certain times. Users of communal standpipes would have to queue for water and would store it in outside butts, which might be subject to mold growth. The shortage of water discouraged personal cleanliness

The Verona line of wash basins from the 1901 Twyfords catalog (The Museum of Science and Industry in Manchester)

and general hygiene. In homes with piped water, the reception point was usually at street level, connected to a tank in the basement. Water still had to be pumped or carried upstairs. From about 1850, major cities began to build reservoirs, often some miles away, which allowed them to provide 24-hour water services. Filtration of water became common practice as a means of increasing water purity. The introduction of more stringent plumbing regulations greatly reduced wastage due to leaks and faulty cisterns.

By 1880, a constant water supply was normal in urban areas, but the extension of a piped supply to working-class homes was still in the early stages in most towns and cities. The adoption of water closets (toilets) as standard indoor accoutrements tended to go hand-in-hand with infrastructure expansion. By 1900, most urban households did have running water, yet the water-borne disease typhoid continued to claim many lives. Therefore, in the early twentieth century, water utilities began to chlorinate the water supply, as chlorine acts as a germicide. By 1930, it was common for middle-class homes to have hot and cold running water throughout the house, although this was still rare in working-class homes until the 1950s and 1960s.

By the 1960s, the last widespread problem that remained was that many older homes still had internal lead piping. As a soft metal, lead was ideal for plumbing, hence the term (from the Latin for lead), but it is also toxic and prone to dissolve in hard water containing magnesium or calcium salts. Today, while we take piped water for granted, we are also more conscious that water is a scarce resource, particularly with the well-publicized threat of increased incidence of drought as a consequence of global warming. In spite of this, worldwide water consumption was increasing at twice the rate of population growth in the late twentieth century, raising fears that critical shortages were imminent in some parts of the world.

See also Showers; Water Closets; Water Heaters

References:

Coley, Noel. "From Sanitary Reform to Social Welfare." In *Science, Technology, and Everyday Life, 1870–1950,* edited by Colin Chant. London: Open University/Routledge, 1990.

Cowan, Ruth Schwartz. *More Work for Mother: The Ironies of Household Technology from the Open Hearth to the Microwave.* London: Free Association Books, 1989.

Wright, Lawrence. *Clean and Decent: The Fascinating History of the Bathroom and the Water Closet.* London: Routledge, 1980.

Plastics

The term "plastics" covers a wide range of materials that can be molded or shaped into different forms under pressure and/or heat. Chemically, plastics are polymers, composed of long chains of repeating molecules. Some derive from natural materials such as amber, rubber, and horn.

The nineteenth century saw the development of semisynthetic plastics such as vulcanized rubber (Charles Goodyear, 1838, U.S.), Parkesine (Alexander Parkes, 1862, UK), and Celluloid (John & Isaiah Hyatt, 1870, U.S.), mainly marketed as imitations of horn and ivory.

The twentieth century saw the first truly synthetic plastics. The most important one for the developing appliance market was phenol-formaldehyde. It is better known as Bakelite, after its inventor Leo Baekeland, a Belgian-born chemist who worked in the United States. He reacted the chemicals phenol and formaldehyde to produce a material that, when combined with fibrous fillers such as paper or cotton flock, could be molded into almost any shape. These fillers gave a mottled effect in dark colors.

Bakelite is hard to ignite and has excellent insulating and molding properties, making it an ideal material for the emerging electrical appliance industries. Bakelite was used for electrical plugs and insulators, radio cases, thermos flasks, hot water bottles, hair dryers, and kettle handles. The later urea-formaldehyde (Edmund Rossiter, 1924, UK) was lighter and came in brighter colors, thanks to the addition of colored powders in the production process. It was also used for electrical appliances and was also popular for picnic wares. In the United Kingdom, brands such as Bandalasta and Beatl were popular.

Plastics and electronics were ideal partners in mass production. Although the set-up costs of producing plastic moldings was high, the profits came from huge production runs that cut out expensive handwork. The 1930s saw plastics being used for products as diverse as radio cases, light switches, electric plug cases, telephone handsets, combs, jewelry, cigarette boxes, coffee grinders, and electric shaver cases.

Developments continued during the 1930s and 1940s, with Du Pont's polyethylene (1933) and Nylon (Wallace Carothers, 1935, U.S.).

Culturally, plastics were plagued by poor consumer responses in the 1950s and early 1960s. The problem was twofold. First, there were many poor-quality plastic products made after 1945 that led to a perception that all plastics were "cheap and nasty." Second, many people did not fully understand the qualities of the material and used them in similar ways to metal objects, leading to a number of melted kitchen products! Nevertheless, their lightness and practicality eventually won the battle in the kitchen. Tupperware helped to sell a better image of the material. A 1955 British Ekco brochure proclaimed

> The modern kitchen, to which Plastics make a valuable contribution, is playing a huge part in easing the tasks of the present day housewife—The range includes jugs, sink bowls and tidies, colanders, hotplates, baths and accessories for the nursery, toilet seats and toilet brush holder, polythene bottles for powders and lotions. All these and many more, flow from the Ekco factories to add to the

Domestic phenolic plastic products from the 1930s and 1940s. (The Museum of Science and Industry in Manchester)

comfort and convenience of life in modern homes throughout the world.

The 1950s saw the plastics market continue to grow in diverse areas, apart from the kitchen. In the United States laminates such as Roanois and Micarta had been introduced in to cafe and bar furniture as well as trains and airplane interiors. In the United Kingdom, Formica had a similar impact, especially when it could be bought and used within the home. The do-it-yourself magazines of the 1950s and 1960s promoted Formica as a modern material that could transform old furniture. By the end of the 1950s, many domestic appliances were beginning to shed their metal skins in favor of plastics. The Italian Kartell Company pioneered the use of quality design in domestic plastics products. Its lemon squeezer of 1958 and Colander KS 1036 of 1960 were both designed by Gino Colombini, the company's technical director. Kartell's plastic bucket was selected for the collections of the Museum of Modern Art in New York.

The 1960s continued this trend as new plastics like ABS and PVC were used in the growing fashion industry. European manufacturers, such as the Italian Comfort and Kartell companies, were also producing stylish furniture in acrylic and ABS. In the United Kingdom, the quality of plastics goods was improving with the likes of Brookes & Adams melamine Fiesta tableware. The increasing use of plastics in appliances continued throughout the 1960s and 1970s. Interestingly, the look of many of these products did not alter radically, and the colors employed tended to be whites and creams.

Changes did take place in the 1980s and 1990s as plastics began to move out of the kitchen and become more visible in other rooms of the house and in the garden. The Sharp QT 50 radio-cassette player of 1984 was produced in a turquoise case with pastel control knobs. Public perception of the material had moved. Plastics were either so ubiquitous that they were "invisible" or "material-neutral" to many people, or they were to be celebrated for their ability to become, as Roland Barthes remarked in 1957, "buckets as well as jewels." The 1990s saw an increasing range of fashionable, inexpensive household plastics.

Further refinement has led to an almost bewildering range of materials that can now out-perform traditional ones. Kettles and toasters can now be made almost entirely of plastics. Now ubiquitous, they continue to be the envelope for most household technologies from computers and televisions to drills and food processors. Plastics are now central to the design and production of almost all domestic appliances and have gained acceptance as being the most appropriate materials for the job.

See also Ekco; Radio; Record Players; Tupperware

References:

Katz, Sylvia. *Classic Plastics.* London: Thames & Hudson, 1984.

Mossman, Susan, ed. *Early Plastics: Perspectives, 1850–1950.* London and Washington, DC: Science Museum/Leicester University Press, 1997.

Sparke, Penny, ed. *The Plastics Age: From Bakelite to Beanbags and Beyond.* Woodstock, NY: Overlook Press, 1993.

Popcorn Poppers

Popcorn, as its name implies, is made from corn kernels. The kernels are heated in oil and expand and "pop" in the pan. Once they stop popping they can be coated with salt or sugar. This simple operation can be done in standard saucepan, but this fact did not deter manufacturers producing specialized equipment for the U.S. home of the 1950s.

Models were either stainless steel or enameled. West Bend produced models in colored aluminum in the 1950s. All had a heated element in the base, which usually stood on short tripod legs. Some had cranks in the lid, which turned the kernels. The Fostoria by the McGraw Electric Company featured a glass lid so that the user could see the corn popping inside.

Contemporary models such as the Prima and Proctor-Silex Popcorn Pumper use a stream of heated air and do not require oil. The kernels are placed into the cylinder and pop out through a chute and into a bowl when they have heated and expanded. Prima also produces Dudley Duck a "children's" model in the shape of a cartoon duck. His yellow feet form the base, a red hat, the lid, and the hot corn pops out of his beak!

Pressure Cookers

The late seventeenth century saw a number of experiments into more efficient cooking methods, many of which became widespread during the second half of the nineteenth century, partly due to the continuing development of the increasingly efficient closed kitchen range. Nevertheless cooking remained a time- and coal-consuming business. Cooking speed could be accelerated if the pressure inside the pan was increased. This principle had been established by the French physicist Denis Papin, working in late seventeenth-century London. Normally, water boils at 100°C (212°F), and no matter how fast it is made to boil it evaporates rather than getting hotter. By putting the container under pressure, the boiling temperature is increased, enabling the food to be cooked faster. Papin invented his "steam digester" in 1679–1680 and demonstrated it at the Royal Society in 1682, using only 8 oz of coal. It was a closed iron vessel with a tightly fitted lid and a safety valve to release the excess pressure.

The use of pressure cookers, or digesters, as they were known, slowly increased during the nineteenth century. They were usually of cast iron with simple weighted safety valves in the center of their domed lids. Working at a pressure of 3 lb per square inch, they were produced by companies such as Kenrick & Sons of Birmingham, England. It is probable that their popularity grew due to the importation of chilled Argentinean beef that, after

Woman in the kitchen using the Presto Pressure Cooker, 1953 (Library of Congress)

the 1860s, supplemented growing home consumption. This beef was tough and needed cooking longer. The "digester" reduced the cooking time and the coal bill as well as tenderizing the meat.

Twentieth-century models developed to be used with more efficient electric and gas stoves that could provide a more constant and even heat. Cast aluminum replaced iron on some models, and these were capable of withstanding pressures of up to 50 lb per square inch. 1930s models such as the Easiwork Health Cooker featured a pressure gauge and a whistle. A bridging clamp held down the lid. Pressure cookers were popular in the United States during the 1930s, with models produced by Presto and West Bend.

The most successful British pressure cookers were the Prestige line. Platers and Stampers of London produced the Prestige 65 in 1948. It featured an aluminum body and black Bakelite handles, with three cooking pressures controlled by three weights that were attached to a central vent in the lid, giving boiling points of 257°F, 272°F, and 284°F. The correct pressure was announced by a continuous hissing. Once the cooking time was completed, the cooker had to be left for the pressure to drop. An improvement on this was the introduction of automatic depressurization, controlled by a lever on the lid. Models of this type came with alloy trivets and divided containers for vegetables. By the late 1970s, Prestige produced "color clad" models and Teflon nonstick surfaces.

Pressure cookers are ideal for steamed puddings, preparing preserves, and cooking vegetables but are less useful for cooking meats. They remained popular during the 1950s and 1960s, as they were a fast and energy-efficient way of cooking food, especially in small apartments. Specialist cook books for pressure cookers were produced to encourage greater usage. Their popularity declined during the 1980s with the rise of the microwave oven that offered even simpler rapid cooking. Nevertheless they continue in production, aided by campaigns for healthy eating that have promoted the steaming of vegetables and seafood. Prestige, Tower, and Tefal are the main producers.

See also Cookware; Presto; West Bend

References:

Brears, Peter. *The Kitchen Catalogue.* York, UK: York Castle Museum, 1979.

Campbell, Susan. *The Cook's Companion.* London: Chancellor Press, 1980.

Feild, Rachael. *Irons in the Fire: A History of Cooking Equipment.* Marlborough, UK: Crowood Press, 1984.

McDermott, Catherine. *The Design Museum Book of Twentieth-Century Design.* London: Carlton Books, 1998.

Presto

Presto was the name given to appliances manufactured by the Northwest Steel and Iron Works of Wisconsin, which was established in 1905. Their products were originally made under the trade name National. One of their early successes was a pressure cooker, and in 1939 the company changed its name to the National Pressure Cooker Company; hence the brand name of Presto.

The company diversified after 1945 and in 1949 produced the first steam iron capable of using tap water. The company continues to produce small appliances.

See also Pressure Cookers

Proctor-Silex

Proctor-Schwartz was founded in 1920 by the Philadelphia Textile Machinery Company (founded in 1885). It moved in to the expanding electrical market and by 1928 had purchased the Liberty Gauge Company of Cleveland, Ohio. Liberty had just marketed the Liberty electric iron, which featured an adjustable temperature control. The company expanded its line to include toasters and coffeemakers. The Silex Company was a manufacturer of coffeemakers and irons, which had acquired a controlling interest in the Chicago Electric Company, which produced sandwich toasters, waffle irons, juicers, portable clothes washers, and electric ice cream makers.

These two companies merged on 1 March 1960 as the Proctor-Silex Corporation, a successful union. After six years of high sales, the company was purchased by the SCM Corporation. It was purchased by Wesray Appliances Inc. in 1983 and in 1986 merged into Wear-Ever Aluminum as a cookware and appliance company. In 1989, the WearEver division was sold and Proctor-Silex continued as a part of NACCO Industries. In 1990, NAACO bought Hamilton Beach and merged the two companies to create the largest manufacturer of small appliances in the United States.

See also Irons; Hamilton Beach

References:

www.hambeach.com

Pyrex

Ordinary glass suffers from low tolerance to rapid heating and cooling, but the introduction of boric oxide solves this problem. The American Corning Glass Works successfully manufactured chemical and heat-resistant borosilicate glass, building on the pioneering work of the German chemist Otto Schott. Corning introduced this in 1912 as Nonex. The wife of one of Corning's researchers first made domestic use of the glass when she baked a cake in a Nonex battery jar! Corning began to develop the glass for domestic use with the brand name of Pyrex in 1915.

Wishing to capitalize on the new material, Corning looked to overseas markets and licensed manufacturers in Europe. They approached the Wear Flint Glass Works in Sunderland after larger concerns had proved to be cautious. This company, later known as James A. Jobling & Co., produced all the British Pyrex ware from 1921. Corning and Saint Gobain formed the Société Le Pyrex to manufacture and market Pyrex in France, Belgium, Holland, Spain, and Italy. Licenses to manufacture in Germany were granted in 1927 and in Japan in 1930.

The early wares took traditional forms and the line sold well, being functional and inexpensive. Most were clear glass, sometimes with engraved decoration, transfer prints, or color sprayed. Colorware was introduced in 1952 and were sprayed with enamel colors such as Jade Green, Powder Blue, and Canary Yellow. After 1945, their patents began to expire and Corning began to invest in design to make the line more attractive and to market it as an "oven-to-table" product. In 1953, Corning introduced Pyroceram, a glass that looked like a white ceramic but was capable of withstanding great variations of temperature. These were marketed in the United States as Corning Ware and were the base of the popular oven-to-table wares of the 1960s and 1970s.

Corning acquired a 40 percent share in Jobling & Co. in 1954 and took complete ownership in 1973. Corning remains one of the leading specialist glass producers in the world. Pyrex continues to be produced today, and although it produces some "fashion" lines, its simplest designs have remained virtually unchanged from the 1930s.

See also Cookware

References:

Baker, John C., Stuart Evans, and Cindy Shaw. *Pyrex: 60 Years of Design.* 1983.

Heskett, John. *Industrial Design.* New York: Oxford University Press, 1980.

R

Radio

Radio communication is the transmission and reception of sound in the form of electromagnetic waves, which travel over much longer distances than sound waves. At first, it was more commonly known as wireless communication to distinguish it from the earlier forms of electrical communication—telephony and telegraphy—which transmitted sound by wire. Whilst the scientific principles and basic technology of radio communication were established in the late nineteenth century, radio had little impact on the home until the onset of public radio broadcasting in the 1920s.

Early Experiments

James Clerk Maxwell, a British physicist, published his mathematical theory of the existence of electromagnetic waves in 1873. German scientist Heinrich Hertz devised the first convincing proof of this theory, through a series of laboratory demonstrations. In acknowledgement of his role, Hertz's name was later given to the unit of radio wave frequency. Oliver Lodge, an English physicist, gave the world's first public demonstration of wireless communication at the Royal Institution in London in 1894. However, the person who pioneered the practical and commercial application of wireless communication was Guglielmo Marconi, a young Italian amateur scientist. Although much of his work lay in improving earlier discoveries rather than innovating, Marconi is generally credited as the inventor of wireless communication.

Wireless Telegraphy

What Marconi envisaged was a system of wireless telegraphy for communicating with ships. In 1896, in pursuit of this vision, he traveled to England, center of the international shipping trade. He filed his first patent in June 1896 and continued to develop his system with funding from the General Post Office. Commercial wireless telegraphy was launched in Britain in 1898. Two years later, Marconi expanded his operations internationally by setting up new companies, including one in the United States.

Wireless Telephony

Wireless telegraphy gained momentum rapidly, but wireless telephony, as it was known, developed more slowly. Wireless telephony was initially conceived as a form of one-to-one communication, hence the name "telephony." The first transmitters were spark oscillators that produced intermittent waves. These worked adequately for wireless teleg-

"ACF" console model 333, c. 1929 (Library of Congress)

raphy, which only required the transmission and reception of long and short pulses (representing Morse dashes and dots), but not for more complex sounds such as the human voice. In 1899, Nikola Tesla, a U.S.-based Croatian inventor, developed a continuous wave high-frequency alternator. This type of transmitter produced a narrow band of frequencies, allowing easier tuning of the receiver. However, early versions only had a transmission range of 20 miles. In 1900, Reginald Fessenden, a Canadian electrical engineer working in the United States, used a spark transmitter to send the first wireless telephone message to a receiver a mile away. Six years later, Fessenden was more successful using a high-frequency alternator and delivered a musical broadcast over a distance of 11 miles. Fessenden was also the inventor of the heterodyne design of receiver circuit, patented in 1901.

Diodes and Triodes

In 1904, the British scientist John Ambrose Fleming patented the diode, an evacuated glass tube with two electrodes that detected and rectified electromagnetic waves. The diode and its successors, collectively known as thermionic valves in Britain and vacuum tubes in the United States, were the basis of the first electronic circuits. In 1906, scientists in Germany, Britain, and the United States patented detectors based on natural and synthetic crystals, which behaved as rectifiers. The next advance was the development of the Audion triode, patented by the American engineer Lee De Forest, in 1907. The triode acted as a rectifier, detector, and amplifier. In 1913, Irving Langmuir, an employee of General Electric in the United States, developed the first high vacuum ("hard") valve. The importance of wireless equipment for military communication during World War I stimulated advances in radio technology, including the mass production of high vacuum valves, which performed better and had a longer life than their predecessors. In 1915, a message was sent from Arlington, Virginia, to Paris by wireless telephony, the first transatlantic transmission of its kind.

Public Radio Broadcasting

By 1919, wireless broadcasting was technically achievable and there was a growing number of amateur wireless enthusiasts. The public radio broadcasting era began in November 1919 when a wireless station at The Hague, in Holland, transmitted the world's first scheduled radio broadcast, consisting of speech and music. The station was owned by the Nederlandsche Radio Industry, manufacturer of radio equipment, which used broadcasting as a strategy to boost sales.

In Britain, the Marconi's Wireless Telegraph Company built a high power transmitter at Chelmsford and, in 1920, began experimental broadcasts, culminating with a concert given by Nellie Melba in July. However, in spite of their popularity, these broadcasts were suspended in November because they were causing interference to telegram services. In the same month, U.S. radio broadcasting was launched with the transmission of the presidential election results from a Pittsburgh radio station owned by the Westinghouse Electric & Manufacturing Company. The Pittsburgh station followed up its initial success with a program of daily broadcasts featuring advertising, thus becoming the first commercial radio station. By the end of 1922, the United States had over 500 radio stations. In Britain, Marconi's Wireless Telegraph Company was granted two broadcasting licenses in 1922 as an interim measure to satisfy public demand while the concept of a single licensed broadcasting company was being explored. In October 1922, 300 manufacturers and dealers formed the British Broadcasting Company, funded by a new license fee levied on all owners of radio receivers. The BBC had a short life as a limited liability company, as, in 1927, it became a public corporation by royal charter and remains so today. In contrast, by 1938, the United States had four national broadcasting networks, which were commercially funded.

Indeed, popular drama serials became known as soap operas because they were associated with advertising by soap and detergent manufacturers.

Modulation

Modulation of the radio signal is the process by which separate radio signals can be broadcast simultaneously. A wave has two interrelated properties: amplitude (depth) and frequency. The properties of a radio wave can be changed by the sound waves that it carries. Radio stations are assigned particular broadcast frequencies, which they achieve through modulation. In the 1920s, all radio broadcasts used amplitude modulation (AM), expressed as a frequency in kilohertz. In 1933, the American electrical engineer Edwin Harold Armstrong invented the technique of wide-band frequency modulation (FM), expressed in megahertz, used to produce FM radio broadcasts. FM broadcasting improved radio reception because, unlike AM, the signal was not affected by local interference from electrical equipment such as transformers.

Radio Receivers in the 1920s

As expected, the introduction of public broadcasting in Europe, Canada, and the United States boosted the sales of radio receivers. While enthusiasts might build a receiver from scratch or assemble it from a kit, most listeners in the 1920s bought a manufactured radio, and the choice of receiver lay between a crystal set and a valve radio. Initially, the crystal set was the more popular because it was significantly cheaper—less than the average weekly wage. Neither type of receiver required mains (wired) electricity. The crystal set, also known as the cat's whisker radio, could receive broadcasts, via an outdoor aerial, within about 12 miles of a transmitter. At its most basic, it consists of a crystal detector activated by a tiny wire coil (the cat's whisker) that the user moves over the crystal to find the best signal and a tuning device, such as a variometer. The first crystal detector, psilomeian, had been discovered by the German scientist Karl Ferdinand Braun in 1901. The only energy supply comes from electromagnetic waves. The signal strength was only sufficient to enable one person or maybe two people to listen via headphones, unless the crystal set was attached to a separate amplifier or combined amplifier and loudspeaker. Crystal sets came in various simple forms, including components mounted on a baseboard with no casing, cylindrical cases, and rectangular boxes with hinged lids. They declined in popularity as the price of valve radios dropped and by 1930 were virtually obsolete.

Early valve radios gave better reception and were suitable for group listening via loudspeaker but were expensive and cumbersome. For example, early valve sets required two types of bulky battery: an accumulator, a low tension wet battery encased in glass that had to be recharged every three to four months, and high tension batteries (dry cells) that were not rechargeable and also lasted about three to four months. The valve set also required an aerial—preferably an outdoor one, but indoor aerials were usually adequate—and had to be grounded. Valve sets could contain as few as one valve or as many as eight. More valves meant better performance but also more expense, so basic two-valve sets were most common. A two-valve set might be capable of driving a loudspeaker, depending on proximity to the transmitter, but was more likely to suit headphone use. The early valve set generally took the form of a simple wooden cabinet on the front or top of which the valves and other components were exposed for ease of replacement.

Fortunately for the consumer, the manufacturers soon realized that improvements were necessary. The key problem was power supply, and this required changes in valve design. Valves were developed that removed the need for use of a wet battery. This reduced the bulk and weight of a valve set, and created the possibility of portable valve sets.

Portable sets duly appeared, with built-in aerials and loudspeakers. The next major step was the development of valves designed to take mains electricity. This was a real benefit to householders with a mains supply as mains power cost about a tenth of the equivalent battery power. The introduction of basic mains radio sets was soon followed by the first models with integrated loudspeakers, although the cabinet design remained austere. Other technical advances that prepared the ground for the application of a more user-friendly design ethos were improvements in amplifier and power output valves, the adoption of the supersonic heterodyne or superheterodyne circuit design, and the use of a metal chassis assembly that allowed easy insertion into the cabinet.

Radio Receiver Design in the 1930s

By 1930, just over 3 million households in Britain and almost 14 million households in the United States had a radio receiver. Manufacturers began to pay more attention to the styling of radio sets. The radio had become a focal point of the household so cabinet design was seen as a way of influencing consumer choice in an increasingly competitive market. As radio cabinet design developed, no single prevailing design aesthetic emerged. Instead, there was a broad spectrum of styles, reflecting the manufacturers' views on the range of public taste. The spectrum ran from the more traditional furniture-like wooden cabinets to the more innovative molded Bakelite cabinets. The use of new materials such as plastics and more structurally versatile older materials such as plywood and metal in place of wood gave designers more freedom of expression. The only functional constraints on the imagination of the designer were the need to accommodate the loudspeaker grille, the control knobs and the waveband display.

Art deco influences enlivened wooden radio cabinets. In the late 1920s, loudspeakers with decorative grilles came on the market and these grilles were incorporated into radio cabinets, usually placed centrally to create symmetry. The geometric effect was enhanced by the use of plywood with contrasting veneers. Plywood can be bent so its use enabled the incorporation of curves to soften the cabinet shape. One of the most influential designers of this type of radio cabinet was Gordon Russell, a British furniture designer who produced cabinet designs for the British company Murphy Radio.

The most adventurous British radio manufacturer was E. K. Cole (Ekco), which commissioned three architects associated with the Modern Movement—Serge Chermayeff, Misha Black, and Wells Coates—to produce designs for Bakelite cabinets. Chermayeff and Black were both born in Azerbaijan and both emigrated to Britain with their parents before World War I. Coates was a Canadian citizen, born in Japan, who moved to Britain in the early 1920s. The Ekco AD65 model of 1934, with a circular Bakelite cabinet designed by Coates, was the first radio design to fully capitalize on the moldability of Bakelite. Perhaps the most spectacular U.S. radio of this period was the Air King, a Bakelite skyscraper-style cabinet designed by the American designer Harold van Doren. In the late 1930s, more elongated, asymmetrical cabinets appeared, first in the United States and then in Europe. These were products of the influence of American streamlining. Leading American advocates of streamlining, including Norman Bel Geddes and Walter Dorwin Teague, entered the field of radio-cabinet design. Companies such as the U.S. mail-order giant Montgomery Ward, with an in-house design team led by Anne Swainson, followed suit.

Radio and Wartime Propaganda

World War II interrupted the smooth progress of radio development. In Europe, the need to conserve materials for the war effort limited production of all consumer goods. Nevertheless, radio production was not completely suspended because radio was viewed as a vital means of public communication. In 1944, the British government approved production of the Wartime Civilian

Receiver, or "Utility set," a simple, cheap radio receiver that economized on materials. This concept was similar to that of a much earlier German radio, the VE301 People's Radio (Volksempfänger), designed by Karl Maria Kersting in 1928. Over 12 million VE 301 radios, which could only receive transmissions from local stations, had been sold by 1939, and radio broadcasting was used as a Nazi propaganda weapon.

Miniaturization of the Radio

After the war, full-scale radio production resumed. As most homes already had radios, the manufacturers' strategy was to persuade consumers of their need for a second radio, perhaps a portable model or a combined alarm clock and radio. The radio industry became increasingly focused on technical improvements that would make radios smaller and cheaper. New types of miniature valves, which had been rare before the war, were introduced and experimentation with printed circuits began. In the United States, "midget radios" had been available from the early 1930s at reasonable prices, but in Britain, "compact radios" were a postwar phenomenon. At the opposite end of the size scale came radiograms, a composite radio and gramophone first seen in the 1930s, which became more popular in the late 1940s and 1950s. Radiogram sales were boosted by the introduction of long playing records by CBS (Columbia Broadcasting System) in 1948.

In December 1947, John Bardeen, Walter Brattain, and William Shockley at the Bell Laboratories in the United States invented the transistor, a major development that later transformed radio miniaturization. The transistor (or transfer resistor) was a return to crystal technology. Bardeen and Brattain had discovered that placing two tiny wires (similar to the cat's whisker) close together on a germanium crystal created amplification, making the transistor the equivalent of the triode valve. It took about five years of further research and development to engineer reliable transistors. A number of companies had transistor radios on the market in the mid-1950s, but it was the Japanese company Tokyo Tsushin Kogyo (TTK) that produced the world's first pocket-size transistor radio, under the brand name Sony, in 1957. TTK had managed to reduce the failure rate of transistor production dramatically, thus slashing the cost of transistors. This marked the beginning of the rise of Japanese companies to the position of supremacy in the manufacture of electronic consumer goods that they hold today. Low labor costs in Japan and nearby countries such as Hong Kong and Taiwan provided a competitive edge in the international market.

Valve radios continued to be produced until the late 1960s, but after the mid-1950s they were gradually supplanted by transistor radios, first in the portable sector and then in table-top radios and hi-fi tuners. Styling of radio receivers became more minimal and functionalist, epitomized by the radios produced by the German company Braun after the mid-1950s. Light colored plastics superseded dark Bakelite. Portable radios, which had resembled small suitcases in the 1930s, became more like colorful vanity cases or were reduced to the essentials of radio (with built-in aerial) plus carrying handle. By 1960, Japanese companies were marketing pocket transistor radios. These changes were in tune with a changing society, in which a new youth culture was emerging and car ownership was increasing personal mobility.

Radio and Lifestyle

Two basic trends of 1950s radio production have continued to the present day: the integration of the radio with recorded music players and the demand for portability. The radiogram developed into the hi-fi system, which has the advantage of allowing various audio components—radio tuner, record deck, CD deck, and cassette deck—to share an amplifier and loudspeakers. The same principle applies to the portable radio incorporated in the "boom box" with a cassette player, although portability suffers. In theory, miniaturization has been assisted by the growth of microelectronics since 1970, which has seen the transistorized printed circuit succeeded by the single-chip integrated circuit and the microprocessor. However, in practice, size has not been reduced because the reduction in size of other components, such as speakers, also reduces listening quality. Pocket radios, which tend to suffer from distorted sound at louder volumes, have been replaced by the personal stereo, which is designed for use with headphones, although it can be used with miniature external speakers.

Since the 1970s, high-tech styling has been dominant in the design of radios and other audio equipment. The main exceptions to this are novelty products, such as radios disguised as Coca Cola cans, and retro designs. High-tech design follows functionalist and minimalist principles and favors clean lines and monochromatic palettes of matte black, gray and silver/chrome. This styling reveals how the selling of objects through design may change as objects become familiar. When radio was new and strange to users, traditional design was a way of introducing familiarity; now that radio is ubiquitous, futuristic design is a way of reintroducing a note of mystique.

Radio has had a huge influence on home life, not least in paving the way for television. Live broadcasts enabled a mass audience to share in major events and to receive up-to-date news coverage. Music, comedy, and drama broadcasts brought low-cost entertainment into the home and gave access to a breadth of culture. In the same way that printing revolutionized written communication, radio has revolutionized spoken communication this century.

See also Bang & Olufsen; Ekco; Loudspeakers; Marconi Company; RCA; Record Players; Sony Corporation; Tape Recorders

References:

Hill, Jonathan. *Radio! Radio!* Bampton, UK: Sunrise Press, 1986.

Radio Corporation of America

See RCA

Rawlplugs

The traditional method of fixing screws into masonry was to drill the hole and pack it with a wooden plug that would receive the screw. The Rawlplug, invented in 1919, gave a much more secure fit. It was invented by an English builder, John J. Rawlings, who was carrying out renovation work at the British Museum. The museum was concerned that the work should cause as little damage as possible. Rawlings's plug was made of jute and pig's blood. The use of blood as a bonding material was not unique; the early plastic bois durci was made from blood and sawdust. The fibrous quality of Rawlings's plug allowed it to expand as the screw went in, resulting in a tighter, neater fit.

Rawlings formed the Rawlplug Company and invested in advertising in the daily newspapers. Demand was such that during the 1920s the company was able to expand into a large factory in London. It developed the larger, metal Rawlbolt in 1934. This was for heavy-duty bolts in masonry and concrete.

The fiber plugs were on sale as late as the 1970s but were being replaced by plastic ones. The Rawlplug was a boon for the growing numbers of do-it-yourself enthusiasts armed with their electric drills. They enabled householders to fix shelves and wall-mount small appliances. Most wall-mounted appliances now come supplied with their fixing screws and Rawlplugs. Similar items are sold in the United States as wall toggles or anchors.

See also Do-It-Yourself

References:

Hillman, David, and David Gibbs. *Century Makers.* London: Seven Dials, 1999.

Razors

For over 200 years most Western men have, in accordance with the dictates of fashion, been clean-shaven, thus necessitating the need for shaving. Until the late nineteenth century this meant a morning encounter with shaving soap and a "cutthroat" razor. The 1880s saw some new developments with "safety" razors such as the Rolls (United Kingdom) the Kampfe and the Home Safety Razor (United States).

All these razors required regular sharpening. The American King Camp Gillette developed a disposable blade in 1895. Gillette designed a wafer thin two-edged blade that could be fitted into a simple holder and, once used, thrown away. He had to wait until 1903 for steel technology to improve enough to deliver his safety razor, and then for the public to get used to the idea of a disposable product. In 1903, Gillette sold 51 razor sets and 168 blades. By 1904, the year in which Gillette patented his razor in the United States, 90,000 razors and 12,400,000 blades had been sold. Spurred on by this, Gillette opened a sales office in London and began manufacturing in Paris. The Gillette razor blade was one of the first truly disposable products of the twentieth century. By 1910, the Gillette Safety Razor Company had rivals such as Wilkinson's (United Kingdom) and Sicherheits (Germany). Some companies, such as Wilkinson's and Drew's, also developed devices for sharpening safety razor blades. Gillette had sold 3.5 million razors by 1918 when it received an order for 3.5 million razors and 36 million blades for the United States Army.

Traditional shaving required hot water, and it was here that electricity made its first incursion. The Merryweather Company (a British company whose main business was fire engines) produced an electric shaving pot in 1898. This was a simple urn with a lid, the water being heated by a separated element similar to those of early kettles.

The first successful electric razor was patented by a retired American army officer, Lt. Col. Jacob Schick of Stamford, Connecticut, in 1928, going into production in 1931. Schick first thought of the idea when gold prospecting in Alaska, where hot shaving water was a rarity! After serving in World War I, he returned to this idea. He was already in the razor business with another of his inventions, the safety razor blade dispenser. He sold his Magazine Repeating Razor Company to finance the production of the Schick dry electric razor. Running off the electrical socket, a small electric motor activated a series of parallel blades below a metal foil. In 1933, a German model electrified a safety razor, utilizing standard blades. The German Braun Company had developed a more sophisticated and stylish model in 1938, which was successfully reintroduced as the S50 in 1950. This had an oscillating cutter block covered by a very thin but strong steel screen. Prof. Alexander Horowitz, of the Dutch Philips Company, designed a circular cutting head in 1937. The result was the Philishave, which has two or three bladed discs. These spin, trap, and cut the facial hair.

The electric razor became popular during the 1960s and 1970s, the market being dominated by companies such as Braun, Philips, Remington, Schick, and Sunbeam, which introduced their Shavemaster in 1947. The Braun Sixtant of 1962 was a technical and design innovator, featuring a "honeycomb" shaving foil and a distinctive matte finish. Portable battery-operated models were introduced. Women's electric razors, designed for shaving body hair, were first introduced in 1940 with the Remington Duchess. The Philips Ladyshave followed in 1950 and the Sunbeam Lady Sunbeam in 1958.

Nevertheless, the "wet" shave also remained popular. Many men felt it gave a closer, more refreshing shave than the dry electric razor. The Gillette Company introduced the twin-blade razor in 1971, swivel heads in 1975, and lubricating strips in 1986. Its rival Wilkinson introduced the Protector in 1992. Designed by Kenneth Grange, it featured protecting wires to further minimize the risk of cuts. The first truly disposable razor was produced by Bic in 1975, using a simple thin blade and a plastic handle.

Gillette responded with its own version within a year. Wet razors have continued to market themselves on the concept of the close, refreshing shave. This has led to the introduction of dual and triple bladed shaving heads.

The 1980s saw the range of razors increase, with more models being introduced for women, most being battery operated and useable on wet or dry skin. They are consciously designed in pastel colors, as opposed to men's razors, which usually have black plastic or aluminum cases. Most men's razors are now cordless socket/battery combinations. The Philishave Cool Skin has electric blades and a cartridge for shaving gel, combining the features of wet and dry shaving.

See also Braun; Philips Electronics

References:

Druitt, Sylvia. *Antique Personal Possessions.* Poole, UK: Blandford Press, 1980.

Tambini, Michael. *The Look of the Century.* London: Dorling Kindersley, 1996.

RCA (Radio Corporation of America)

The Radio Corporation of America (RCA) was formed in 1919 to acquire the assets of the Marconi Wireless Telegraph Company of America. It was initially owned by corporate investors, including General Electric and Westinghouse, and became an independent company in 1932. Since 1988, RCA has been owned by the French multinational corporation now known as Thomson Multimedia, which is the world's fourth largest consumer electronics company. The RCA division is the United States' leading television, video, and digital decoder manufacturer and is ranked second in audio and communications equipment.

Guglielmo Marconi, the British-based Italian inventor of radio, had formed the Marconi Wireless Telegraph Company of America in 1900 in order to exploit his radio patents in the United States. In 1917, the U.S. government entered World War I and took control of all wireless stations in order to control strategic communications for the duration of the war. After the war, convinced that wireless communications should continue to be nationally controlled, the government engineered the acquisition of Marconi's U.S. subsidiary by a consortium of U.S. companies, led by General Electric and including Westinghouse and American Telephone and Telegraph. In 1919, RCA was born and began to restructure its operations. RCA did not manufacture radios but instead marketed radios made by General Electric and Westinghouse under the Radiola brand name. In 1922, David Sarnoff, previously Marconi's commercial manager and an enthusiastic advocate of radio broadcasting, became vice-president of RCA. Sarnoff had been born in Russia and emigrated to the United States with his parents as a child.

In the early 1920s, radio stations were purely local, but Sarnoff had a vision of national radio broadcasting. Largely as a result of Sarnoff's drive, RCA, General Electric, and Westinghouse combined to form the National Broadcasting Corporation (NBC) in 1926. A year later, NBC had a coast-to-coast network of 25 radio stations. In 1929, RCA diversified by taking over the Victor Talking Machine Company, which manufactured record players and produced records under the His Master's Voice label. Sarnoff was promoted to the position of president of RCA Victor in 1930. RCA Victor set up a subsidiary, the RCA Radiotron Company, to manufacture radio valves. However, at this point, the federal government became concerned that the close ties between RCA, General Electric, and Westinghouse amounted to a virtual monopoly of the radio industry. General Electric and Westinghouse were forced to divest their interests in RCA in 1932. In spite of the uncertain economic climate of the Depression, RCA Victor and NBC set up a new joint headquarters, Radio City, in New York's Rockefeller Center. The move to Radio City took place in 1933.

Meanwhile, Sarnoff had developed a new obsession—television. In 1929, he had at-

Painting of Nipper, which became an RCA trademark in 1929 (Library of Congress)

tended a convention of radio engineers at Rochester, New York, where Vladimir Kosma Zworykin had demonstrated his electrical television system. Zworykin had also been born in Russia, and he emigrated to the United States in 1919. He developed his television system while working for Westinghouse as a research engineer. The two components of the system were the iconoscope (the television camera), developed in 1923, and the kinescope (the picture tube), prototyped in 1924. As Westinghouse was losing interest in the possibilities of television, Sarnoff hired Zworykin as director of electronic research for RCA. It took ten years before the system was ready to be launched for public broadcasting. The RCA television demonstrations were the major talking point of the 1939 New York World's Fair. The opening ceremonies, including an address by President Franklin Delano Roosevelt, were broadcast by RCA to the 200 television owners within a 50-mile radius of New York.

The outbreak of World War II suspended RCA's normal operations, as factories were required to support the war effort by producing military equipment such as mine detectors and sonar devices. After the war,

RCA resumed television broadcasts and introduced the 630TS table television set, which had a 10-inch screen. As RCA produced a range of electronic tubes during the war, its core television technology advanced. RCA expanded its television production facilities to cope with the rapid growth of demand. While CBS was the first television network to develop a color television system, it was RCA's color television system that became the U.S. standard in 1953. RCA color television sets went on the market in 1954, but the high price meant that sales were slow initially. The next major advance in RCA color televisions was the launch of solid-state color televisions in 1970.

The 1970s saw RCA take an active role in the new video age. Unfortunately, this proved to be an unwise move financially and led the company into a downslide. RCA had been involved in the professional application of videotape recording since the 1950s because of its television broadcasting interests. However, it initially took a backseat in the rivalry that developed in the mid-1970s over formats for consumer videocassette recording. In 1977, when Matsushita's VHS format had become the clear front-runner, RCA announced that it would produce VHS recorders and scored a small coup by introducing the first four-hour videocassettes. RCA's mistake was its decision, taken in the early 1960s, to develop a videodisc format. Three rival video disc formats were announced in 1975, all play only and nonrecordable, and appeared on the market a few years later. RCA's Selectavision system, launched in 1978, was competitively priced but unattractive to consumers in comparison with recordable videocassette systems. After five years, when it ceased Selectavision production, RCA had sold half a million players and 10 million discs, but had spent $300 million on research and development and lost approximately the same amount on production.

Perhaps not surprisingly, following the commercial disaster of Selectavision, RCA ceased to be an independent company in 1986 when it was taken over by General Electric. Less predictably, little more than a year later, General Electric sold off not only RCA, but also its own consumer electronics operations, to the French electronics multinational Thomson Grand Public. The enlarged company was renamed Thomson Consumer Electronics. Ironically, Thomson began life as the French subsidiary of the U.S. Thomson-Houston Electric Company, which merged with Edison Electric Light Company to form General Electric in 1892. This completed a series of prestigious acquisitions by Thomson in the 1980s, which include the German companies SABA and Telefunken and the British company Ferguson. Under new ownership, RCA began to flourish again. It reached a major milestone in 1989 when its 50 millionth color television set came off the assembly line at its plant in Bloomington, Indiana, which is the world's largest television assembly plant.

In 1993, Thomson became a founding member of the Digital HDTV Grand Alliance, an international body formed to agree on global standards. A year later, the RCA Digital Satellite System introduced digital satellite television broadcasting in the United States. The parent company, Thomson Consumer Electronics, was renamed Thomson Multimedia in 1995 to signal its growing interest in digital home-entertainment products. Moreover, in 1998, it made equity holdings available to four companies that were considered to be suitable partners for new digital developments. These companies, with an aggregate 25 percent shareholding, are Microsoft, DirecTV, Alcatel, and NEC.

See also Electrical and Musical Industries; General Electric; Marconi Company; Radio; Record Players; Television

References:

www.rca.com

www.thomson-multimedia.com

Record Players

Until the late nineteenth century, the only form of musical entertainment available in

Scientific American. [DECEMBER 22, 1877.

Scientific American.

ESTABLISHED 1845.

MUNN & CO., Editors and Proprietors.

PUBLISHED WEEKLY AT
NO. 37 PARK ROW, NEW YORK.

O. D. MUNN. A. E. BEACH.

TERMS FOR THE SCIENTIFIC AMERICAN.

One copy, one year, postage included.................................. **$3 20**
One copy, six months, postage included **1 60**

Clubs.—One extra copy of THE SCIENTIFIC AMERICAN will be supplied gratis for every club of five subscribers at $3.20 each; additional copies at same proportionate rate. Postage prepaid.

The Scientific American Supplement

is a distinct paper from the SCIENTIFIC AMERICAN. THE SUPPLEMENT is issued weekly; every number contains 16 octavo pages, with handsome cover, uniform in size with SCIENTIFIC AMERICAN. Terms of subscription for SUPPLEMENT, $5.00 a year, postage paid, to subscribers. Single copies 10 cents. Sold by all news dealers throughout the country.

Combined Rates. —The SCIENTIFIC AMERICAN and SUPPLEMENT will be sent for one year, postage free, on receipt of *seven dollars*. Both papers to one address or different addresses, as desired.

The safest way to remit is by draft, postal order, or registered letter.

Address MUNN & CO., 37 Park Row, N. Y.

☞ Subscriptions received and single copies of either paper sold by all the news agents.

Publishers' Notice to Mail Subscribers.

Mail subscribers will observe on the printed address of each paper the time for which they have prepaid. Before the time indicated expires, to insure a continuity of numbers, subscribers should remit for another year. For the convenience of the mail clerks, they will please also state when their subscriptions expire.

New subscriptions will be entered from the time the order is received; but the back numbers of either the SCIENTIFIC AMERICAN or the SCIENTIFIC AMERICAN SUPPLEMENT will be sent from January when desired. In this case, the subscription will date from the commencement of the volume, and the latter will be complete for preservation or binding.

VOL. XXXVII., No. 25. [NEW SERIES.] *Thirty-second Year.*

NEW YORK, SATURDAY, DECEMBER 22, 1877.

Contents.

(Illustrated articles are marked with an asterisk.)

THE TALKING PHONOGRAPH.

Mr. Thomas A. Edison recently came into this office, placed a little machine on our desk, turned a crank, and the machine inquired as to our health, asked how we liked the phonograph, informed us that *it* was very well, and bid us a cordial good night. These remarks were not only perfectly audible to ourselves, but to a dozen or more persons gathered around, and they were produced by the aid of no other mechanism than the simple little contrivance explained and illustrated below.

The principle on which the machine operates we recently explained quite fully in announcing the discovery. There is, first, a mouth piece, A, Fig. 1, across the inner orifice of which is a metal diaphragm, and to the center of this diaphragm is attached a point, also of metal. B is a brass cylinder supported on a shaft which is screw-threaded and turns in a nut for a bearing, so that when the cylinder is caused to revolve by the crank, C, it also has a horizontal travel in front of the mouthpiece, A. It will be clear that the point

Fig. 1.

on the metal diaphragm must, therefore, describe a spiral trace over the surface of the cylinder. On the latter is cut a spiral groove of like pitch to that on the shaft, and around the cylinder is attached a strip of tinfoil. When sounds are uttered in the mouthpiece, A, the diaphragm is caused to vibrate and the point thereon is caused to make contacts with the tinfoil at the portion where the latter crosses the spiral groove. Hence, the foil, not being there backed by the solid metal of the cylinder, becomes indented, and these indentations are necessarily an exact record of the sounds which produced them.

"The Talking Phonograph"; article in Scientific American, *December 22, 1877 (Library of Congress)*

the home was live performance. While wealthy householders could afford to hire professional singers and musicians to provide entertainment at social gatherings, most people had to rely on their own musical abilities. The invention of the phonograph in 1877, followed by the gramophone in 1888, introduced recorded music to the home. Although a third of British homes had a gramophone by 1913, it was only after the advent of vinyl discs in the late 1940s that recorded music became a huge money earner.

The Phonograph and Gramophone

The concept of sound recording dates from 1857, when the French artist Éduoard-Léon Scott de Martinville developed a cylinder recording machine called the phonautograph. As the phonautograph did not provide a playback facility, it had no commercial potential. This omission was rectified by the American inventor Thomas Alva Edison in 1877. In Edison's phonograph, sound waves entering a speaking cone caused a thin, metal diaphragm at the base to vibrate. The vibrations were transferred to a steel stylus, which impressed a groove in tinfoil or waxed paper covering a rotating brass cylinder. The recording was played back through a separate stylus and diaphragm, attached to a set of ear tubes. Because the cylinder was rotated by turning a handle or working a foot treadle, it was difficult to maintain the consistent rhythm required for good sound reproduction. The phonograph in its original form, as patented in 1878, could only create a single recording that could not be copied. This meant that the phonograph was more suitable as a dictating machine for office use than as a machine for playing music. The other limitation of the original phonograph was that the tinfoil or waxed paper sheets were fragile and easily damaged during replaying.

The durability of phonographic recordings was improved in 1886, when the graphophone, a machine that used a cylinder coated with hard wax, was patented. The graphophone's inventors were the American telephone pioneer Alexander Graham Bell, his cousin Chichester Bell, and Charles Sumner Tainter. For group listening, an acoustic horn provided amplified sound, but there was no way of controlling the volume other than by muffling the horn. However, the problem of nonreproducibility remained. Only two years later, Emile Berliner, a German immigrant, solved this problem. Berliner took the idea of a recording disc mooted by the French poet Charles Cros in 1877 and came up with the gramophone. He substituted the phonographic cylinder with a glass disc coated with a mixture of lampblack and linseed oil. As the horizontal disc rotated on a turntable, the stylus vibrated laterally, cutting a continuous spiral groove from the circumference toward to the center. After recording, the disc was electroplated to create a copper or zinc "master" disc. Multiple copies could be created by pressing discs of shellac hardened by the addition of ground slate or vulcanite (rubber hardened by the addition of sulfur) against the negative master disc.

In 1894, Berliner's Gramophone Company sold 1,000 gramophones and 25,000 records. These early 7-inch records played at 30 rpm (revolutions per minute). In 1897, Berliner introduced the 10-inch 78-rpm records that soon became standard for the next fifty years. While Berliner's gramophones were hand-cranked, another American inventor, Eldridge R. Johnson, had the idea of fitting a spring motor to the gramophone and filed a patent for this design in 1897. The spring motor, also known as the clockwork motor, had to be wound up periodically and lost speed when it needed rewinding. Johnson's Consolidated Talking Machine Company adopted the "His Master's Voice" trademark for its record label. This trademark, featuring the fox terrier Nipper, was retained when the Johnson and Berliner companies merged in 1901 to form the Victor Talking Machine Company. The typical tracking weight of the steel stylus at this time was 255 g (9 oz), which caused rapid wear to both the stylus and the record. Manufactur-

Bush music center with vertical record deck and tangential tone arm, c. 1980 (The Museum of Science and Industry in Manchester)

ers also produced thorn styli, which were gentler on records but needed to be sharpened frequently.

Although the gramophone ultimately triumphed over the phonograph, the development of techniques for mass-producing wax cylinders in the first years of the twentieth century extended the phonograph's life for another twenty-five years. Both gramophones and phonographs tended to conform to a similar design, with polished wood bases and ornate metal horns. Hand-cranked and spring-driven models coexisted until the electric record player became standard.

The Electric Record Player

In 1915, American Telephone and Telegraph's Bell Laboratories pioneered an electric sound recording technique that used the same kind of carbon microphone that had become standard in telephone mouthpieces. The microphone incorporated an electromagnet so that the vibrations in the diaphragm were converted to electric pulses. Over the next ten years, Bell Laboratories improved the microphone for use in sound recording and developed powerful valve amplifiers to boost the electric signal. In the mid-1920s, electric sound recording using equipment manufactured by AT&T's Western Electric subsidiary was widely adopted by American recording studios. This greatly improved the frequency range of recorded sound.

The development of the dynamic moving coil loudspeaker in 1924 meant that the same improvement in sound quality could be achieved when playing records by fitting an electric pickup (or cartridge) to the stylus. The pickup translated the vibration of the stylus to an electric signal that was amplified and fed to the dynamic loudspeaker. The first pickups were small electromagnets. The vibration of the stylus moved the magnet, inducing pulses of electric current in the surrounding coil. An alternative form of pickup, introduced in the mid-1930s, was based on the piezoelectric properties of crystals, such as quartz. When vibrations compress a piezoelectric crystal, a change in voltage results. Lightweight magnetic pickups appeared in the late 1930s and became the preferred form. As the pickup increases the responsiveness of the stylus, the tracking weight of the stylus could be reduced, resulting in lower friction and lower rates of wear of both the stylus and the record. Diamond or sapphire styli replaced steel styli, and the tracking weight dropped from about 85 g (3 oz) in the mid-1920s to a few grams in the 1950s.

In the 1920s, electric motors began to supplant spring motors, although wind-up record players continued to be made both as portable models for outdoor use and to accommodate the needs of homes without electricity. The first fully electric record player, with electric pickup and electric motor, was the Panatrope, introduced by the U.S. company Brunswick in 1925. At first, the drive mechanism was a direct-drive worm gear, a shaft with a helical screw end connecting the motor to the cog at the base of the turntable spindle. The disadvantage of this format was that the motor noise tended to be quite audible. After World War II, the drive mechanism was changed to a format that isolated the turntable from the motor more effectively by the insertion of an intermediary drive. The first kind of intermediary drive to be used was a pulley with a rubber wheel; the later format, still used today, connects the motor shaft to a pulley with a rubber belt and the pulley to turntable base by a second rubber belt. The typical record player of this period was fitted in a simple wooden case with a hinged lid.

Changes in Disc Technology

The original rigid shellac discs required a heavy steel stylus, which resulted in audible surface noise. In 1906, the Columbia Phonograph Company, one of the leading U.S. record companies, introduced a flexible record, created by laminating shellac around a paper core, which was more sensitive to stylus pressure. It permitted a lower stylus tracking weight, thus reducing surface noise. The Radio Corporation of America (RCA), which had taken over the Victor Talking Machine Company in 1929, introduced a 33⅓ rpm vinyl disc in 1931, but the type of vinyl used, called Vitrolac, was not the same kind that became that the standard record medium after World War II. There were a number of reasons why record companies were interested in the possibilities of synthetic thermoplastic resins. One was supply: shellac is resin secreted by a scale insect that is native to the jungles of the Far East and Indian subcontinent, so its availability was subject to natural restrictions. Another reason was that radio broadcasting companies had found that discs coated with cellulose acetate provided improved sound. However, that technology was not economic for mass production.

A particular vinyl, polyvinyl chloride (PVC), proved to be a highly suitable recording medium. In particular, it had the advantage that the groove could be cut in a tighter spiral, increasing the length of track on a disc of a given diameter. This became known as the microgroove technique and was developed in 1947 by the American sound engineer Peter Goldmark. In 1948, Columbia introduced its first vinyl long-playing records (LPs), which were 12-inch discs to be played at 33⅓ rpm. In the following year, RCA Victor introduced another vinyl format, the 7-inch, 45 rpm extended play record (EP). In the late 1940s, the concept of popular music was transformed by the arrival of youth-oriented rock and roll, and the 7-inch format was also adopted for single-play records. As with the original 78 rpm records, the short playing time of 45s meant users liked the convenience of being able to stack records for automatic consecutive play. However, the problem with the use of automatic record changers was that the friction between stacked records increased wear and tear. This led to record changers being discontinued as a feature of higher-quality record players.

Stereophonic Sound

Until the late 1950s, all commercial records and record players were monophonic or monaural. This means that the constituent parts of the sound were recorded as a single track or channel and played back through a single loudspeaker. Sound engineers recognized that this meant that recordings tended to sound unnatural and lacking in depth. Because humans are binaural (having two ears), we are used to picking up live sound from all directions, favoring the ear nearest to any particular sound. Therefore, stereophonic or binaural recording, whereby sound is recorded as two distinct, synchronized sound tracks that are played back through a two-channel pickup and separate loudspeakers, is more natural to our ears. This is achieved by using the opposite walls of the groove for separate sound tracks. The original sound recording may consist of more separate tracks that are then combined for the optimal stereo effect.

The first stereophonic recording system was developed by the British sound engineer Alan D. Blumlein at Electrical and Musical Industries (EMI) and patented in 1931. Blumlein's system cut two lateral tracks, each at 45° to the perpendicular center of the groove, thus creating a groove with a 90° angle. Two years later, engineers at the Bell Laboratories also developed a stereophonic recording system, which began to be used in the United States in 1940. At first stereo recording was restricted to professional environments, such as radio broadcasting. In 1958, the recording industry agreed to adopt the Blumlein stereophonic technique as the global standard and the first stereo LPs became available. However, monophonic sound persisted into the 1970s, as many people were initially reluctant or unable to invest in expensive stereo equipment. As teenagers and young adults began to emerge as the dominant group of record purchasers in the 1950s, cheap, semiportable mono record players were ideal for the bedroom. These record players were typically in colorful vinyl-covered cabinets, appealing to youthful tastes. In the 1970s, quadrophonic sound, involving four sound tracks and four speakers, made a brief appearance. Quadrophonic music systems only achieved a very small market because the equipment was expensive, there was a limited choice of quadrophonic recordings, and many people did not feel that the sound was significantly better than stereo.

The Hi-Fi System

While the term "high fidelity" or "hi-fi" was coined in the mid-1930s, referring to the improved sound quality from electrical sound recording, true hi-fi only came with the introduction of stereophonic sound and solid-state electronic equipment. The first type of integrated audio system was the radiogram, a floor-standing cabinet incorporating a radio, record player, and record storage space. Radiograms were produced from the early 1930s to the 1960s. In general, they conformed to a conventional wooden cabinet form so that they harmonized with other household furniture. The Hyperbo radiogram, which was introduced by the Danish company Bang & Olufsen in 1934, was an exception in that it had a modernistic, Bauhaus-influenced tubular metal frame. Stereo radiograms became available in the early 1960s, but by that time, modular audio systems were taking over at the top end of the audio market.

Although modular audio systems only became common in the late 1960s, they first became available thirty years earlier. One of the pioneers of such systems was Avery Fisher, who set up his hi-fi business in the United States in 1937. The advantage of the modular hi-fi system was that the consumer had more control over the specification and could combine the technical strengths of different manufacturers and, as with the radiogram, have a radio tuner and record deck sharing the same amplifier and loudspeakers. However, there was only a minority market for such systems because, aside from the cost factors, most consumers did not have such sophisticated expectations. Another problem was that valve amplifiers were bulky and unattractive.

In the 1950s and 1960s, perceiving a growing market for high-quality audio equipment, a number of European manufacturers began to develop integrated audio systems for the top end of the market. One of the most influential designs was the Phonosuper SK4, introduced in 1956 by the German company Braun. The Phonosuper was a combined radio and record player in a compact tabletop form. Designed by Dieter Rams and Hans Gugelot, its salient features were a predominantly metal casing (with wooden side panels) and a clear acrylic lid. In the following year, Rams carried over the basic elements of the Phonosuper design into the Braun Atelier I radiogram. Minimalist design became the essence of modern hi-fi systems.

The Italian company BrionVega took a functionalistic approach in its RR126 stereo hi-fi system of 1965. Designed by the brothers Achille and Pier Giacomo Castiglioni, the RR126 was mounted on a castor-wheeled stand and had cube-shape speakers that were hinged so that they could be folded up to save space when not in use. Bang & Olufsen became a particularly notable hi-fi manufacturer in the 1960s because not only were its systems strikingly styled but they also introduced technical features that later became widespread. Jakob Jensen's designs highlighted the technical sophistication of B&O's products. The Beolab 5000 music system of 1965 introduced user-friendly sliding controls while the Beogram 4000 stereo turntable of 1972 featured the world's first tangential pickup arm. The straight double tone arm was electronically controlled by a light spot and its tangential path eliminated wandering of the stylus.

In terms of the mass market, the stereo hi-fi age has been characterized by the rise to ascendancy of Japanese electronics companies such as Sony, Matsushita, and Toshiba. Matsushita is not a household name as its products are marketed under the JVC (Japanese Victor Company), Panasonic, and Technics brand names. Japanese companies were quick to see the potential of the new solid-state transistor technology in the 1950s. Their focus on the mass market meant that they embraced minimalist design, not so much for its aesthetic qualities but for its economic advantages in terms of standardization. By the 1970s, they had evolved a version of the modular system whereby matched units of standard size were designed to stack within purpose-built hi-fi cabinets with glass doors and lids. The units had metal cases, typically in silver or black, expressing their high-tech performance. Although consumers could still buy component units separately and assemble a system to their own specifications, it was generally cheaper to buy a complete system. Manufacturers began to incorporate some of the features used in professional recording equipment. For example, amplifiers were fitted with graphic equalizers, rather than just bass and treble controls. Dolby Laboratories, a company formed by American physicist Ray Dolby in London in 1965, has been responsible for many advances in sound reproduction. For example, Dolby Pro Logic "surround sound" technology, launched in 1987, is now incorporated in the amplifiers of more expensive stereo systems. In the 1970s and 1980s, the integrated equivalent of the stacking system was the music center, but music centers were generally conceived as lower performance, budget purchases. As the width of the stacking system was primarily dictated by the width of the record turntable, the introduction of the compact disc player in 1983 paved the way for smaller audio systems. While the record player is not yet an extinct technology, it is no longer seen as a critical element of an audio system. By the end of the twentieth century, most hi-fi systems—those designated as midi, mini, and micro—omit the record deck.

See also Bang & Olufsen; Compact Disc Players; Loudspeakers; RCA; Radio; Sony Corporation; Tape Recorders

References:

Gardner, Robert, and Dennis Shortelle. *From Talking Drums to the Internet: An Encyclopedia of Communications Technology.* Santa Barbara, CA: ABC-CLIO, 1997.

The Origins of Everyday Things. London: Reader's Digest Association, 1998.

Schoenherr, Steve. "Recording Technology History." On http://ac.acusd.edu/history/recording/notes.html.

Refrigerators

A refrigerator is an artificially cooled cabinet for storage of perishable foods. Cooling occurs when the refrigerant, preferably a substance with a low boiling point, is forced to change from a liquid to a gas by the application of pressure or heat. As the liquid evaporates, it draws heat from its surroundings, thus chilling food. The gas is then caused to reliquify either by being passed outside the

1930s Electrolux electric refrigerator, sold in Britain through the General Electric Company (The Museum of Science and Industry in Manchester)

cabinet to a condenser, where it is able to expand and give off heat to the surrounding air, or by gravity. This cycle operates continuously. The basic principles of both methods of refrigeration—compression and (heat) absorption—were established in the nineteenth century and applied in commercial contexts such as brewing and shipment of meat. Refrigerators on a smaller scale, suitable for household use, did not appear until the early twentieth century.

Before domestic refrigerators became available, for many households the only way to keep food cool was by storing it in a naturally cool place, such as a cellar or a larder. A more effective method was to pack blocks of ice around food. Ice became more widespread as a commercial commodity in Europe, Canada, and the United States during the nineteenth century. By 1900, department stores were stocking ice boxes, which were well-insulated wooden cabinets with one compartment for ice, another for food, and a tray to collect water when the ice began to melt.

Compression Refrigerators

In 1913, a Chicago company brought out the Domelre, the commercial version of the first domestic electric refrigerator invented by A. H. Goss. By this time, small electric motors had been used in various domestic appliances, including vacuum cleaners and washing machines. Within five years, the U.S. companies Kelvinator and Frigidaire had entered the market. These early refrigerators all used the compression method, which was first patented in 1834 by Jacob Perkins, an American then working in Britain. These early fridges had little to commend them to the average householder. First, few could afford them—even by the early 1920s, the price was equivalent to a quarter to a third of average annual wages. Second, they were large and heavy because, in addition to the wooden cabinet of the passive ice box, there was a bulky mechanism in the form of the motor and compressor pump. Third, because they were water-cooled rather than air-cooled, they had to be permanently connected to a water supply.

Absorption Refrigerators

In Britain, there was little interest until Frigidaire began exporting smaller models, marketed as "electric ice boxes," in 1923, but these were still expensive. At about the same time, the Swedish company Electrolux was developing an absorption-type refrigerator on the premise that the mechanical simplicity of the absorption method meant more compact, cheaper refrigerators. The absorption method was conceived by the Frenchman Ferdinand Carré in 1857, when he designed an ice-making machine that used heat to evaporate ammonia. In 1922, Swedes Baltzar von Platen and Carl Munters presented their design for an absorption refrigerator. Electrolux acquired the manufacturing rights in 1925 and launched the D-fridge, the world's first absorption refrigerator. Electrolux fridges were made in Britain from 1927. In the United States, the patent rights were acquired by Servel in 1925 and production began in 1926. Electrolux also pioneered the built-in refrigerator, which first appeared in 1930.

Absorption refrigerators are often referred to as gas refrigerators. Although the heat source could be gas or electricity, in practice electrical appliance manufacturers favored the compression method. Gravity cooling slowed down the cycle of the absorption fridge. However, the absence of moving parts gave absorption refrigerators two advantages over compression models: they were less prone to mechanical failure and considerably quieter. The pumps on early compression refrigerators were quite noisy. However, gas fridges failed to achieve more than a small share of the market in Britain or the United States. The main reason for this was that the refrigerator market in the United States was soon dominated by four powerful companies, able to command economies of scale and invest heavily in research, development, and marketing. These companies—General Electric, General Motors (owner of Frigidaire), American Motors (owner of Kelvinator), and, later, Westinghouse—backed the compression method. Kelvinator and then General Electric developed air-cooled compression fridges in 1925, six years before Electrolux's first air-cooled gas fridge appeared. General Motors and American Motors were able to draw on their sheet metalworking experience to remodel and lighten fridge cabinets.

Refrigerants

Ammonia was the first substance to be recognized as a refrigerant, in the wake of experiments conducted by the British scientist Michael Faraday in 1823. Ammonia and other early refrigerants such as sulfur dioxide and methyl chloride are toxic, which led to concerns about possible leakages. In the early 1920s, Charles F. Kettering developed a nontoxic refrigerant in Britain, but this was not widely adopted. In 1930 and 1931, the American industrial chemist Thomas Midgley Jr. and colleagues at General Motors developed a new class of compounds that they named Freons, now more commonly known as CFCs (chlorofluorocarbons). Freons are halogenated hydrocarbons, produced by a reaction between the hydrocarbons methane or ethane with the halogens fluorine and chlorine, whereby the halogen atoms replace hydrogen atoms. Freon 12 (dichlorodifluoromethane) became a popular refrigerant because Freons are stable, nontoxic, nonflammable, and noncorrosive. The negative qualities of CFCs were only recognized much later.

Design Issues

The first all-steel refrigerators were produced in the late 1920s. At first, the makers just focused on thermal performance. Steel cabinets featured better insulation in the form of a fiberglass filling that was lighter and more effective than the cork lining of wooden cabinets. Bimetallic thermostats were fitted from about 1930, improving temperature control. The first refrigerator to make a design statement was also the first one to make real market impact. This was the General Electric Monitor Top, an air-cooled fridge with the mechanism encased in a squat cylinder placed above the cabinet. It was designed in 1925, adapted for mass production in 1926, launched in 1927, and had achieved sales of 50,000 units by 1929. In Britain, a model of similar appearance made by BTH (British Thomson-Houston) was known as the "beehive" refrigerator.

In the 1930s, the widespread adoption by refrigerator manufacturers of the techniques of pressing and stamping metal used in the automobile industry cut production costs. It also provided the opportunity for leading U.S. designers to introduce the streamlined style. An early design commission went to Lurelle Guild in 1933, when he designed a refrigerator for Norge. A year later, Raymond Loewy produced what is generally regarded as the classic design of the 1930s, the

Coldspot Super Six model for the U.S. retail giant Sears, Roebuck. The hallmarks of the Coldspot were the white enameled, pressed steel body with rounded corners, a flush door, and chrome hardware (hinge, handle, and maker's badge). The look emphasized hygiene. The Coldspot also featured a well-equipped interior with rust-proof aluminum shelving, a semiautomatic defroster, and accessories such as a glass rolling pin and ice box. In five years, Sears, Roebuck sold 275,000 Coldspot fridges. In 1935, the American designer Walter Dorwin Teague added a new interior feature—storage space in the door—when he designed the Shelvador fridge for Crosley. This design sequence took a further step in shaping the fridge of today when another American designer, Henry Dreyfuss, designed a refrigerator with a full-length cabinet (and no legs) for General Electric in 1938. The final development of the 1930s came in 1939 when Westinghouse pioneered a pressing technique that eliminated the need for an internal frame.

Refrigerators have changed little since 1940 except cosmetically, in response to changing fashions, or functionally in response to changing lifestyles and consumer concerns. In the 1950s, further developments in sheet metal technology enabled rounded edges and curves to be replaced with sharp edges and corners, more suited to the seamless, uniform ideals of the modular fitted kitchen. Two-tone "harlequin" kitchen units were complemented by colored fridges. In the 1960s and 1970s, as the growth in female employment was accompanied by increased sales of convenience foods, including frozen food, the small ice compartment of the fridge developed into a freezer section, followed by the combined fridge-freezer. The use of expanded plastic insulation meant that full-height fridges or fridge-freezers were light enough to be maneuverable. Magnetic door contacts removed the need for locking mechanisms, allowing discreet hand-holds to replace protruding handles for a flush front. The only major chore associated with fridges—periodic defrosting—was simplified by the introduction of push button and self-programmed defrosting mechanisms and frost-free refrigerators.

In the 1980s and 1990s, there was a revival in the designer fridge, albeit with limited commercial impact. In 1987, the Italian company Zanussi produced the Wizard refrigerator, designed by Roberto Pezzetta. Available in gray or black and further distinguished by its pointed "roof" and a protruding plinth that wasted space, it was a flop. Another Italian company, Smeg, has been more successful with its range of large fridges and fridge-freezers, which combine U.S. 1950s-style shapes with bold colors. Meanwhile, in the United States, the emphasis has been on size and accessories rather than styling, with companies such as Frigidaire and General Electric producing giant double-door fridges with features including external chilled water and crushed ice dispensers.

Growth in Ownership

In the United States, refrigerator sales rose sharply in the 1930s. Even without persuasive marketing, the climatic range and the remoteness of rural areas meant that there was a sizeable proportion of the population for whom the refrigerator, once affordably priced, was an obvious benefit. The number of American homes with a refrigerator doubled between 1933 and 1936, and by 1941, home refrigerator ownership had exceeded 50 percent. In Britain, the temperate climate cast refrigerators firmly in the luxury class. The result was that in 1939, only 3 percent of British homes had refrigerators. After World War II, refrigerator production picked up in Britain with new manufacturers, such as Lec, setting up. While refrigerators in the United States got bigger and bigger, in Britain compact models were the norm. Postwar food rationing ceased in 1956 when restrictions on meat sales were removed, theoretically creating a market for larger fridges. Advertising of refrigerators in Britain reflected the upper-middle-class nature of the target market by playing on social snobbery ("cocktails on tap") and personal pride ("keep germs at bay"). The combination of rent-purchase options, falling prices, and rising wages produced an increase in home ownership of refrigerators from 8 percent in 1956 to 69 percent in 1971.

Environmental Issues

By the late 1980s, reducing energy consumption had become an issue for refrigerator manufacturers in common with manufacturers of all white goods (stoves, washing machines, and refrigerators). This led to the introduction of energy-saving fridges such as the 1991 Electrolux Low-Energy Refrigerator that used only half the energy of a conventional model. However, the major environmental concern affecting the refrigerator industry relates to the choice of refrigerants. By the 1970s, scientists had recognized that CFCs, including Freons, the miracle refrigerant of the 1930s, were causing depletion of the ozone layer, which filters the sun's harmful ultraviolet radiation. The United Nations Environmental Programme, established in 1972, convened a meeting in Vienna in 1985 to discuss protection of the ozone layer. The next convention in 1987 produced the Montreal Protocol on the control of ozone depleters. The signatories to the protocol included the United States and the countries of the European Economic Community. The Montreal Protocol has since been amended four times, most recently in 1997. As a result, the production and use of CFCs is being phased out. In the case of refrigerators, the CFCs are sealed in, and the issue is one of safe disposal when refrigerators are scrapped. The solution was to suck out the CFCs for recycling. Chemical companies developed HFCs (hydro-fluoro-carbons) as an alternative refrigerant. HFCs are currently the preferred refrigerants of most refrigerator manufacturers, but as "greenhouse gases" that contribute to global warming, their use is strictly regulated. The environmental-pressure group Greenpeace has advocated the use of environmentally friendly refrigerants through its Greenfreeze campaign. Greenfreeze fridges are now produced in a number of countries, and the British company Iceland Frozen Foods became the first company ever allowed to use the Greenpeace name to promote its range of fridges and freezers.

The Fridge of the Future?

In 1999, Electrolux previewed its Screenfridge, a new concept in refrigerator design. The name refers to the touch screen mounted in the outer face of the door, the concept being that the refrigerator is an ideal appliance to double as a communication device because of its frequency of use. The Screenfridge is equipped to receive television and radio signals, connect with the Internet and act as a monitor for security cameras. It can thereby act as a home entertainment unit, a household message board, and a household management tool. By the end of 1999, there was no news on the commercial release of the Screenfridge.

See also Electrolux; Freezers; Whirlpool Corporation; Zanussi

References:

Cowan, Ruth Schwartz. *More Work for Mother: The Ironies of Household Technology from the Open Hearth to the Microwave.* London: Free Association Books, 1989.

Giedion, Siegfried. *Mechanization Takes Command.* New York: Oxford University Press, 1948.

Heskett, John. *Industrial Design.* New York: Oxford University Press, 1980.

Relationship Technology

Relationship technology combines previously separate elements of personal and household technology, a logical development of microelectronics, changing lifestyles, and market research. An example is the combination of a microwave oven and an on-line Internet banking service, created by the Knowledge Lab, an international technology research consortium attached to NCR. The combination of these elements is based on market research that reveals that many people look at their statements and bills in the kitchen as they cook food.

The welding of technologies into single appliances is not new, but the elements have usually been clearly associated. The audio industry has successfully merged record players, radios, and cassette recorders, first as radiograms in the 1950s, then as music centers in the 1970s, and more recently as portable CD, radio, and cassette combinations.

The technology now exists to combine diverse appliances, but to be commercially successful they need to fulfill or create a need with consumers. An early attempt was the Phonolamp, which was displayed at the 1923 British Ideal Home Show. It was an electric table lamp on a circular pedestal base, which had doors that opened to reveal a gramophone player. It was not a success. The German Braun Company met with more success in 1929 when it was one of the first companies to incorporate a radio receiver and speaker into one set.

Less successful examples followed. The American Hurley Machine Company produced the Thor brand Electric Servant (1936) that was a washing machine with attachments for food mixing, blending, and juicing. Thor's core products were, not surprisingly, washing machines. The British Atmos Mechanical Housemaid was a clothes washer, rinser, wringer, drier, iron, and vacuum cleaner. The advertisement stated, "It sounds to good to be true!" It probably was, and like the Thor, it was not a success.

A truly successful example is the Sony Walkman, which created a new market for personal stereos by removing a traditional component, the speakers, and replacing them with a headset. The resulting technology was engineered around its relationship with a moving human being rather than another technology. It was a fortunate mix of a pioneering product meeting a receptive audience.

The advances in Internet and mobile phone technology have led to experiments to combine them, and also to link them to other appliances. The Merloni Elettrodomestici (owner of Ariston and Indesit) has launched a washing machine that can be controlled over the Internet by a mobile phone. Time will tell how useful such a thing will be; do people really need to control their washing over a distance? Whatever the outcome, the mobile phone is certainly a target for future development. There are already models that are combination personal electronic organizers, faxes, and phones with Internet connections. Mobile banking is probably the growth area. Oko, a Danish Bank, has been operating a mobile-based service since 1997. Another Scandinavian initiative is Bluetooth, named after Harald Blåtand, King Harold II of Denmark (c.950–c.985). It is a globally available shortwave radio frequency that utilizes a chip. Bluetooth is currently being supported by Ericsson, IBM, Intel, Nokia, Toshiba, Microsoft, and Motorola and has the potential to carry out automated transactions between a user's "smart" card (a card containing the working electronics of a mobile phone), their bank, and the shop they are buying goods from. A logical step from this increasingly sophisticated miniaturization is the concept of wearable computing, with the hardware integrated into jewelry or even clothing.

The development of Internet and cable technology could also prompt a growth in hybrid communication and entertainment systems where a single "appliance" could combine the functions of television, telephone, fax, home computer, and stereo system. Pieces of music and films could be ordered from a central source. Modern computers already have the ability to do most of these functions via the Internet and CD/DVDs.

The key to the mass acceptance of this developing technology is cultural as much as it is technological. The example of the mobile phone illustrates that if the product fulfils a real or a constructed need it will, if allied to falling prices, become popular.

See also Microwave Ovens; Refrigerators; Mobile Phones

Remote Controls

In the home, the remote control has become a familiar device for operating various items of home entertainment equipment. The first remote control devices were developed in the early twentieth century for military use and used radio waves to transmit control instructions to motorized weapons. Civilian applications only emerged after World War II, initially in the form of controls for automatic garage doors.

The television remote control first appeared in 1951 when the U.S. company Zenith Radio launched its Lazy Bones device. Unlike its descendants, it was not a wireless device. A cable connected the remote control to the television set. Users found the cable inconvenient so Zenith went on to develop a wireless remote control, the Flashmatic, in 1955. The Flashmatic, as the name implies, used photoelectric technology. The remote control emitted a light beam, which had to be directed at photoelectric cells located at each corner of the television screen. The disadvantage of this method of remote control was that a chance beam of sunlight could accidentally activate the cells.

Zenith continued its quest to develop a foolproof television remote control. The solution was the use of ultrasonics. The Space Command remote control of 1956 contained four aluminum rods of different lengths that, when struck by a trigger mechanism, produced distinct high-frequency sounds in the 35 to 45 kilohertz range. The main drawback of this model was financial: the remote control system, consisting of the remote control handset (the transmitter) and the transducer inside the television (the receiver), added about a third to the price of a television. A lesser disadvantage was that appliances could be accidentally activated by stray ultrasonic signals. Ultrasonic remote controls were also limited in the number of functions that they could control, as each needed a separate frequency. More compact, battery-powered remote controls accompanied the solid-state televisions of the 1960s, and remote control technology became a less expensive addition. Later Zenith models produced eight frequencies, and the U.S. company Magnavox developed a variant of the ultrasonic remote control that emitted ultrasonic pulses, increasing the number of control functions.

Developments in microelectronics paved the way for the modern infrared remote control technology that became available in the early 1980s. The infrared remote control is a digital device that uses an infrared diode to emit pulses of infrared radiation. The integrated circuit within the remote control handset is programmed so that the depression of a particular function key sends an electric signal to the diode, which converts the electric signal to a specific infrared pulse code. In the recipient appliance, the infrared signal is detected by a sensor, then decoded to produce an electric signal that is interpreted by a microprocessor that controls the voltage supply to the particular function in the appliance circuit.

The infrared remote control became available at a time when the range of home electronic goods was expanding. Infrared remote controls became standard not just for televisions, but also for videocassette recorders, compact disc players, cable television boxes, and satellite television decoders. In general, manufacturers program remote controls so that they can only operate a specific appliance. While this has the benefit of preventing accidental operation of another appliance, it becomes a drawback when the householder needs an array of remote controls to operate closely related appliances. The most common combined remote controls are those designed to operate both the television and the videocassette recorder. Another is the so-called universal remote control. In practice, these are multi-appliance rather than strictly universal, being preprogrammed with the codes for popular models of a small number of appliances, usually television-related. The user selects their particular models from a menu.

The restriction of infrared technology is that there must be a clear path between the remote control and the appliance, as infrared waves cannot travel through solid objects. This restriction may be overcome by the addition of a remote control extender, which converts the infrared signal to a radio signal. However, as yet, this technology has limited appeal, because there is little point in operating the common remotely controlled appliances from another room.

See also Philips Electronics

References:

Goodman, Robert. *How Electronic Things Work . . . and What to Do When They Don't.* New York: McGraw-Hill, 1999.

www.zenith.com

Rice Cookers

See Cookware

S

Sandwich Toasters

Toasted (grilled) sandwiches became popular in cafes and diners with fillings such as cheese, cheese and ham, and tuna. The sandwich toaster consists of two hinged hollow plates. The bread is buttered on the outside; one slice is placed on the bottom plate, then the filling and the top slice added and the toaster closed. The toaster is then heated, the bread toasts, and the filling is baked.

Domestic models are either simple types, which can be heated on a stove, or electric ones. Electrically heated models became popular in the 1970s with the Breville Company being a market leader. Leading U.S. manufacturers included Hamilton Beach and Sunbeam. Rectangular with a hinged lid and nonstick plates, they could toast, cut, and seal the sandwich.

Sandwich toasters provided a quick and easy snack. Their only drawback was a leaking filling that often led to a sticky mess on the plates. Some models now have removable plates.

Sears, Roebuck and Company

Chicago-based Sears, Roebuck is one of the world's most prominent retail groups. It operates more than 800 department stores in the United States and Canada, and has about 50 million mail-order customers.

The company began life in 1886 when Richard Sears, a station agent in North Redwood, Minnesota, purchased a consignment of watches from a Chicago jewelry company and sold them to other railway employees. This venture was so successful that Sears set up a watch company in Minneapolis. In 1887, Sears relocated to Chicago and recruited a young Indiana watchmaker, Alvah Roebuck. The first Sears mail-order catalogue was launched in 1888. Five years later, the company took the name Sears, Roebuck and Company.

In 1895, a Chicago clothing manufacturer Julius Rosenwald became a partner in the company. The business branched out into general merchandise, including clothing and household goods. It prospered because it met the needs of the rural market in the Midwest. Rural stores charged high prices to compensate for low turnover, whereas mail-order companies could buy in bulk and offer both lower prices and a wider selection. Richard Sears conceived the idea of enlisting customers as unofficial agents by offering free gifts for recruiting new customers. In 1906, the rapid growth of the business necessitated a number of changes. In order to raise investment funds, Sears, Roebuck stock was traded, bringing the company into public ownership. The company also moved into a new mail-order plant, which, with 300,000 square meters (3 million square feet) of floor space, was the world's largest business premises. Sears, Roebuck also opened an office in Dallas, Texas, to improve services in the southwestern states.

Service and product quality became crucial to maintaining competitive advantage in the mail-order business. Sears, Roebuck adapted assembly-line production techniques to the processing and shipping of orders. A communication system of belts and chutes dispatched order dockets, prioritized by shipping time, to the correct assembly point. Meanwhile, the creation of a testing laboratory in 1911 provided the means of ensuring that products met defined quality thresholds.

By the early 1920s, urbanization had caused significant changes within the retail sector. Chain stores were proliferating. For example, F. W. Woolworth, pioneer of the "five-and-dime" store, had almost 700 stores in 1913 when it opened its new New York offices in the world's tallest building. Growing car ownership meant that rural customers could also benefit from the urban chain stores. Sears opened its first store at the Chicago mail-order plant in 1925. Its success encouraged the company to embark on an expansion program that resulted in the launch of 400 stores by 1933. Sears was followed into the chain store field by its big mail-order rival Montgomery Ward, also based in Chicago.

The competitive environment meant that there was a need to develop distinctive and exclusive products. Sears introduced brand names such as Kenmore and Craftsman for various product categories. It also engaged the services of leading designers to style products. For example, leading American designer Raymond Loewy was employed to design the highly successful Sears Coldspot Super Six refrigerator of 1934. Another strategy was to improve the design of the store itself. Sears set up a department to plan store layouts and displays in 1932. The first Sears store built to a predefined layout opened in Glendale, California, in 1935. Mail-order was promoted in the department stores through the introduction of catalogue desks, while new catalogue offices were opened in smaller towns. A new venture launched in 1931 was insurance. The Allstate Insurance subsidiary sold policies both through mail and through in-store sales points.

In the 1940s, Sears began to expand overseas. It opened stores and sales offices in Central and South America and Europe. More successful in the long term was the creation in 1953 of a Canadian subsidiary Simpsons-Sears. Business confidence was still high in the late 1960s when Sears decided to build a new corporate headquarters in downtown Chicago. The Sears Tower, designed by Skidmore, Owings and Merrill, was completed in 1973. The 110-story skyscraper was the world's tallest building until it was surpassed in 1996 by the Petronas Twin Towers in Malaysia.

At the beginning of the 1970s, Sears seemed to have a position of unassailable supremacy in the U.S. retail market. It was the fifth-largest employer in the United States, and department stores were still the major type of retail outlet for household goods. However, retailing was already in transition, with the spread of giant suburban shopping malls providing extensive car parking. For example, in Sears's home city of Chicago, the Woodfield Center was the world's largest mall when it opened in 1971. At first, department stores flourished as the keystones of such malls. By 1980, the sales share of department stores had declined sharply owing to the dramatic growth of discount chains and the influx of new specialty stores. In the U.S. retail rankings, Sears was overtaken by the discount chains Wal-Mart and Kmart.

Sears responded to this challenge by undertaking a program of major restructuring and streamlining. By 1995, it had divested itself of its peripheral businesses, such as insurance and financial services, and reduced its foreign retail activities. Perhaps the most symbolic act was the relocation of its headquarters from the Sears Tower to a low-rise campus in the north suburbs of Chicago. Meanwhile, remote shopping was no longer restricted to conventional mail order, having been revitalized through television shopping channels and electronic shopping via the Internet. Sears now provides an Internet shopping service and describes its appliance section as the largest on-line appliance site.

References:

Worthy, James C. *Shaping an American Institution.* Urbana: University of Illinois Press, 1984.

www.sears.com

Security Systems

Crimes against property are by no means a twentieth-century phenomenon. English law developed with draconian punishments for burglary and theft. Technically the thieving of goods valued at over 40 shillings from a house or on a highway was a capital offense, but one that was impossible to fully uphold. The result led to overflowing jails and transportation to Australia.

The propertied classes relied on physical security and night watchmen to protect their property. In a country with a far-from-effective police force, Georgian and Victorian town houses were protected by curtain walls and railings. Internally, wooden shutters were also a deterrent to the opportunistic burglar. Silver and other valuables were protected in butler's strong rooms with steel doors. Servants were in the house practically twenty-four hours a day and also a form of alarm. Paradoxically, they could also become accomplices to burglaries. Dismissed servants did supply thieves with inside information as a way of enriching themselves and repaying their former employers.

The twentieth century saw more effective policing throughout Europe and America and the rise of an affluent suburban middle class. Home security remained an issue of physical protection rather than one of electronic surveillance. By the 1970s the problems of urban decline within inner cities led to an increase in property crimes and an equal increase in the fear of crime. In the United States the "doughnut" effect began to take hold, with the urban centers becoming abandoned by the middle classes as places to live. There was a flight to the suburbs, leaving business centers empty in the evening and surrounded by decaying suburbs or project housing. Those in the suburbs began to develop a laager or defensive mentality, resulting in the growth of housing associations that lobbied hard with developers and corporations for low-growth and low-rise single-family housing. Such approaches had the dual effect of maintaining property values and thereby deterring inner city dwellers from joining the suburban good life. Ironically, well-maintained apartment blocks with twenty-four-hour guards offer less temptation to burglars than the stand-alone family house.

The electronics industry had developed alarm systems for businesses and offices in the 1960s. Linked to the rising fear of actual and potential crime, the home security market grew rapidly during the 1980s, using variants of detection devices developed for commercial buildings.

By the end of the century, the standard intruder alarm featured the following devices. The system is controlled by a keypad panel that sets the alarm system, which is either hard-wired or radio-control linked to the detectors. These can be set to cover a number of zones, and more sophisticated models have ward settings to allow sections of the house to be occupied while others remain alarmed. The most popular detectors are passive infrared (PIR). They receive an amount of invisible infrared radiation that is naturally emitted from the area under protection. If a body moves within that area the amount of radiation changes, causing the detector to set off an alarm. An improvement on this is a combined infrared and microwave detector. Vibration detectors detect the energy transmitted above a preset level by force being exerted in an area close to where it is fitted. They are often used near windows, possibly alongside breakglass detectors, which are mounted either side of the window frame. The alarm is activated in response to the generation of noise created by breaking glass. Another popular device is the magnetic reed contact alarm, often fitted to windows and doors. When the contact is broken the alarm is activated.

Some home alarm systems only sound an external bell, but others are linked by telephone to a central control room that will alert the emergency services. Infrared fire alarms can also be included within such systems, although many homes rely on simpler battery-operated smoke alarms.

The physical side of security was also upgraded with the introduction of strong saw-proof locks, door chains, and window locks. Some elite U.S. suburbs have linked the electronic with the human in the form of the discreet sign on the lawn warning "Armed Response." The extreme form has developed in gated estates that employ state-of-the-art electronic security, closed circuit television, and armed guards.

Continuing social problems and the ever-present poverty gap signal that the concept of security will continue to be a domestic priority and one that will be increasingly designed into houses and estates.

See also Houses; Servants

References:

Chesney, Kellow. *The Victorian Underworld.* London: Pelican, 1976.

Davis, Mike. *City of Quartz: Excavating the Future in Los Angeles.* London: Pimlico, 1998.

Servants

Many middle- and upper-class European households entered the twentieth century with one of the most adaptable of domestic aids, the servant. The nineteenth century had seen the servant population in the United Kingdom rise as the population and the wealth of the country increased. This wealth was the result of initial preeminence in industrial production and the captive markets of a huge empire. The working classes were therefore subject to the booms and slumps of a laissez-faire capitalist economy, and the life of a domestic servant, although subject to long hours and little freedom, offered steady employment in an uncertain world.

Large households could afford a retinue of specialized servants, from butlers, cooks, and maids to grooms, coachmen, and gardeners. The British upper classes also had full-time nannies to bring up their children until they reached school age. Middle-class suburban households often had a single female maid, the "maid-of-all-work," who either came in on a daily basis or "lived-in." A huge number of women worked in this way. In 1881, it was estimated that the United Kingdom employed 1,230,406 female domestic servants. This number grew slowly to 1,330,783 in 1901 and 1,359,359 in 1911. These figures indicate a slowing down in the expansion of servants during the early years of the twentieth century and a decline in their numbers as a percentage of a rising population. In 1911, they represented 3.6 percent of the population as opposed to 6.1 percent in 1861.

The life of a maid-of-all-work was generally very hard. Even in 1861, Mrs. Beeton had commented in her *Book of Household Management* that the general servant was "perhaps the only one of her class deserving of commiseration: her life is a solitary one, and, in some places, her work is never done . . . the mistress's commands are the measure of the maid-of-all work's duties."

Those duties in a modest suburban house were generally as follows:

Daily Duties:

Dust and arrange the sitting rooms and bedrooms, hall, and passages

Attend to the grates

Sweep the carpets and stairs

Wash the front doorsteps

Answer all bells

Lay the cloth and serve all meals

Wash up china, glass, and plate on use

A housemaid sweeps a kitchen floor in 1937 (Hulton-Deutsch Collection/Corbis)

Attend to lamps, candlesticks, and lights generally
Clean ladies' boots and shoes
Weekly Duties:
Thoroughly turn out and clean each room
Polish furniture and stained boards
Clean and polish plate and brasses
Prepare soiled linen for the wash; air and distribute fresh linen
Occasional Work:
Wash brooms and brushes
Take up stair carpets and wash stairs beneath
Clean out the pantry

Given that the work was unpopular and that women voted with their feet if given other opportunities, it is not surprising that the employers of the early twentieth century saw the getting of good servants as a real problem. This "servant problem" was hardly new, as employers had long had trouble keeping reliable servants. The 1904 *Book of the Home,* aimed at the middle classes, summed up the problem:

> The public depreciation of service is growing apace, born of unhealthy intentions and a feverish desire to ape the manners of the class just above . . . in America it is notorious that no free citizen would willingly undertake what he calls menial service . . . so great is the demand now for decent servants across the water, that Americans on their travels often engage English, Swiss and French girls to take back with them to the States.
>
> The increasing difficulty of getting good servants in England has led to an importation of Swiss, German and Belgian maids.

In fact, few houses employed servants from overseas and simply had to adjust routines and activities to the fact that the servant population was declining and that they could make do without them. Female servants were aspiring to a better way of life; they often wanted to leave and marry, or work in the increasing number of clerical posts that were opening up to them.

World War I delivered an economic and social blow to domestic service. The necessity of war work opened up many new opportunities for women. Many who previously had little alternative to domestic service were able to become factory workers, nurses, bus conductors, and the like, usually with the double bonus of more freedom and better wages. The war reduced female servants by around a quarter.

Many of these opportunities closed after the peace, but others opened up, especially in the electrical and light engineering fields. Some servants returned to service but many did not. By 1931, only 5 percent of the population employed resident domestic servants. This figure was to fall even more dramatically after World War II, which accelerated many of the social changes begun in the first one. By 1951, only 1 percent of households had a resident servant. The idea of the "servantless household" was an incentive for domestic appliance manufacturer's to market themselves at the middle-class housewife of the 1930s and 1940s. It is also possible that improvements in controls and the introduction of easier to clean enameled panels on cookers were made because the manufacturers realized that the "lady of the house" was now using it rather than the maid. Certainly the Regulo thermostatic control of gas cookers (1923) allowed the cook to leave food unattended for longer. A 1930s advertisement for New World cookers illustrated a young housewife making a bed and carried the copy "You can forget the dinner . . . it's cooking itself with the 'Regulo' control." The servantless house was reflected in the Ideal Home Shows of the 1920s and 1930s and also in the design of suburban houses, which had smaller kitchens and fewer bedrooms, but always a bathroom and hot water supply.

If servants were "human domestic appliances" for cleaning, washing, cooking, and lighting fires, what was their relationship with the appliances that were growing at the same time that they were declining? It is tempting to think that the servant classes would have welcomed such things as vacuum cleaners and washing machines with open arms. They worked long hours and certainly must have developed effective short cuts in some areas. They also moved from post to post quite frequently so could have picked up new ideas and techniques. Despite this, the evidence suggests that the only area where servants innovated was in the field of cookery, but not in other areas of domestic technology. There are no appliances whose invention can be attributed to servants, and there is no evidence of them campaigning for the introduction of piped water or labor-saving devices.

This was probably due to a combination of poor education and ignorance of what was available. In fact, there was hostility to the introduction of labor-saving devices that were seen as a threat to jobs. In 1923, a British Ministry of Labour committee investigating the condition of female domestic servants reported that there was "a curious and quite unreasoning hostility among maids themselves to the use of such appliances, due presumably to that conservatism of which the British race is not infrequently accused." The date of the report is significant because servant numbers were falling and the committee had suggested that employers purchase better equipment in order to attract and retain staff. The real reason for the continuing decline in servant numbers was a combination of economic and social factors: simply, there were better-paid jobs elsewhere that offered more opportunity and improved social standing.

Up to the mid-1920s there is no direct relationship between the decline of servants and innovation in domestic appliance and house design. By the 1930s, the servantless household was becoming the norm, and manufacturers were beginning to realize that the market was changing. It is important to note that many of the early applications for small electric motors were developed in the United States, which did not have such a strong tradition of employing servants. Many Americans lived in apartments in densely populated Eastern cities or on self-sufficient farms and in small towns. The ratio of servants and households was small compared to that of the United Kingdom. The number of people employed as servants fell from 1,851,000 in 1910 to 1,411,000 in 1920, at a time when the number of households was rising. The growth of the U.S. suburbs continued from the 1920s and rapidly expanded in the 1940s and 1950s, due to rising prosperity. As people had more wealth and leisure, then they chose to invest in appliances. In the United Kingdom, the suburbs had begun to grow again in the 1930s and the new houses were aimed at the servantless.

The concept of the servant working for one master or mistress had gone, but the cultural assumption that domestic chores were women's work remained until the advances of feminism in the 1960s and 1970s, when the time taken to do housework and its social status were questioned. The 1980s and 1990s witnessed a growing service industry that served a number of houses providing a range of services such as laundry work, gardening, swimming-pool cleaning, and house cleaning. In the United States, this type of low-paid work has traditionally been done by each successive wave of newcomers to the country. Today the "cleaning crew" or the "pool crew" are often Mexicans or Hispanics working for a contract cleaning company. This trend has grown through increasing wealth and the numbers of working women. Child-minding has also become a profession that has replaced the concept of the full-time nanny. In Europe, the concept of the au pair, a foreign student living in the family house in return for cleaning and child-care work, has been seen as an alternative.

In affluent Western societies many people have become cash-rich and time-poor and find it more convenient to pay others to do their chores. In 2000, the Unilever Company proposed an on-line venture to provide domestic help via the Internet. The employer may be able to find a person to do the work by electronic means, but the nature of that work is the same. Domestic service remains low-paid, low-status work, but has been modernized and contractualized.

References:

Cowan, Ruth Schwartz. "The Industrial Revolution in the Home." In *Material Culture Studies in America,* ed. Thomas J. Schlereth. Nashville, TN: American Association for State and Local History, 1984.

Davidson, Caroline. *A Woman's Work Is Never Done: A History of Housework in the British Isles, 1650–1950.* London: Chatto & Windus, 1982.

Davidson, H. C., ed. *The Book of the Home.* London: Gresham Publishing, 1904.

Gathorne-Hardy, Jonathan. *The Rise and Fall of the British Nanny.* London: Arrow Books, 1974.

Horn, Pamela. *The Rise and Fall of the Victorian Servant.* Gloucester, UK: Alan Sutton, 1986.

Sewing Machines

The mechanization of sewing was a natural progression from the mechanization of yarn and cloth production. Hence, the sewing machine is almost wholly a nineteenth-century household innovation. However, the social context of the sewing machine in the home changed in the twentieth century. The sewing machine was conceived primarily as a professional tool, providing the means of increasing productivity eight-fold. This had an impact on the home because many dressmakers and tailors worked at home. The amateur market was limited in the nineteenth century at the one end by lack of financial means and at the other by snobbery. Accordingly, the first half of the twentieth century was the period when the sewing machine became a truly widespread domestic appliance. More recently, as ready-made clothes have declined in relative cost, the sewing machine has declined in popularity.

The main challenge for early sewing machine inventors was to devise a successful method of continuous stitching. Unlike hand sewing, with machine sewing there was no

Daguerreotype of a woman working at a sewing machine, c.1853 (Library of Congress)

way that the whole needle could pass through the cloth. The first sewing machines, beginning with Thomas Saint's British patent of 1790, were experimental rather than commercial. In 1830, Bartholomey Thimmonier, a French tailor, devised a chain-stitch machine, operated by a foot treadle, that inserted a hook-type needle through the cloth to pull up loops of thread from an underlying thread holder. Two features of this machine—the horizontal cloth plate and presser foot—appear in modern machines, but there was no mechanism for feeding cloth. Thimmonier's sewing machine was the first to enjoy commercial success. However, although there were further patents for chain-stitch machines, lockstitching soon surpassed chain-stitching.

The lockstitch was the invention of the American Walter Hunt, who only made one machine and never patented it. Hunt's machine of 1834 used two threads to form the stitch, with the upper thread held by an eye-pointed needle (a needle with the eye in the pointed end rather than the blunt end) and the lower thread held in a shuttle that passed through the loop formed by the upper thread. Ten years later, John Fisher and James Gibbons obtained a British patent for a sewing machine using the eye-pointed needle and shuttle action but for decorative, rather than straight, stitching. The first patent for a lockstitch machine for straight seams was granted to the American Elias Howe in 1846. The main design weakness of Howe's machine was that the cloth was held vertically and was only fed in short runs. Over the next ten years, a number of rival lockstitch machines were marketed. In 1854, Howe successfully sued for patent infringement, which meant that he could charge royalties for licensing his design.

One of the American manufacturers affected by the 1854 judgment was Isaac Merritt Singer, the man who made the greatest single contribution to the mass-marketing of the sewing machine. Singer's first patent of 1851 introduced several design improvements, most notably a straight vertical needle (rather than the earlier curved needles) allowing horizontal placement of the cloth and a grooved feed. Like Thimmonier's machine, it was operated by a foot treadle rather than a hand wheel so that both hands were free to guide the cloth. I. M. Singer and Company used extensive advertising and introduced rent-purchase in 1856. It also targeted the household market by developing a range of hand- or treadle-operated, lightweight Family sewing machines, employing door-to-door selling, and offering part exchange and after sales service. By the late 1860s, over 100,000 Singer sewing machines has been sold in the United States, and by 1890, Singer had sold more than 9 million sewing machines worldwide.

Other manufacturers were also refining and improving the design and technical performance of sewing machines. In the United States, Allen Wilson and Nathaniel Wheeler formed a company in 1853 and launched the rotary-hook lockstitch machine. This machine, like current sewing machines, featured a stationary disc bobbin to hold the lower thread, and a rotary hook around the bobbin caught the thread from the needle and fed it under the bobbin thread. The replacement of the moving shuttle with the stationary bobbin created a quieter and more mechanically efficient machine. Another of Wilson's innovations still in use today is the four-motion feed, a set of rotating metal teeth, patented in 1854.

Wheeler and Wilson became the dominant U.S. manufacturer of the 1850s and 1860s because its compact sewing machine was more suited to home use than the Singer. It was less successful than Singer in the export market and was eventually acquired by Singer in 1905. The first portable sewing machine was designed by the U.S. company Grover and Baker in 1856. Between 1856 and 1877, the four major American patent holders—Howe, Singer, Wheeler and Wilcox, and Grover and Baker—protected their collective interests by forming the Singer Machine Combination, which coordinated and controlled the licensing of their patents. For example, the most successful British sewing machine company, Jones, entered the industry in 1859, with a license to manufacture sewing machines based on the Howe and Wilson patents. From 1869, Jones began to establish its own patents. One company that avoided paying any royalties to the Combination was Willcox and Gibbs. Their simple chain-stitch machine of 1857 was half the price of lockstitch machines, making it an attractive option for home use.

While there was still room for improving technical performance, for example by better tension control, manufacturers began to pay more attention to extending the range of functions. By 1865, attachments such as the hemmer, binder, and braiding foot were standard. The first sewing machine capable of reverse stitching appeared in 1861, but this feature was uncommon until the 1920s. Zigzag stitching, as required for buttonholing, was conceived in the 1850s, but except in specialist buttonholing machines for industrial use, it did not become common until the late twentieth century.

Today's electronic swing-needle sewing machines are routinely capable of straight stitching and a range of zigzag stitches, including buttonholing. Small versions of special sewing machines for overlocking raw edges are retailed commercially. The sewing machine market is now dominated by Japanese companies such as Brother, which took over the British manufacturer Jones in 1968, and Toyota. The chain-stitch machine has reemerged in the form of small, battery-operated, hand-held models, designed for doing in situ emergency repairs, for example to curtains.

Hand-operated sewing machines were initially more popular for home use because they were cheaper and easier to accommodate. From the late nineteenth century, treadle-operated machines grew in popularity. They became more useful and more harmonious with the home with the introduction of the drop-down design, whereby the baseplate of the sewing machine was hinged so it could be hidden beneath a table top, and the incorporation of drawers for storage. The application of electricity to the sewing machine for industrial use was tested as early as 1867 in France. However, the electrification of the sewing machine for domestic use only became feasible after the development of the small AC (alternating current) motor by Nikola Tesla in 1888. In the following year, Singer adapted a treadle sewing machine to be driven by an electric motor. The treadle remained functional, acting as both an on-off switch and a speed control. For the next thirty years, the electric motor was an optional extra because the market did not justify the development of specifically electric models. In 1921, Singer introduced a portable, electric domestic model. The full treadle was replaced by a small foot pedal, although some early electric sewing machines featured the alternative of a curved bar to be operated by the knee.

The approach to the aesthetic design of sewing machines has come full circle. As industrial tools, early sewing machines were plain and functional, relying on performance as the main selling point. As makers strove to increase sales by targeting the household market, more attention was given to styling, mainly through surface ornamentation, as the basic shape of the sewing machine was functionally determined. The metal sewing machine body was embellished with gold lacquering or mother of pearl inlays or japanning, while the wood cases and stands might feature decorative veneering. Marble bases were attractive and gave small hand-operated sewing machines greater stability. This emphasis on appearance declined early in the twentieth century and styling has been minimal since the introduction of plastic casings in the 1960s.

See also Electricity

References:

Godfrey, Frank P. *An International History of the Sewing Machine.* London: Robert Hale, 1982.

Jewell, F. Brian. *Veteran Sewing Machines: A Collectors' Guide.* Newton Abbot, UK: David & Charles, 1975.

Showers

The ancient Greeks and Romans believed that bathing was essential to physical regeneration and well-being. The cleansing effect was secondary. Bathing was a communal activity, encouraged by the provision of public baths. Two types of bath were distinguished: the plunge bath involved the immersion of the body in a pool of water, while the shower bath involved water cascading over the body from overhead outlets. In Western Europe,

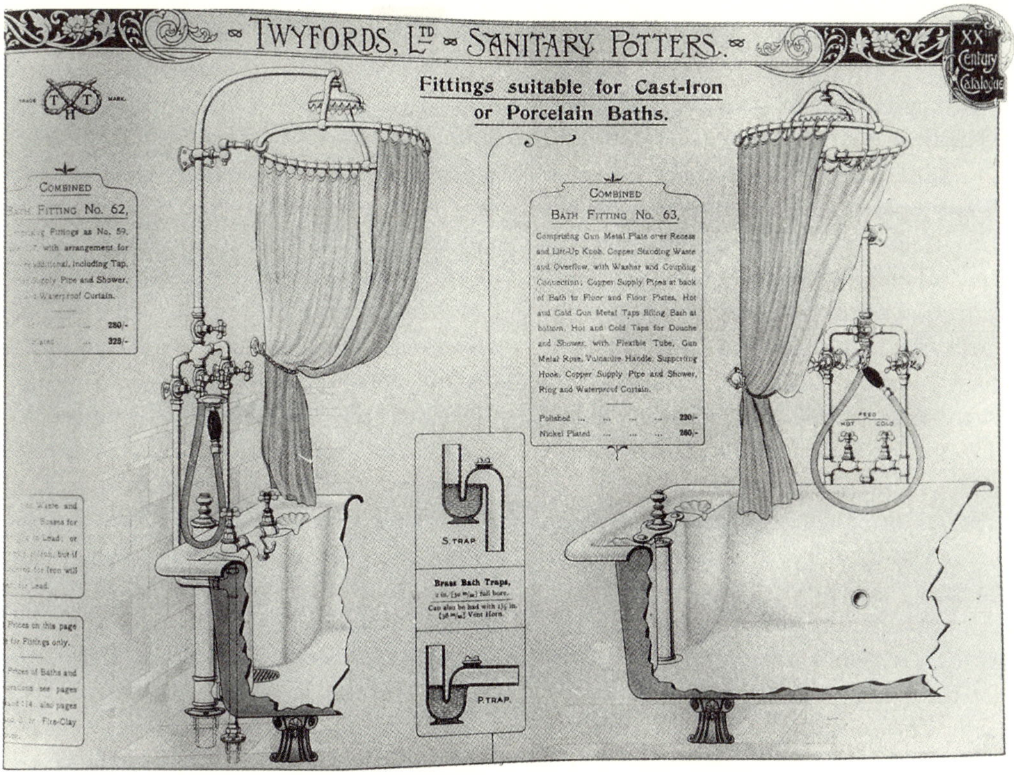

Shower fittings from the 1901 Twyfords catalog (The Museum of Science and Industry in Manchester)

the ritual of bathing faded with the end of the Roman Empire. Bathing for cleaning purposes became an infrequent activity and took the form of the plunge bath. There was a revival in showering as therapeutic activity in the mid-nineteenth century, but the popularity of the shower in the late twentieth century hinged on its convenience value as a quick and effective way of washing.

The earliest patent for a shower was granted to the Englishman William Feetham, in 1767. However, as even wealthy homes had only limited plumbing at this time and regular bathing was still an alien notion, showers did not take hold. Bathing as a health measure, known as "taking the cure," at spa towns or seaside towns became fashionable in the eighteenth century, and the modern concept of hydropathy emerged in the early nineteenth century. Vincent Priessnitz, a Silesian peasant, became convinced of the healing powers of water through successfully treating his own sprains and injuries with cold compresses. He opened a clinic offering water cure treatments in 1829. Hydropathic clinics soon became widespread. Cold water shower treatments could be gentle or intense. The gentle type, the rain bath, involved nothing more than the patient receiving an all-over shower of water from a rose-shaped shower head. The douche bath, on the other hand, involved the localized application of a high-pressure jet of water to the injured or aching part of the body.

As an appreciation of the value of cleanliness increased, a range of showers became available. Because water shortages were common throughout the nineteenth century, water economy was an advantage of the shower over the bath. The rose-type showerhead, which could be fed from an overhead tank or by pressurized piped water, is simple and effective. An American design, patented in 1879 by Warren Wasson and Charles Harris of Carson City, Nevada, featured a rocker-type foot pump that circulated a fixed supply of water from the shower trough through a long, flexible hose with a rose-type shower head. For homes without piped water, the most convenient type of shower was the tank with perforations in the base, which released water when a handle was pulled. Another option was the rubber shower hose that was connected to the taps. For example, Dr. Melcher's Shower Yoke of 1902 was a perforated rubber pipe that rested on the bather's shoulders. The most elaborate type of shower was the needle shower. This consisted of a semicylindrical frame of perforated horizontal pipes supported by metal uprights, usually brass or copper. A German needle shower was designed as early as 1832. From the 1870s, the most luxurious type of shower fitting was the hooded bath with its built-in shower cubicle, a forerunner of today's multifunction baths.

By the end of the nineteenth century, the expansion of piped water services in urban areas and the availability of efficient water heaters meant greater convenience for regular bathing. Initially, most houses were provided with only one cold-water tap, usually in the kitchen or the scullery. Many houses did not have bathrooms, which were introduced slowly in larger houses, often by the conversion of an existing bedroom. This encouraged shower manufacturers to tailor their products to a wider market. For example, simpler versions of needle showers appeared, such as the Kennedy Needle Shower of 1916, which consisted of four jets on one horizontal chromed pipe connected to the taps. After 1900 most of the new houses built in London had upstairs bathrooms, and all the suburban developments of the 1920s and 1930s had bathrooms and plumbing for hot and cold water. The water was heated by either gas water heaters or a combination kitchen range.

Despite the relatively simple technology of showers, they were not a standard fitting in British bathrooms from the 1920s to the 1970s. The large bath, whether made of cast iron, pressed steel, or molded plastic, remained the primary bathroom fitting. Older houses, which often were converted into flats and bedsits (studio apartments), continued to have only the most rudimentary plumbing and shared bathrooms. The shower was far more popular elsewhere in Europe, which had a longer and more advanced system of apartment living, and in the United States. In Britain, showers started to become more popular in the 1960s through a combination of technical, social, and cultural factors. By then, almost all houses had a reliable piped water supply and there had been improvements in instantaneous water heaters and boilers, the latter being linked to central heating systems. Compact electric shower units were ideal for smaller homes and homes without instantaneous water heating for general use. Socially, British people had begun to travel more and enjoyed showers on holiday in the Mediterranean. Constant piped water encouraged more frequent washing, which, combined with changing lifestyles, cultivated the habit of the quick daily shower rather than the leisurely twice-weekly bath.

Showers were installed throughout the 1970s and 1980s, linked to rising standards of cleanliness and increased consciousness of the need to conserve energy and water resources. In the 1980s, the main shower companies were Mira, Triton, and Aqualisa, which introduced the "power shower." Incorporating a pump to increase the force of water, the power shower was designed to massage and invigorate the body of the user through an adjustable water force, providing the perfect bathing accessory for the thriving exercise culture. The 1980s also brought designer showers with the Uno series of shower heads by Frogdesign for the German manufacturer Hansrohe.

Showers are now more popular than baths as they are quick, refreshing, and economical. Separate shower cubicles are regular fixtures within modern homes, especially those with en suite bathrooms. With daily washing established as a cultural norm, one bathroom is no longer enough for many family homes and the space-efficient shower room is a popular solution.

See also Jacuzzi; Piped Water; Water Heaters

References:

Muthesius, Stefan. *The English Terraced House.* New Haven, CT, and London: Yale University Press, 1982.

Wright, Lawrence. *Clean and Decent: The Fascinating History of the Bathroom and the Water Closet.* London: Routledge, 1980.

Sinclair Radionics and Sinclair Research

The British inventor and entrepreneur Clive Sinclair founded his first company, Sinclair Radionics, in 1961, three years after leaving school. He went on to found a succession of other companies producing consumer electronic products and computers, of which Sinclair Radionics and Sinclair Research were the longest-lived. Knighted in 1983, Sir Clive Sinclair was more successful as an inventor than as an entrepreneur, hence the prevailing public perception of him as an eccentric boffin.

Sinclair's interest in electronic engineering was initially expressed through editing electronics magazines and writing articles and books. Sinclair Radionics stalled early on through lack of financial backing, a recurrent theme in Sinclair's businesses. Its first product, marketed in 1962, was a micro-amplifier, followed in 1963 by the Slimline pocket transistor radio. Amplifiers and radios formed the core of the business for the first ten years, but Sinclair also developed the Microvision miniature television in 1966, and the System 2000 hi-fi system in 1968. Sinclair's first major success came in 1972, when he launched the Sinclair Executive pocket calculator, the world's first pocket calculator, which weighed only 2.5 oz (71 g). The Executive calculator contained a Texas Instruments GLS 1802 integrated circuit, 22 transistors, and an 8-digit light-emitting diode (LED) display. Its designer, Richard Torrens, was awarded the Design Council Award for Electronics in 1973. However, its technical weaknesses were that it performed only four arithmetic functions, was heavy on battery power consumption, and overheated dangerously if left on continuously. Moreover, retailing at almost £100, it was too expensive for the average consumer. Nevertheless, the Executive calculator netted £1.8 million in profits for the company, enabling Sinclair to develop models that both were cheaper and performed a wider range of calculations.

Sinclair also continued to develop other electronic products, such as the Project 80 hi-fi system of 1973. Only four years after the financial success of the Executive calculator came Sinclair's first financial disaster. He launched a digital watch, the Black Watch, in 1976 that resulted in losses of £335,000. The company's liquidity was restored in 1977 by the sale of a 43 percent stake and the withdrawal of Black Watch production. New products included the TV1A pocket television and the Wrist Calculator, but profits were hit by increasing competition in the calculator market from U.S. and Japanese companies, resulting in losses of £2 million in 1978.

Sinclair turned to the new field of home computers. He developed his first microcomputer in 1978 and set up Science of Cambridge in partnership with Chris Curry to market the Mk. 14 microcomputer that came only as a kit. Sinclair and Curry parted in 1979, with Curry going on to found Acorn Computers and Sinclair forming Sinclair Research. Sinclair's first commercial microcomputer was the ZX80, which was available both as a kit and pre-assembled. At £100, it was the cheapest microcomputer on the market and an immediate success. The next model, the ZX81, was cheaper and had a more extensive programming language. Sinclair's triumph was the Spectrum computer, launched in 1982, which featured color graphics and, for the time, generous memory, but still sold at less than £200. In spite of technical problems and mail-order delivery delays, it became the top-selling home microcomputer in Britain. Sales of 300,000 units in 1982 meant revenues of £30 million, and by Easter 1983, sales were running at 12,000 a week.

Problems began to mount up when the ZX Interface tape-loop and ROM (read-only memory) peripherals proved faulty and unpopular. Matters worsened in 1984, when in an attempt to beat the much-heralded launch of the Apple Macintosh, Sinclair rushed the release of the new QL computer, which was not fully debugged. Nevertheless, annual sales of £77 million and early orders worth £5 million for the QL saw the company valued at £134 million. A dramatic downturn in the home computer market in early 1985 left the company in a precarious position, with its creditors becoming increasingly anxious. Faith in Sir Clive Sinclair's entrepreneurial abilities was further shaken by the launch of the Sinclair C5 electric road vehicle. While the C5 was the product of a separate company, Sinclair Vehicles, the creditors of Sinclair Research saw it as confirmation of Sinclair's flawed commercial vision. Sinclair staved off the creditors temporarily by negotiating a £10 million bulk purchase of Spectrum Plus computers by the electrical retail group Dixons. With the income from this sale the company was able to repay most of its £15 million in debts. However, in April 1986, he admitted defeat by selling off his computer production, marketing, and distribution rights to the rival British computer company Amstrad for £5 million.

Sinclair Research remained in business as a research and consultancy service, focusing on mobile phones and wafer-chips. In 1988, Sir Clive Sinclair reentered the computer business by setting up Cambridge Computers to produce the Z88 notebook computer, but a majority interest was sold off in 1991. Returning to his interest in electric road vehicles, he began to develop the lightweight Zike electric bicycle for subcontracted manufacture. The Zike failed to recoup its research and development costs of £500,000. Since then, Sinclair Research has continued to trade at a loss and has had modest sales success with products such as the Zeta, a bolt-on unit to convert any bike to electric power, and the X1 Button Radio, billed as the world's smallest radio and fitting inside the ear.

See also Calculators; Computers; Design Council; Radio; Television

References:

Adamson, Ian, and Richard Kennedy. *Sinclair and the Sunrise Technology*. Harmondsworth, U.K: Penguin Books, 1986.

www.nvg.ntnu.no/sinclair/

Singer Corporation

The name Singer is almost synonymous with that of the domestic sewing machine. Isaac Merritt Singer was born in 1811. He had trained as a mechanic and cabinetmaker before becoming an actor and inventor. He had already patented an excavator and was making wooden printing type when his attention was drawn to the commercial possibilities of improving sewing machines. He collaborated with Orson C. Phelps of Boston to produce such a machine. Phelps Company was manufacturing a type of lockstitch machine designed by John A. Lerow and Sherburne C. Blodgett in 1849.

Singer founded I. M. Singer and Company in 1850 and began producing sewing machines for the commercial market in the same year. His machine, which was patented in 1851, was an improvement on an earlier machine designed by Elias Howe. By 1858, the company had a Broadway headquarters and unit sales of over 3,000 a year with three manufacturing plants in New York City. Commercial machines cost around $100, and in 1856 the company introduced a very early form of rent-purchase scheme.

In 1858, Singer launched the Family, its famous lightweight domestic model, which was also sold on rent-purchase. The Family was not only a highly successful product but also one of the first industrially inspired ap-

Singer electric sewing machine, c. 1945 (The Museum of Science and Industry in Manchester)

pliances designed specifically for the home. Singer machines had a standard design that was to continue into the twentieth century; a metal casing set on a wooden base. Treadle machines sat on a cast-iron frame and were powered by a foot-operated treadle for rapid sewing and a hand crank for finer work. Tabletop models relied on the crank only. The more robust New Family followed in 1865. It assisted the company's expansion into world markets, and a factory was opened in Glasgow, Scotland, in 1867. In 1870, the company adopted its red "S" girl trademark, a scantily clad girl blowing a trumpet in front of a sewing machine entwined by a red "S." By 1880, world unit sales exceeded 500,000.

Singer, one of the first truly multinational companies, expanded in Asia as well as the United States and Western Europe. It established an office in India in 1871 and Singer Thailand in 1888.

Being relatively large, sewing machines were easy to electrify, and Singer produced the first electric model in 1889. The company entered the twentieth century as the Singer Manufacturing Company, with annual global sales in excess of 1 million. The Singer Sewing Machine Company was established to handle sales and distribution. The new 47-story Broadway headquarters was then the tallest building in the world, at 612 feet. Although the core business was sewing machines, the company did begin to branch out into other appliances, producing the Featherweight vacuum cleaner in 1929. Further diversification followed in the 1950s and 1960s, with the acquisition of office-equipment and electronic-systems manufacturers.

Singer continued to innovate with sewing machines, introducing the Athena 2000, the first domestic model with a microchip, in 1975 and the 2010 Touchtronic in 1978, which incorporated a touch-sensitive membrane panel. Microchips continued to be used in the Unlimited 9900 of 1989, a machine that featured an LED message center offering hints to the sewer. The Quantum line of the 1990s continued this increasing sophistication.

Singer has expanded its role as a global company and also produces both large and small domestic appliances. It has a distribution network in over 150 countries and retail outlets in Asia, Latin America, Africa, and the Middle East. It has manufacturing bases in Brazil, Taiwan, and Japan. Nevertheless, after 150 years it remains the world's leading manufacturer of sewing machines.

See also Sewing Machines; Vacuum Cleaners

References:

Jewell, F. Brian. *Veteran Sewing Machines: A Collectors' Guide.* Newton Abbot, UK: David & Charles, 1975.

www.singerco.com

Slicers

The slicing, chopping, and shredding of vegetables are basic requirements for almost every form of cooking. For many, the oldest method, using a knife, remains the fastest and most economical, and a variety of knives and cutters developed long before the twentieth century. The nineteenth century saw numerous hand-operated gadgets appear for such tasks as chopping and slicing vegetables, peeling apples, and cutting oranges for marmalade. Eventually these functions were electrified by the development of mixers, blenders, and food processors, but a number of mechanical appliances continued in use throughout the century.

The mandoline has remained popular throughout Europe as an efficient handheld slicer. Essentially it comprises two fixed blades set into a rectangular frame, one for fine slicing, the other for thick slicing. More sophisticated models have additional blades and stands to support the frame at 45 degrees. A variant is the spin slicer for producing long shreds or spirals of vegetables. It has two blades, one toothed and one straight, set in a plastic "A" frame. The vegetable to be shredded is secured on spikes fixed to a hand-operated slide. As the handle is turned the vegetable is pressed against the blades and shredded. The hand-operated rotary vegetable shredder also remains popular in Europe and is mainly manufactured in Italy and France, either in tinned steel or plastic. The French Mouli-legume is one of the most popular and has three cutting discs.

One problem vegetable has been the onion because of the tears associated with chopping it. A late twentieth-century solution is a hand-operated cutter with a lidded hopper above the rotary blades. The onion is placed into the hopper, and chopped onion falls into a clear plastic container below. An earlier 1970s variant was the "bouncing chopper" that had a stainless steel blade bent to form six edges inside a plastic dome. The blades were connected to a spring-loaded plunger. As it was depressed the blade bounced up and down and rotated to cut the onion or vegetable.

Bread slicers with a guillotine blade were introduced in 1880 (the Danish Raadvad) and are still used in France for slicing crusty, thin loaves (batons). Electric bread and meat slicers were initially developed for restaurants and shops to deal with the fine slicing of meats and volume sandwich production. Domestic versions appeared in the 1960s and 1970s. Some early models were hand operated. Domestic slicers have rotary blades that cut as the bread or meat is presented to them on a slide set at right angles to the blade. Most of the midprice models have plastic bodies. The most stylish contemporary model is the Futura F-10 produced by the German company Graef. Made of stainless steel, it has two blades; one toothed for bread, the other plain for meats. The popularity of bagels has led to the introduction of specialist electric slicers such as the Chef's Choice Bagel Pro. Bagels, croissants, muffins, and rolls are dropped into a hopper to be halved by rapidly reciprocating stainless steel blades.

See also Food Mixers; Food Processors

Slow Cookers/Crock-Pots

Slow cookers, or Crock-Pots, are electric casserole or stew pots. They were introduced in the late 1970s and feature a glass or ceramic crock that sits inside either a metal or a plastic casing and is used as a pot when it is heated by thermostatically controlled elements. Most have two heat settings and are designed to allow food to cook slowly and energy efficiently over a few hours.

A U.S. forerunner was the Bean Pot produced by Naxon Utilities Corporation. They were taken over by the Rival Company (established in 1932) of Kansas City in 1970, which manufactured slow cookers in the 1970s.

A more versatile close relative is the flatter, rectangular multicooker, which can be used to stew, braise, roast, steam, boil, fry, and act as a slow cooker with an accompanying ceramic dish. The temperature is again thermostatically controlled and can range from simmering to 424° F.

See also Cookware

Sony Corporation

The company was founded in 1946 by Masaru Ibuku and Akio Morita as Tokyo Tsushin Kogyo (Tokyo Telecommunications Engineering) and first used the name Sony in 1955 as a brand name for marketing its first transistor radio. In 1958, it was renamed the Sony Corporation. By 1970, Sony was established as one of the world's leading and most innovative manufacturers of consumer electronic products. Sony is particularly known for its miniaturization of a range of electronic goods and for its modern, minimalist approach to design.

Sony entered the consumer electronics market with its launch in 1950 of the G-type tape recorder, the first commercially available Japanese model. A more significant early milestone was the acquisition of a license to produce transistors in 1954, the same year that the world's first transistor radio was manufactured in the United States. A year later, Sony launched its first transistor radio, the TR-55. Sony's great contribution to transistor production was to reduce the failure rate in the production process, thus reducing the costs. Sony's predilection for miniaturization became evident in 1957 when it introduced the world's first pocket-size transistor radio, the TR-63. The still smaller and lighter TR-610 pocket transistor radio of 1958 featured a hinged wire stand and came in a range of colors.

Having mastered radio production, Sony next applied its skills to television sets and video recording. In 1960, it launched the world's first fully transistorized (solid-state) television, the TV8–301. This was a portable model, weighing 6 kg (13 lb), with a rigid carrying handle. The Trinitron color television, with its squarer, flatter screen, was launched in 1968 and has continued to evolve. One of Sony's few commercial failures has been its consumer videocassette recorder format. Sony entered the video market very early with the launch of world's first compact reel-to-reel videotape recorder in 1963. However, this would not qualify as compact by today's standards and was an expensive luxury item for the home consumer. Sony continued to lead the way in video technology, introducing the world's first videocassette recorder in 1971. While its U-Matic videocassette recorders became popular with professional users, the scaled-down Betamax format, launched in 1975 as a would-be mass-market format, lost out to the rival VHS tape format. Similar issues of tape format arose in the consumer camcorder market in the mid-1980s. However, in this case, Sony's Video8 format became more popular than the rival VHS-C format.

Another significant milestone in Sony's development was the creation of the Product Planning Center in 1978. The function of this center was to analyze lifestyle trends and develop matching products. This strategy soon paid off with the launch in 1979 of the Sony TPS-L2 Walkman personal stereo player, a pocket-size cassette player designed for use with headphones. It proved to be a successful format for Sony, spawning a series of personal entertainment devices: the Watchman "Voyager" pocket television in 1982; the D-88 DiscMan compact disc player in 1988; the DAT (Digital Audio Tape) Walkman in 1991; the digital DiscMan in 1999. The MiniDisc format, a recordable two-inch floppy disc launched in 1992, is another Sony innovation, but one that has only achieved low market penetration.

The audio compact disc was developed as a joint venture between Sony and Philips Electronics of the Netherlands. Launched in 1982, the compact disc player became one of the most instantly successful new consumer products ever. Sony made a late entry into the electronic games market with its 1996 PlayStation, but quickly became a market leader in that field too. In 1999, Sony again showed its inventive spark with the introduction of the world's first home entertainment robot, the dog-like AIBO.

See also Camcorders; Compact Disc Players; Electronic Games; Radio; Tape Recorders; Television; Videotape Recorders

References:

Lyons, Nick. *The Sony Vision.* New York: Crown Publishers, 1976.

Morita, Akio, Edwin M. Reingold, and Mitsuko Shimomura. *Made in Japan: Akio Morita and Sony.* New York: Dutton, 1986.

www.world.sony.com/

Steamers

See Cookware

Stoves

See Cookers

Sunbeam

The Sunbeam Corporation began life in 1897 as the Chicago Flexible Shaft Company, founded by John K. Stewart and Thomas J. Clark to manufacture mechanical horse clippers. It later added sheep-shearing equipment. Its first electrical appliance was the Princess electric iron, produced in 1910 as a response to the seasonal sales fluctuations associated with the livestock trade.

The line expanded with toasters in 1922 and coffeemakers in 1929. The great success of the Sunbeam Mixmaster food mixer in 1930 sealed their success. The company continued to develop and greatly benefited from the design skills of the Swedish-born Ivar Jepson, who redesigned the Mixmaster and went on to design a wide range of new products. Jepson was born in 1903 in Sweden. He studied engineering in Heslehom and Berlin, moving to the United States to take up the promise of a job in North Dakota. He stopped off in Chicago and stayed. By 1925, he was working as a draughtsman for the Chicago Flexible Shaft Company and by 1932 was chief engineer. He became vice president of research and development in 1956. Jepson made an important contribution to appliance design. His approach was design- rather than market-led, and he did not respond to the increasing use of market research and corporate management that were being introduced in the late 1950s. He retired in 1963 and died of a heart attack in 1965.

The Chicago Flexible Shaft Company remained profitable during the Depression. It introduced the pop-up toaster and the Coffeemaster and by 1946 was in a prime position to take advantage of postwar economic growth and rising prosperity. Such was the visibility of the Sunbeam brand that the company changed its name to the Sunbeam Corporation in the same year. It also began to introduce personal shavers for men and women.

Sunbeam prospered during the 1950s and 1960s, continuing to introduce products such as an egg cooker (1950), a Jepson-designed electric frying pan (1953), and the first electric blanket (1955). In 1960, Sunbeam acquired the John Oster Manufacturing Corporation, which had been founded in 1924 to manufacture hair clippers for barbers and domestic use. In 1935, Oster expanded the line to include an electric massager, the Oster

Massage. During World War II, the company continued to make motors and hair clippers. After the war it acquired the Stevens Electric Motor Company and refined their drinks mixer into a domestic blender. The Osterizer was introduced in 1946 and followed by the hand-held Osterett in 1948. Sunbeam has continued to use the Oster brand, marketing it as a premium line of appliance. The Sunbeam Corporation now incorporates a number of brands, including Mr. Coffee coffeemakers and Grillmaster barbecues. It has also expanded into the outdoor-leisure market with brands such as Peak outdoor equipment and Campingaz.

See also Food Mixers; Razors; Toasters

References:

Goldberg, Michael J. *Groovy Kitchen Designs for Collectors, 1935–1965.* Atglen, PA: Schiffer Publishing, 1996.

www.sunbeam.com

Sunray Lamps

The warmth of the sun was seen by Northern Europeans as a restorer of health, and many eighteenth- and nineteenth-century doctors advocated a period in Southern France or Italy as a cure. The warmth was seen as beneficial, but the side effects of tanning were not. The popularity of sunbathing grew in the 1920s and 1930s, influenced by fashion leaders like Coco Chanel and Hollywood film stars. A tanned complexion was increasingly seen as a symbol of health and the good life.

"Health lamps," which produced infrared heat, were introduced in the 1920s and 1930s and marketed as a relief to those with muscular or rheumatic pains but without the means and leisure to take a cure. Later sunray lamps were either ultraviolet or capable of delivering both types of heat. Such lamps deliver intense heat, and users need to wear protective goggles and to limit the time their skin is exposed. The lamps helped to build up a tan prior to a holiday.

Variants are the sun-bed or home solarium that allow the sunbather to recline while developing a tan. These were developed for health clubs, but are now marketed for home use, although they are rather large and expensive. Philips is a major manufacturer. The sunbather lies below a curved canopy fitted with six to ten ultraviolet tubes. The more expensive "double" models have twenty tubes, ten fitted above and ten below.

Sunbathing is no longer seen as a symbol of health due to the results of research on skin cancer in the 1980s, although the suntanned look can be achieved through cosmetic fake-tan creams.

Synthetic Fabrics

In the case of natural fibers, intrinsic properties dictate what kind of use particular fibers are most suited for. The key measurable properties include dry and wet strength, thermal performance, and elasticity, but look and feel are also important to consumers. Properties are also the starting point when new artificial fibers are synthesized, although production cost is another important factor. The first synthetic fibers, rayon and viscose, were based on plant fibers and were mainly inspired by the perceived market for a cheaper artificial version of silk. Later synthetic fibers were based wholly on chemical raw materials and were engineered to improve on, rather than merely imitate, the properties of natural fibers. In the home, the easy-care properties of synthetic fabrics meant that they could be washed at lower temperatures and dried quickly.

The British inventor Joseph Wilson Swan developed a form of artificial silk by accident in the course of trying to produce a more durable lamp filament. In the early 1880s, he tried soaking cotton thread in sulfuric acid and created nitrocellulose. At about the same time, a French chemist, Louis-Marie Hilaire Bernigaud, Comte de Chardonnet, was consciously developing a method for producing artificial silk. He soaked mulberry leaves, the food of the silkworm, in nitric and sulfuric acid to create a nitrocellulose pulp from which filaments could be drawn. In 1890, he set up the first factory for manufacturing rayon, as it became known. In 1892, the British chemist Charles F. Cross developed an alternative method for producing rayon, known as the viscose method. The main advantage of this method was that it used a widely available plant matter, wood, as the source of cellulose. It was described as regenerative because the reagents, sodium hydroxide and carbon disulfide, only had the role of breaking down the cellulose into a suitable pulp and the end fiber was pure cellulose. By the 1905, a number of variant methods for producing rayon had been patented. Early rayons had relatively poor wet strength and tended to loose shape after repeated washing. In terms of appearance, rayon fabrics fell some way short of real silk, as they were uniformly shiny, with none of the natural variation of silk. They became more popular after 1920, when a technique for cutting filaments into short fibers emerged, allowing rayon to be blended with natural fibers to produce hybrid yarn.

In 1918, Camille Edward Dreyfus founded the Celanese Corporation in Cumberland, Maryland, to manufacture a new synthetic fiber, cellulose acetate, which was water repellent. Like the other early synthetic fibers, it was based on material of plant origin, and it was not until the late 1930s that the first wholly synthetic fiber, nylon, appeared. The raw materials used to produce nylon were petroleum, natural gas, air, and water. Nylon was one of the results of a concerted research program into synthetic polymers (long-chain molecules) by the U.S. chemical company Du Pont in the 1920s and 1930s. Wallace Hume Carothers, the head of the Du Pont research team, perfected the synthesis of nylon shortly before his death in 1937. The key properties of nylon are its lightness, strength, and elasticity. Nylon is particularly versatile because filament thickness can be varied and its low melting point makes it easy to mold. Commercial production began in early 1938, and it was introduced to the public in the form of toothbrush bristles. In 1940, the first nylon stockings appeared on the market in the United States and were such an instant success that 5 million pairs were sold on the first day. Nylon dries very quickly, so it became popular for easy-wash shirts and bed sheets, although its susceptibility to becoming electrostatically charged was a disadvantage and it also produces clamminess of the skin.

Nylon is one of the few synthetic materials that were not the result of a conscious effort to simulate a natural material. As the dominant natural fiber, cotton was a prime candidate for imitation. In 1941, the British chemists John Rex Whinfield and James T. Dickson developed the synthetic fiber polyester by polymerizing ethylene glycol and terephthalic acid. The manufacturing rights were sold to ICI (Imperial Chemical Industries) in Britain and to Du Pont in the United States. Wartime constraints meant that polyester was not commercially manufactured until World War II had ended. In the early 1950s, Du Pont began to market polyester as Dacron and ICI adopted the brand name Terylene. Like cotton, polyester fabrics are durable and keep their shape well, but they have the added advantages of being lighter in weight than cotton and crease-resistant. However, as with rayon and silk, polyester was considered inferior to cotton in terms of look and feel. Blends of cotton and polyester (polycottons) became a popular compromise, particularly for bed clothes and shirts, as they were easy to wash and dry. Since 1970, polyester has been the leading synthetic fiber in terms of production volume.

The last of the major synthetic fibers, acrylic, or acrylonitrile, was developed by William Hale Church for Du Pont in 1950 and was marketed as Orlon from 1951. Acrylic fibers were developed as a cheaper substitute for wool and have mainly been used for the same kind of end products as wool. Like wool, acrylic fabrics need to washed carefully at lower temperatures in order to reduce stretching, but like nylon, acrylic fabrics are also fast-drying. Other well-known brand names for acrylic fibers are Courtelle, Acrilan, and Dralon. Later synthetic textiles include Moygashel (an artificial linen), Lurex (a plastic-coated aluminum thread), and Gortex (a breathable fabric used for waterproof clothing), but the most significant is Lycra, introduced by Du Pont in 1959. Generically described as spandex or elastomeric fiber, Lycra is a highly elastic filament, which is blended in low proportions with other fibers to make warm, stretchy fabrics suitable for underwear and sports clothes.

See also Plastics; Washing Machines

T

Tape Recorders

The audiotape recorder makes and plays back recordings on magnetic tape. In the recording process, the tape is wound past an electromagnet that acts as a recording head. The sound is transmitted to the recording head as electric signals. When current flows through the recording head, magnetic crystals in the tape are scattered, producing a pattern corresponding to the depth and frequency of the signals. Stereophonic sound is recorded as two parallel tracks. During playback, the varying magnetism of the sound track induces weak currents in the playback head. When amplified and fed through loudspeakers, this output reproduces the original sound input. Tape recorders also have an erase head that produces a strong magnetic field to remove any magnetic pattern prior to recording.

The inventor of magnetic sound recording was Valdemar Poulsen, a Danish engineer. In 1898, he developed a machine for recording sound on steel wire or tape. Poulsen's recording machine was named the telegraphone. Like Edison's phonograph, the telegraphone was primarily conceived as a dictating machine. It was another thirty years before tape was used as a recording medium for music. The next major developments took place in Germany. In 1928, Fritz Pfleumer invented a method for applying a magnetic coating to paper or plastic film. Two German companies, BASF and AEG-Telefunken, collaborated on the commercial development of this process. Their magnetophone, which used 12 mm (½ in) tapes of plastic coated with magnetic ferrous oxide, was first demonstrated in Berlin in 1935, but the new technology had little impact outside Germany.

Interest in tape recording intensified in the late 1940s. An American Signal Corps captain, John Mullin, came across magnetophones at a radio station in Germany and brought two machines plus reels of tape back to the United States. A magnetophone demonstration at the National Broadcasting Corporation (NBC)/American Broadcasting Corporation (ABC) studios in Hollywood aroused the interest of Bing Crosby, who wanted the freedom to prerecord his radio programs instead of performing live. Prerecording radio programs using gramophone disc technology was costly and inefficient. Fortuitously, the Minnesota Mining and Manufacturing Company (3M) had been working on the development of magnetic tape since 1944 for the U.S. Navy. The company that undertook the task of building the

BSR Sound Riviera portable reel-to-reel tape recorder, c. 1960 (The Museum of Science and Industry in Manchester)

tape recorders was the Ampex Corporation, founded by the Russian emigrant Alexander M. Poniatoff and based in California. In 1948, Ampex delivered its first Model 200 tape recorders to ABC. The recorders took large reels of 3M Scotch 111 acetate tape. A full reel was placed on one spindle and the end of the tape was threaded through guides in front of the tape heads and onto an empty reel on another spindle. The tape was transferred from reel to reel as the recording progressed.

By the mid-1950s, portable reel-to-reel tape recorders were available. A hinged lid protected the top-loading tape deck. Although portable, these machines were bulky and heavy. They were only found in a minority of homes, partly because of expense, but also because few people had record players of sufficient quality to make recording worthwhile. Tape recording became a more attractive option in the 1960s, when hi-fi record players became more widespread and a new compact tape format was launched. In 1963, the Dutch company Philips introduced its compact audiocassette, a 10 cm (4 in) long plastic case contained a spool of 4 mm (1/8 in) polyester tape, and the complementary EL3300 cassette recorder. This battery-operated tape recorder was genuinely portable, a major selling point. While the pocket transistor radio had provided the means to listen to music whilst on the move, the portable cassette recorder meant that the consumer could listen to music of their own choice.

A disadvantage of the audiocassette in the early years was that the sound quality, even of commercial prerecorded tapes, was inferior to that of gramophone records because there was more surface noise, also described as background hiss. Cassette sound quality was improved when Dolby Laboratories, a company formed by American physicist Ray Dolby in London, introduced its B-type noise reduction system in 1968. From 1970, the Dolby noise reduction system was used in the production of prerecorded cassettes and incorporated in cassette recorders. Cassette recorders became available as hi-fi cassette deck modules, designed to form part of an integrated stacking sound system. Emphasis on sound quality also influenced tape technology. In addition to standard ferrous oxide tape, now commonly referred to as normal tape, tape manufacturers marketed more expensive tape types, including chromium dioxide and metal, which offered a better frequency response. Meanwhile, front-loading reel-to-reel tape decks, which offered superior sound quality, were pitched at the hi-fi enthusiast.

In 1979, the convenience appeal of the audiocassette was boosted by the highly portable Walkman, produced by the Japanese company Sony. The Walkman, a pocket-size stereo cassette player with compact headphones, was an immediate success. The personal stereo, as the product type became known, was a triumph of lifestyle design and marketing, reflecting the highly mobile nature of modern life. Because the personal stereo was used as an accessory and not as a substitute for a cassette deck, the lack of a recording function was no loss to the average consumer. Personal stereos have proliferated since the launch of the Walkman, and Sony alone has sold 50 million. Today, personal stereos may also incorporate radios, and cassette recorder models such as the Sony Pro Walkman are available, although primarily intended for voice recording rather than music recording.

The cassette recorder was also twinned with the radio to create the ghetto blaster, or boom box, another example of the lifestyle product. These names were derived from the ghetto blaster's association with urban black music of the early 1980s, music that demanded to be played at high volume with good bass response. Incorporating stereo loudspeakers, the early ghetto blaster was typically an elongated black plastic box with a large folding top handle. The prevailing design has since been updated to give a softer outline, and the speakers may be fixed or detachable. By the mid-1980s, prerecorded audiocassettes were outselling vinyl discs. This generated a demand for tape-to-tape recording, satisfied by the production of dual cassette decks and dual cassette ghetto blasters.

The instant popularity of the compact disc (CD) player, introduced in 1982, paved the way for the development of digital tape technology. In 1986, Sony and Philips, the same partnership that produced the compact disc, collaborated on the development of the digital audiotape (DAT) format. The main difference between an analog tape recorder and a digital tape recorder is that the digital recorder has rotary heads, like the video head of a videotape recorder, rather than fixed heads. This allows the sound track to be created as a series of diagonal tracks across the tape, greatly increasing the tape's storage capacity for the same playing time. DAT decks came on the market in 1987, and Sony launched the DAT Walkman in 1991. However, while the manufacturers may have assumed that consumers would wish to match their CD players with tape recorders of equivalent sound quality, they seem to have overestimated the elasticity of the market and high-priced DAT recorders have not emulated the commercial success of CD players.

See also Camcorders; MiniDisc Recorders; Philips Electronics; Record Players; Sony Corporation; Videotape Recorders

References:

Rushin, Don. "The History of Magnetic Tape." On www.technicalpress.com/Articles/History/History_tape.htm.

Schoenherr, Steve. "Recording Technology History." On http://ac.acusd.edu/history/recording/notes.html

Tea Makers

Given the British preference for a cup of tea with breakfast, it is hardly surprising that automatic tea makers appeared there. The earliest example was patented by Frank Clarke, a Birmingham gunsmith, in 1902. It operated through springs and levers connected to an alarm clock. When set it would ignite a match by running it across emery paper, thus lighting a spirit lamp that would heat the kettle above it.

The Goblin Company, which was well known for its vacuum cleaners, produced the first electric machine in 1933, the Goblin Teasmade. Designed by Brenner Thornton, it was also linked to an alarm clock but had a special kettle that could be set to boil before the alarm went off. The boiling water decanted into the teapot, which sat on a stand.

Pye "Polly Put the Kettle On" tea maker, 1952 (The Museum of Science and Industry in Manchester)

The weight of the water slightly tipped the teapot, engaging a switch that lit a bedside lamp attached to the machine. Despite folktales of scalded sleepers who had forgotten to replace the teapot, the Teasmade gradually became relatively popular.

Meanwhile, the tea bag was introduced in America. Designed by Thomas Sullivan in New York, its intended use was for sampling tea. It went into commercial production for caterers and had become a popular domestic item by the 1930s. The New York–based Tetley introduced tea bags into Britain in 1953.

Tea bag sales now dominate the market and offer improved blends and convenience, including "one-cup" bags. Despite this, a restyled and electronically more sophisticated Teasmade remains in production today. Swan also produces a model where the kettle and pot are hidden behind a panel that can accommodate a family photograph. The U.S. Chef's Choice Company also produces the TeaMate, an appliance that looks similar to a coffee percolator. Based on the samovar method, it steams the tealeaves and then introduces boiled water to create a concentrate that is blended with the rest of the water.

See also Convenience Foods

Tefal

The French Tefal company is a major manufacturer of cookware and small electric appliances such as toasters, irons, and kettles. It made its name in the cookware market with the first "non-stick" frying pan of 1956 and takes its name from the substance (polytetrafluoroethylene) that gives such pans their properties. It is now a subsidiary of the French SEB group and operates in over 120 countries.

See also Cookware; Irons; Kettles; Teflon; Toasters

Teflon

Teflon was an accidentally discovered in Du Pont's laboratory in New Jersey in 1938. Researcher Roy Plunkett's task was to develop a nontoxic gas for use in refrigerator compressors. Most refrigerators of the time used toxic sulfur dioxide gas as a refrigerant, which was potentially dangerous if a pipe broke or leaked.

On 6 April 1938, an assistant named Jack Rebok opened a cylinder of Freon that Plunkett had prepared, but no gas emerged and both men believed it was empty, the gas having leaked out. They weighed the cylinder and found that it was still full. Curious, they sawed it in half to find out what was wrong. They discovered that the gas, when compressed, had formed into long chains. It was a white, inert substance, and subsequent tests found it to be unaffected by heat, corrosion, electricity, acids, and solvents. It also had the lowest coefficient of friction known at the time, essentially most things slid off it!

Such a material was developed under secrecy as war enveloped Europe. It is also known as PTFE (polytetrafluoroethylene) The U.S. military used it in the construction of the first atomic bomb. The existence of Teflon was announced in 1946. Its properties were of interest to kitchenware manufacturers who saw the possibility of nonstick pots and pans. In the United States, Teflon was being used on some commercial cooking and baking wares, but not for domestic cookware. Although a U.S. invention, it was the French Tefal Company who produced the first Teflon-coated wares in 1956, four years before they appeared in America. The early products suffered from the very properties that made Teflon unique, it was difficult to bond the Teflon coating to the metal body of the pan.

This bonding problem was overcome and nonstick Teflon pots and pans were marketed with considerable success in the 1960s. Tefal coatings were applied to steel and aluminum cooking utensils as well as electric appliances such as sandwich toasters. The nonstick frying pan remains one of the most popular items.

Teflon has also been used in medicine, the NASA space program, and the construction industry, but for most of us, this "miracle material" is best known for helping us cook a better breakfast!

See also Cookware; Tefal

References:

Hillman, David, and David Gibbs. *Century Makers.* London: Seven Dials, 1999.

Telephones

Derived succinctly from the Greek words for far (tele) and sound (phone), the term "telephone" was coined in the early nineteenth century to describe any instrument that could extend the distance over which sound could be sent. Today, the term is more narrowly applied to an instrument that can send and receive sound messages by converting them to pulses of electric current, electromagnetic waves, or light pulses. Invented in 1876 by Alexander Graham Bell, the modern telephone was only used by a minority of businesses and homes by the end of the nineteenth century. Even by the mid-twentieth century, the telephone had still not become the essential tool for personal communication and standard piece of household equipment that it is today.

Early Telephony Experiments

In the late seventeenth century, the English scientist Robert Hooke conducted experiments in transmitting sound along a taut wire and concluded that it might be possible for sound to travel by wire for more than a mile (1.5 kilometers). The first electrical telephone was demonstrated by the German scientist J. Philipp Reis in 1861, but there was no immediate recognition of the commercial potential of telephony. Alexander Graham Bell was a specialist in teaching deaf people to communicate who emigrated with his parents from Scotland to North America in 1870, settling first in Ontario, Canada. In 1872, Bell moved to Boston, Massachusetts, and continued his work in the education of deaf people and speech therapy. Some years earlier, Bell had become interested in the acoustic resonance experiments of the German scientist Hermann Ludwig Ferdinand von Helmholtz. Bell thought that von Helmholtz's work suggested that the telegraph could be adapted to send sound and wanted to devote himself to developing this idea.

Bell was fortunate to secure financial backing to support his experiments from the American financier Gardiner Greene Hubbard, who later became Bell's father-in-law. Hubbard's daughter Mabel was one of Bell's deaf pupils. Thomas Sanders, the father of another pupil, also provided backing. Bell was able to take on a skilled mechanic,

"Future calling": woman inside a modern telephone booth, 1958 (Library of Congress)

Thomas Watson, to assist him. In 1876, Bell succeeded in sending a telephone message from one room to another and quickly filed a patent application, granted later that year. Bell's timing was perfect because a rival Chicago-based inventor, Elisha Gray, had achieved the same breakthrough almost simultaneously and filed his patent application on the same day as Bell, but some hours later.

In Bell's original telephone handset, the mouthpiece (the transmitter) and earpiece (the receiver) were virtually identical, each containing an electromagnet and a metal diaphragm. The complexity of the sound waves produced by the human voice suggested to Bell that accurate reproduction of the sound waves required the transmitting and receiving devices to be closely matched. When sound waves entered the mouthpiece, they cause a thin, flexible diaphragm to vibrate against an electromagnet, which translated the vibrations into pulses of electric current. At the receiving end, the electric pulses traveled down the wire to the more powerful electromagnet in the telephone earpiece, which set up fluctuating magnetic fields, causing the metal diaphragm to vibrate and reproduce the original sound waves. A stronger electromagnet was required for the earpiece because the strength of the electric signal became weaker over distance, unless it was boosted at intermediate points.

Although Bell's earpiece, effectively a small loudspeaker, was satisfactory and became standard, the carbon microphone type of transmitter soon superseded his mouthpiece. Bell had used acidified water, an electrolyte that assists the flow of electric current, to improve the strength of the transmitted signal. The German-born American inventor Emile Berliner improved Bell's design in 1877 by adding an induction coil. Bell acquired the rights to Berliner's patent in 1878. However, Berliner's transmitter was inferior to the carbon transmitter patented by the American inventor Thomas Alva Edison in 1877. In Edison's mouthpiece, which became standard, the resistance of tightly packed carbon granules behind the diaphragm varied according to pressure from the vibrating diaphragm. Greater pressure, corresponding to greater sound volume, reduced the resistance of the carbon, thus increasing the signal strength.

Early Telephone Networks

Bell, Watson, Hubbard, and Sanders set up the National Bell Telephone Company in 1877 to exploit Bell's patent commercially. They faced competition from the powerful Western Union telegraph company, which had acquired the rights to Gray's telephone and Edison's transmitter. The ensuing patent dispute between National Bell and Western Union's American Speaking Telephone Company was settled in 1879, when Western Union ceded its claim. National Bell, renamed American Bell, thereby gained a virtual monopoly on telephone services and equipment manufacture in the United States. In 1882, the Western Electric Manufacturing Company was set up in Boston to manufacture telephones for American Bell. In 1885, a reorganization created clear divisional responsibilities: the American Telephone and Telegraph Company (AT&T) was formed to operate long-distance services and Bell System took responsibility for local services. The expiry of Bell's patents in 1894 paved the way for renewed competition, but, in practice, American Bell's control of the long-distance trunk lines made it virtually impregnable. In 1899, AT&T became the name of the parent company and New York became the corporate headquarters. In 1921, AT&T's effective monopoly was endorsed by the Graham Act, which made telephone services exempt from the Sherman Antitrust Act.

Edison had retained the foreign rights to his transmitter patent and joined with National Bell to form the United Telephone Company in Britain. However, in 1880, the state-owned General Post Office, which had a monopoly of telegraph services, gained licensing control over all telephone services. In 1889, a merger of the competing private companies created the National Telephone Company, with the Post Office taking over long-distance trunk lines. Finally, in 1912, the mixed ownership situation was resolved by the nationalization of telephone services, excluding a few municipally owned services.

A small number of private telephone switchboards were installed in both the United States and Britain in 1877. The first commercial telephone exchange was set up in New Haven, Connecticut, in 1878, while the first British telephone exchange was set up in London in 1879. These early telephone exchanges were manually operated and created new jobs for women as telephone operators. When a call was received at the exchange, the operator would call the intended recipient and connect the two circuit points on the switchboard with a cable. The short burst of current from the exchange activated the ringer in the receiving telephone set. Manual exchanges continued to be used up to the 1950s, but were increasingly replaced by automatic exchanges. The increase in telephone subscribers led to the development of the system of identifying subscribers by a series of numbers. An American undertaker, Almon Brown Strowger, patented an automatic exchange in 1889 and a dial telephone in 1891. He set up the Automatic Electric Company in Kansas City to manufacture automatic telephone equipment. The first Strowger automatic exchange came into operation in the United States in 1897. It used a step-by-step series of on-off switches to direct calls to the correct number. In Britain, the first automatic telephone exchange was installed in 1912.

A strike by telephone operators in 1919 led AT&T to adopt dial telephones, but automatic exchanges only became more widespread in the 1920s, when the levels of telephone use began to warrant it. The number of telephones in the United States doubled from 10 million to 20 million between 1918 and 1929, representing two-thirds of the world's telephones. In Europe, Britain had led the way in telephone use in the early decades, but by 1920 it had been overtaken by the Scandinavian countries and Germany. Long-distance telephone services were slow to develop, even in the United States, because of the problems of maintaining signal strength. The first long-distance trunk line between New York and Chicago came into operation in 1892. Transcontinental telephone services had to wait until the completion of the New York–San Francisco trunk line in 1915. These early trunk lines relied on periodic inductance coils and electromechanical repeaters to boost the signal. Triode valves, which were more effective signal amplifiers, were adopted in the 1920s.

Telephone Design 1876–1950

Early telephones were very utilitarian in appearance, in keeping with their image as a serious communications tool and not a casual toy. The construction and design were inhibited by the fact that early transmitters were sensitive to movement. Therefore, in the 1880s, telephone sets tended to be wall-mounted wooden boxes. The mouthpiece was set into the box and the separate tubular earpiece, connected by a twisted flex, was hooked onto the side when not in use. From the 1890s, the tabletop candlestick phone became more popular. The mouthpiece formed the top of a slim column, with the separate earpiece suspended parallel with the column. Candlestick phones were made from wood and metal until Bakelite became the favored material for telephone bodies in the 1920s. A technically superior, but more cumbersome, tabletop alternative was the skeleton phone, which introduced the convenience of the combined mouthpiece and earpiece handset. The name referred to the fact that the skeleton phone had no shell covering the mechanism. The sturdy cast-iron frame, typically finished with black lacquering and gold painted detailing, formed the magnet bars, and the handset lay across the receiver hooks on top. The Swedish manufacturer Ericsson was the leading maker of skeleton phones.

At first, telephone design was not major consideration for two reasons: first, telephone subscribers bought a service that included a rented telephone, so the phone itself was not regarded as a selling point; second, the service monopolies meant that there was no choice of provider. Moreover, telephone companies were more interested in business users than household users until the 1920s. As the capacity of telephone networks increased, telephone companies began to promote their services more actively to private consumers. For example, in the United States, the introduction of "measured-rate" charges, instead of flat-rate charges, gave subscribers more control over telephone costs. Social use of the telephone suggested the need for a more friendly design. Bell introduced a telephone with a combined handset in 1927. The small rounded footprint and waisted body was attractive, but it was easy to accidentally dislodge the handset.

The first completely molded plastic phone was the Bakelite Neophone of 1929, made by the German company Siemens. Ericsson became particularly noted for its innovative telephone designs. In 1930, Ericsson commissioned the Norwegian engineer Johann Christian Bjerknes and the Norwegian artist Jean Heiberg to develop a handset phone. Heiberg's design, a sturdier version of the Neophone's basic shape, went into production in 1932 as the DHB 1001 model. Shaped loosely like a concave-sided, truncated pyramid, this telephone design was one of the most long-lived and influential. It visibly inspired the Bell 300 series, designed by the U.S. design practice Henry Dreyfuss Associates in 1937. Dreyfuss also designed the Bell 500 series in 1949, which featured a lower, slightly convex-sided telephone.

This early 1980s Ferranti Flip-phone is a variation on the Siemens Grillo phone of 1965 (The Museum of Science and Industry in Manchester)

Expansion of Telephone Networks

The growing number of telephone subscribers in the 1930s placed pressure on the telephone infrastructure. The multiplexing technique, developed for radio transmission, enabled coaxial cable to be used to carry more than one telephone call. The multiplexer allocated a different frequency band to each call. AT&T had held the U.S. rights to the multiplexer system, which was developed in France, since 1912. The need to provide increased capacity for phone calls, particularly in urban areas, was greatest in the United States, which continued to dominate world telephone use, albeit less so from 1930 onward. The level of telephone ownership in the United States grew from 13 percent of the population in 1934 to 23 percent in 1947, whereas in Britain the respective levels were only 5 percent and 9 percent. However, Britain was quicker to appreciate the value of the telephone as a means of improving public emergency services and introduced the 999 emergency number in 1937. In the United States, the equivalent 911 number was only introduced in 1968 in New York and subsequently adopted nationally.

In 1927, Harold S. Black, an AT&T employee, developed a technique called negative feedback for improving the quality of long-distance telephone calls. This entailed the sampling of output signal from amplifiers for comparison with the input signal, followed by the automatic correction of any distortion. By the early 1950s, transistorization was assisting the development of improved automatic exchanges and faster trunk lines. AT&T's "Direct Distance Dialing" service was introduced in Englewood, New Jersey, in 1951, and was available to 90 percent of phone users by 1965. Microwave technology, which had advanced greatly during World War II owing to the military importance of radar, offered a new means of long-distance telephone transmission. AT&T introduced a microwave telephone relay between New York and Chicago in 1950.

In Britain, new telephone technologies were slower to take hold. While the director exchange system permitted direct dialing in major urban areas, the more primitive satellite system persisted elsewhere. The director system assigned an area code to each of the local exchanges so that calls could be automatically routed from one exchange to another. The satellite system used a manual central exchange to route calls between local exchanges. Automatic distance dialing, known as subscriber trunk dialing (STD), was introduced experimentally in Bristol in 1959. It was then adopted in London in 1961 and gradually extended nationwide.

While AT&T had experimented with intercontinental telephone transmissions since 1915, the first full commercial service, between New York and London, did not come into operation until 1927. Shortwave radio was the method used to convey telephone calls across oceans until the mid-1950s, when submarine cable links were developed. Although submarine cables had been used for telegraph transmission for almost a century, voice transmission was more complex because of the booster requirements. Special amplifiers in highly protective housing were created, and a transatlantic telephone service using submarine cable came into use in 1956. Six years later, the launch of Bell Laboratories' Telstar communications satellite introduced another means of long-distance and international telephone transmission.

Deregulation of Telephone Services

While telephone services remained as state or private monopolies, consumers inevitably had little power of influence. In the United States, choice was defined as one of the four basic rights of consumers in 1962. However, it took another twenty years for that principle to take effect in the provision of telephone services. Since 1934, the Federal Communications Commission had regulated telephone services in the United States. A 1949 antitrust action against AT&T, settled in 1956, limited its activities. The first steps toward liberalization of telephony provision came in the late 1960s when the FCC granted limited licenses to two companies. The government's intentions became more explicit in 1974, when the Justice Department indicted AT&T on antitrust grounds. At that time, the government's preferred solution appeared to be the separation of AT&T's equipment manufacturing and network service functions. For a number of reasons, the case was slow to reach the courtroom, and the trial did not begin until 1981. AT&T confounded expectations by agreeing to a negotiated settlement in 1982, ratified by the courts in 1983. While AT&T kept its manufacturing division and retained the provision of long-distance telephone services, it lost the provision of regional and local services. Seven independent regional telephone operating companies, nicknamed the "Baby Bells," came into being.

In Britain, the existence of a state, rather than private, monopoly meant that deregulation was a decision not for the courts, but for the legislature. The Conservative government of 1979 came to power on a sweeping free-enterprise manifesto. The Post Office lost its exclusive role as a supplier of consumer telecommunications equipment in 1980. It was then restructured to create two independent divisions, mail and telecommunications. The telecommunications division was labeled British Telecom and privatized in 1984. Meanwhile, a second telephone service supplier, Mercury Communications, was granted a license.

Digital Telephone Services

In both the United States and Britain, residential customers were inclined to doubt the supposed benefits of deregulation. They felt that competition and revised charges had done more to benefit business customers than themselves. However, the changes did encourage the faster upgrading of the infrastructure technology.

Scientists in both Britain and the United States had begun to study the use of optical fiber cables for voice transmission in the 1960s. Optical-fiber cables transmit sound, video, and data in the form of digital light pulses, emitted by a laser. British researchers at the Standard Telecommunications Laboratories in Harlow were the first to demonstrate the potential of optical fibers for voice transmission in 1966. However, at this time, optical fibers could only be made in small batches in order to guarantee quality control. In 1970, the American glass company Corning invented the process of vapor deposition, which made it possible to guarantee uniform quality along the fibers. Productivity increased again in the 1977, when Japanese optical-fiber manufacturers refined the vapor deposition process to increase the length of continuous fiber that could be drawn out.

Optical-fiber cable has a higher capacity and is a quicker transmission medium than copper. It is also based on a cheap and plentiful material—sand. AT&T created the world's first optical-fiber network in the Chicago business district in 1977. In Britain, British Telecom installed its first optical fiber network in Milton Keynes. Since 1980, optical-fiber cable has become widely used for long-distance and international telecommunications over land and under the sea. The increasing reliance on digital optical-fiber cables for telephone transmission encouraged the replacement of analog telephone exchanges by digital ones, bringing about fully digital telephone services. The final step in the optical-fiber story in the twentieth century was the development of an international standard for the transmission of voice, video, or data. Japan led the way in this field with the announcement of its forthcoming Information Network System (INS) in 1979. The international standard, the Integrated Services Digital Network (ISDN), offered faster transmission rates than standard telephone optical-fiber cables. This became a more significant advantage in the late 1980s as the number of computer-to-computer data transmissions by modem increased.

Telephone Design 1960–1999

The 1950s was the period when telephone design finally began to reflect the social role of the telephone, rather than its business role. In particular, manufacturers began to style telephones to appeal to female users. They were assisted by the fact that the newer plastics could be produced in bold and pastel colors. Ericsson again led the way with its 1949 Ericofon, designed by the Swedish team of Ralph Lysell, Hugo Blomberg, and Gösta Thames. The Ericofon was the first one-piece telephone. It was shaped like a standard curved telephone handset but the mouthpiece end was splayed to form a flat base with a built-in dial on the underside. Available in a range of colors, including bright ones, from the mid-1950s, the Ericofon design was licensed to other companies, including AT&T. The Princess telephone of 1959 was one of AT&T's own designs. Produced in pastel shades, it had a small, oval footprint, low profile, and an illuminated dial, summed up in the advertising slogan "It's little, it's lovely, it lights." In the United States, with its high levels of telephone ownership, thought was also given to the role of the telephone throughout the house. Hence, in the mid-1950s, AT&T reintroduced the wall-mounted phone as the ideal phone for the kitchen.

The introduction of touch-tone keypads in place of dials enhanced the potential for smaller sleeker phones. AT&T previewed the first touch-tone phone at the 1962 World's Fair in Seattle and introduced it commercially in 1964. The combination of touch-tone keypads and the miniaturization of electronic circuits, in the form of the integrated circuit board, made highly compact phones feasible. The term "trimline phone," now commonly used for compact phones, was coined for the design for a two-piece dial phone produced by Henry Dreyfuss for AT&T in 1965. The dial was set in the middle of the handset. However, a more innovative design of the same year was the Grillo phone, designed for Siemens by the German designer Richard Sapper and the Italian designer Marco Zanuso. It was a flat, one-piece, touch-tone phone, with a hinged mouthpiece that made the call reception contact in the closed position. The Grillo phone proved to be more influential in the design of cellular mobile phones than wired phones.

Since the 1960s, compact telephones, with an emphasis on convenience rather than style, have been the dominant design for home use. Novelty phones and retro phones cater to quirkier tastes. Speakerphones, which allow the user to replace the handset and hold a phone conversation with both hands free, were developed with the office in mind. The growing popularity of mobile phones in the 1980s led telephone companies to develop a way of making the wired phone more mobile to increase its convenience in the home. The cordless phone handset incorporates a shortwave radio antenna, which picks up signals from transmitter/receiver "base station." Users can carry the handset around the home with them, instead of having separate phones or phone jacks or using extension lines. Cordless phones suffer from some of the same problems as mobile phones—the need to keep batteries charged, susceptibility to electrical interference, and in analog models, the loss of guaranteed privacy. Aside from convenience, the other main trend of recent years has been the development of combined function phones, incorporating the answering machine and/or fax machine.

See also Answering Machines; AT&T; British Telecom; Ericsson; Fax Machines; Mobile Phones; Modems; Videophones

References:

Fischer, Claude S. *America Calling: A Social History of the Telephone to 1940.* Chicago and London: University of Chicago Press, 1994.

Gardner, Robert, and Dennis Shortelle. *From Talking Drums to the Internet: An Encyclopedia of Communications Technology.* Santa Barbara, CA: ABC-CLIO, 1997.

Lupton, Ellen. *Mechanical Brides: Women and Machines from Home to Office.* New York: Cooper-Hewitt National Museum of Design, Smithsonian Institution, and Princeton Architectural Press, 1993.

Television

Television, the transmission and reception of synchronized sound and pictures via electromagnetic waves or wire, was one of the most influential new technologies of the twentieth century. The word "television" first appeared in the journal *Scientific American* in 1907. Its invention is commonly attributed to the British engineer John Logie Baird, but Baird's mechanical scanning reception method of the 1920s never achieved commercial viability. Electronic scanning, which has formed the basis of television broadcasting systems since the 1930s, was pioneered in the United

Television set, c. 1950 (Bettmann/Corbis)

States by Vladimir Kosma Zworykin and Philo Taylor Farnsworth. These systems produced black and white images; color television, introduced in the 1950s, only became widespread in the late 1960s. Although television was originally conceived as a wireless broadcast medium, today television channels may use wireless (or terrestrial), cable, or satellite transmission technology. The most recent new phenomenon, the integration of television with newer digital communication systems, has made television an even more powerful medium of communication.

Mechanical Television

By 1920, sound transmission by radio and telephone was well established, and still images had been transmitted by early facsimile methods. However, the transmission of moving images was as yet a dormant technology. In 1923, Baird began experimenting with a mechanical scanning system, based on a design patented by the German engineer Paul Nipkow in 1884. Both the transmitter and receiver contained a rotating disk with spiral perforations through which light waves could travel. During transmission, a selenium light detector was placed behind the disk, and as the disk rotated, an image of the picture in front of the disk was built up as lines of light pulses of varying brightness. The photoelectric selenium cell converted the light pulses to electric pulses that could be transmitted as electromagnetic waves. In the television receiver, the current produced light pulses in an electric lamp behind a Nipkow disk, recreating the picture. Although the image is not really continuous, being a series of lines, it appears so because of the phenomenon of persistence of vision (the way that the brain interprets the images formed on the retina of the eye).

Baird termed his receiver a "televisor." He was able to transmit a simple still image to a receiver within the same room by early 1924. After moving to London, Baird was hired to give television demonstrations in Selfridge's department store. With family financial backing, he set up Television Ltd. and improved his technology by using an array of three disks in the transmitter and placing lenses in the perforations of the front disk. In October 1925, his refined system successfully transmitted the live image of a person. A similar demonstration was laid on for members of the Royal Society in January 1926, with the transmitter and receiver in separate rooms. Encouraged by his progress, Baird successfully applied for a license to transmit television signals and began experiments over a distance of 10 miles. In 1927, Baird demonstrated long-distance telecasting by transmitting from London to Glasgow. In the following year, he made the first transatlantic television broadcast from London to a radio station in Hartsdale, New York.

Baird's successes brought him increased financial backing, and he was able to set up a studio in 1928 near the headquarters of the British Broadcasting Corporation (BBC) at

Crystal Palace, in London. He secured a contract with the BBC to make experimental television broadcasts, initially twice weekly for half an hour, using its Crystal Palace transmitter. By this time, the Baird system delivered a 32-line picture at a rate of 12.5 frames per second. Less than a thousand homes in London received these broadcasts because the price of a Baird televisor represented, on average, almost three months wages. Because the televisor only provided picture reception, it had to be used in conjunction with a radio receiver in order to hear the accompanying sound track.

Electronic Television

Meanwhile, in the United States, television pioneers were experimenting with systems inspired by cathode ray tube (CRT) technology. In 1897, the British physicist John Joseph Thomson established that cathodes gave off negative particles, later termed "electrons." In the same year, the German scientist Karl Ferdinand Braun developed the cathode-ray tube. The cathode-ray tube is a glass vacuum tube with an electron gun at one end and a fluorescent screen, a coating of phosphors, at the other end. The electron gun, consisting of an electrically heated cathode and three anodes, fires a beam of high-energy electrons at the screen, where the phosphors glow in reaction to bombardment by electrons. When the electrons rebound against the sides of the tube, which are coated with conducting black graphite paint, the energy is absorbed as heat and fed back to the electron gun. The electron beam is focused either by pairs of vertical and horizontal plates between the electron gun and the screen or by magnetic coils on the outside. Braun used the cathode-ray tube as a laboratory oscilloscope for measuring changes in voltage. Its suitability for the electronic picture scanning was theorized in 1907 by the Russian engineer Boris Rosing and in 1911 by the British engineer Alan Archibald Campbell Swinton. However, neither of these men actually carried out any practical experiments.

Vladimir Kosma Zworykin was a Russian electrical engineer who immigrated to the United States in 1919. He obtained employment at the Westinghouse Research Laboratories and persuaded his employers to support his experimental research on the electronic scanning method of television. In 1923, he applied for a patent for his iconoscope, the key component of his electronic television camera. The iconoscope was positioned so that light entering a camera lens fell on a silvered target plate coated on one side with photosensitive cesium. The screen is continuously scanned by a beam of electrons emitted by an electron gun, producing pulses of electric current. In 1924, he produced the first version of a complementary picture receiver, which he called the kinescope. In the kinescope, the electric television signal was amplified by triodes and passed to the cathode-ray tube, where the picture was recreated on the screen. In 1929, Zworykin became director of electronic research at the Radio Corporation of America (RCA), whose president, David Sarnoff, had been impressed by a demonstration of the Zworykin television system. Westinghouse agreed to this move, as it had become doubtful that the concept would be commercially viable. It took Zworykin more than a decade to perfect his equipment, and he was only finally granted patents in 1938. By that time, the Zworykin system was capable of delivering a 525-line picture at 30 frames per second.

This delay was partly a result of technical problems and was complicated by rival claims from another American inventor, Philo Taylor Farnsworth, who filed a patent application for his Image Dissector in 1927. In the Image Dissector, light entering the camera lens was directed onto a curved photocathode, a glass plate coated with cesium, from which electrons were emitted. The electron beam was focused and sampled by two scanning magnetic coils. The beam traveled toward a metal plate perforated by one tiny hole, through which electrons could pass to reach a second metal plate. Both plates were electrically charged, with the inner plate at a higher voltage, causing pulses of electric current to be transmitted. Farnsworth's Image Dissector was simpler than the iconoscope but was more dependent on the subject being brightly lit, as it generated a feebler signal. It was capable of delivering a 240-line picture at a rate of 30 frames per second.

In Britain, the picture quality of the Baird system was causing concern by 1930. Electrical and Musical Industries (EMI), the company formed in 1931 by the merger of the Gramophone Company and the Columbia Graphophone Company, began to experiment with television. It started off by improving the resolution of mechanical scanning, by using a technique known as "flying spot scanning," which used bright illumination behind the disk. This method ultimately produced a 243-line picture. However, mechanical scanning was soon abandoned because EMI felt that the size of disk needed to recreate a picture of adequate size was too large for home use. Led by Isaac Schoenberg, a Russian engineer who had emigrated to Britain in 1914, the EMI research team turned to electronic scanning, based on Swinton's proposals. In 1933, EMI produced the Emitron television camera, which delivered a 405-line picture at a rate of 50 half-frames per second. Each half-frame consisted of alternate lines of the picture. In 1934, EMI formed a consortium with Marconi (the Marconi-EMI Television Company) in order to take advantage of Marconi's extensive expertise in transmitter construction. Baird Television became concerned about this new rival and entered into a licensing agreement with Farnsworth for use of his Image Dissector.

Early Television Broadcasting

While the embryonic television industry in the United States was beset by patent squabbles, its British counterpart was edging nearer to public television broadcasting. In Britain, the situation was simplified by the fact that a state corporation, the BBC, controlled radio broadcasting. In 1932, the BBC decided to take control of Baird's broadcasts. A parliamentary committee, the Selsdon Committee, was set up in 1934 to investigate the existing systems and recommend standards of service. In 1935, the Selsdon Committee recommended that the BBC should be given charge of television broadcasting and should implement a service with a "high definition" picture quality of at least 240 lines. Initially, the BBC was to invite the Baird Television Company and Marconi-EMI to carry out trial broadcasts.

The world's first regular television broadcasts began in late 1936 from the BBC's new television station at Alexandra Palace in London. Baird and Marconi-EMI transmitted broadcasts on alternate weeks for a four-month period. The trials were not intended for the public but for a selected group of professionals. Nevertheless, television sets capable of receiving both broadcast systems went on sale at prices equivalent to about half the annual wage of a skilled manual or clerical worker or the price of a small car. In February 1937, the BBC decided in favor of the Marconi-EMI 405-line system and began broadcasting for the public. However, only those living within a ten-mile radius of Alexandra Palace were within reception range. By the outbreak of World War II in September 1939, when television broadcasting was suspended, 20,000 television sets had been sold.

In the United States, the 1939 World's Fair in New York, themed "The World of Tomorrow," provided the perfect opportunity for launching public television broadcasts. The General Electric Pavilion offered RCA televisions for purchase for the first time (with prices starting at $200) and a small studio where visitors were televised for viewing on nearby sets. The National Broadcasting Corporation (NBC), an affiliate of RCA, broadcast the opening ceremonies, including an address by President Franklin Delano Roosevelt. The audience consisted of a mere 200 television owners within a 50-mile radius of New York.

As in Britain, World War II led to the cessation of television broadcasting.

Television Blossoms

British television broadcasting resumed in June 1946, and a combined radio and television license fee was introduced. By 1952, there were three television transmitters in operation and 1.5 million television licenses had been issued. A few years later, 80 percent of the population lay within reach of a transmitter. The televising of the coronation in 1953 prompted a surge in sales of television sets, raising the annual total to 1.4 million. It attracted an estimated television audience of 22 million. While rental was one option for those who could not afford to buy a television, television ownership was stimulated by the removal of restrictions on rent-purchase in 1958. Television broadcasting remained exclusively within the control of the BBC, and funded through the license fee, until 1955. The 1954 Television Act set up the Independent Television Authority (ITA) to award franchises for a new channel to be operated by regionally based commercial television stations, which would be self-funding through advertising. The first independent television (ITV) company began broadcasting in London in 1955. The ITA was later renamed, first as the Independent Broadcasting Authority and then as the Independent Television Commission. It authorized two more independent channels, Channel 4 in 1984 and Channel 5 in 1997.

In the United States, the postwar recovery of manufacturing was much more immediate than in Britain, and the resulting economic boom encouraged purchasing of new appliances. By 1950, just under 9 percent of U.S. homes had televisions. As in Britain, the radio networks, such as NBC, ABC (American Broadcasting Corporation), and CBS (Columbia Broadcasting System), had taken the lead in television broadcasting. However, by 1948, the Federal Communications Commission had issued licenses for about 100 television stations and decided to review its policies in order to encourage more structured growth. Television broadcasting in the United States was a commercial operation from the beginning.

Television Sets

The first television sets sold in Britain were tall consoles with a lid, concealing an angled mirror, which formed the viewing screen. The reason for this awkward arrangement was that the current state of CRT focusing technology meant that in order to achieve a large enough and flat enough screen, the tube had to be very long. Technology soon improved enough for the resulting tubes to be mounted horizontally, allowing a vertical screen, by 1938. Before World War II, the largest available screen was 12 inches (30 cm), but cheaper 5-, 7-, and 9-inch (13-, 18-, and 23- cm) screens were more common. To improve ease of viewing, manufacturers provided either built-in or clip-on magnifying screens. The range of models included table televisions and consoles of various sizes, which might incorporate a radio and record player, forming an early version of the home entertainment system.

Wartime disruption of production meant that television sets had changed little by 1950. As had happened with radios in the 1930s, Bakelite became popular as an alternative to wood, because it could be molded into more varied shapes. In Britain, a popular model was the Bush TV12 Bakelite table television. The most common size of screen in the early 1950s was 9 inches (23 cm). As the rounded edges of the screen caused distortion of the picture, in the late 1940s, manufacturers came up with the idea of projection television. The concept was the same as that of the early televisions with mirror screens, but the picture was projected from a special projection tube via an angled mirror onto a translucent, ground glass screen mounted vertically. This produced a sharper, but dimmer, image. Predictably, the price was significantly higher, because of the complexity of construction. Projection televisions died out when larger and flatter picture tubes became available. However, the projection television system has been revived in recent years in a large-screen color format to create a form of "home cinema." The original system with built-in projection has been joined by a rear projection system, with the screen completely separate from the projection unit. The advantage of rear projection is that the distance of the projection unit from the screen can be varied, with a greater distance producing a larger image. By 1955, 17-inch picture tubes were available, followed, three years later, by the 23-inch picture tube.

Color Television

The principles of color television are based on fundamental properties of light and the human eye. While white light can be separated into seven colors of different wavelengths by refraction through a prism, the retina of the human eye only recognizes three colors—red, green, and blue. These primary colors are combined by the brain in appropriate proportions to produce all other possible colors, a process described as color addition. The same process of color addition is performed when electron beams strike colored phosphors, as in tinted fluorescent lamps. An alternative mixing process, known as color subtraction, occurs when white light is shone through colored filters, which absorb the light wavelengths of colors other then their own.

Baird demonstrated a primitive form of color television using mechanical scanning as early as 1928, but he did not pursue this idea further until the early 1940s. His initial system used the simple technique of putting color filters across the perforations in the disk, a type of color mixing described as a sequential color system. In 1944, he demonstrated his new Telechrome tube, a two-color system that used two electron guns whose beams converged on a translucent screen coated on one side with blue-green phosphors and on the other side with red-orange phosphors. As a two-color system, this inevitably rendered a smaller range of colors than a three-color system. However, the main reason why this did not progress further was that Baird died two years later. In 1948, CBS came up with the first commercial color television system, which used a three-color disk in front of a cathode ray tube. This was poorly received by both consumers and regulators because the CBS color television sets and broadcasts were both incompatible with the existing monochrome system.

The FCC decided that it would specify a single set of common standards that all future color television broadcasting must follow. A key requirement was that the color television system must be compatible with the monochrome system. The National Television Standards Committee (NTSC) issued its report in 1952. It was approved by the FCC in 1953, and NTSC color broadcasting began in January 1954. Meanwhile, RCA had been developing a color television set in readiness for the beginning of color broadcasting. The RCA color picture tube had a fluorescent screen formed by a fine mosaic of triangles of red, blue, and green phosphor dots. A triangle of three electron guns, one for each color, fired electrons through a finely perforated "shadowmask," which was aligned to fit the screen mosaic. Large television cabinets were required to accommodate the bulkier tube and the increased array of valves needed to amplify the signal. Color television cameras were also very large until the 1960s, when the development of a broadcast-quality video camera created a more compact alternative. The NTSC color system was also adopted by Canada and Japan, but dimness and color inconsistency were major drawbacks until the early 1970s, when integrated circuitry provided a means of correcting color balance. The problem with NTSC was that the detailed picture content was supplied as the luminance signal, a monochrome rendering of brightness, with phased intermediate pulses carrying a compressed chrominance (color hue) and synchronization signal. The television set reintegrated the signals, but if the color signal experienced any lapses, owing to interference, color quality suffered.

The BBC carried out the first color television trials in Britain in 1955, using the NSTC system modified for the 405-line picture format, and in 1956, using the PAL (phase alternate transmission) color system developed by the German company Telefunken. The PAL system delivered better color because it transmitted the color signal twice, once in reverse, so that any variations could be detected and evened out. However, as PAL was based on a 625-line picture, it was not compatible with the 405-line monochrome system and sets. There was no further progress until 1961 when the Television Advisory Commission recommended the adoption of the 625-line system and PAL color transmission. The new 625-line service was tested on the new BBC second channel (BBC2) in the London area in 1964 and was gradually extended to all channels and the whole country. During this transitional period, manufacturers produced television sets that accommodated both line formats. Color services began on BBC2 in 1967 and were available for all channels from 1969. A higher license fee was introduced for color television users. The first British solid-state color television was the BRC Thorn 2000 television of 1967. PAL became the most widely used system internationally. All Western European countries adopted PAL, except France, which had developed its own system, SECAM (système electronique couleur avec mémoire), in the mid-1950s. In 1989, France agreed to change over to PAL so that all color television broadcasting within the European Community would be compatible.

Solid-State Television Sets

The first transistor radios appeared in the mid-1950s. Transistorizing the television required more development time, as the circuits were more complex. As with radio, it was the Japanese electronics companies that led the way in solid-state television technology. In 1959, Sony introduced its TV8–301 model, which was the world's first solid-state television and the also the first truly portable television, weighing 13 lb (6 kg), with a built-in aerial and rigid handle on top. It also had a radically modern appearance: the screen had no frame other than the case edging and the case was grey steel. As portable televisions fell in price during the 1960s, they became popular as a second television for use in the kitchen or the bedroom. In 1977, the British company Sinclair Electronics introduced the Microvision MTVI miniature television, which contained a tiny 2-inch picture tube. Sinclair went on to produce a pocket television by using an LCD (liquid crystal display) screen, rather then a tube, but it was Sony that commercialized the pocket TV with its Watchman Voyager of 1982, which reintroduced the magnifier to aid viewing. However, the pocket format has not been as successful for television as for radio and personal stereos, because very small pictures are hard on the eye.

It was common for full-size televisions to be part-valve and part-transistorized during the 1960s. By 1960, 75 percent of British homes and almost 90 percent of U.S. homes had television sets. With the market nearing saturation, manufacturers began to put more effort into using design as a selling point. The mood of the 1960s invited futuristic designs. The classic populist example of this trend is the spherical Videosphere television, made by the Japanese company JVC (Japanese Victor Company) and nicknamed "the Sputnik." A more upmarket example is the minimalist ST/201 television, designed for the Italian company BrionVega, by Marco Zanuso and Richard Sapper. It looked like a stark black box when not in use, as there was a smoked glass panel in front of the picture tube.

As picture quality was the critical area of technical performance, manufacturers could reasonably hope to gain a competitive advantage through genuine advances. Launched in 1968, the Sony Trinitron revolutionized color television. Sony redesigned the color formation system in the picture tube, abandoning the triangular (or delta) formation that had become standard in favor of a sequence of vertical phosphor stripes. The picture tube contained a single electron gun that emitted three beams of electrons and a mask pierced by full-length vertical slots. The Trinitron was the progenitor of the flatter, squarer screen that has since dominated television design. It also introduced the black or charcoal plastic shell that has become the dominant form of television housing. In the mid-1970s, the U.S. company General Electric introduced another type of color formation. This was similar to the Trinitron system in that it used the linear phosphor stripe formation, but it used three in-line electron guns and a mask pierced by short vertical slots. This proved to be better for creating line images than continuous images, and thus more suited to the demands of color computer monitors than television. The Danish company Bang & Olufsen has favored slim, large screen televisions since its BeoVision MX 2000 television was introduced in 1984. Aside from projection systems, the ultimate in flat screen technology is the LCD screen. However, LCDs of the size and resolution necessary for full-size televisions are still extremely expensive.

Until quite recently, when people became accustomed to the high quality of digitally recorded sound, sound quality received far less attention than picture quality. In 1990, the BBC introduced television broadcasts with Nicam (near instantaneous companded audio modulation) digital stereo sound. "Near instantaneous" refers to the fact that the stereo tracks are transmitted sequentially, like the phased color picture delivery. "Companded audio modulation" refers to the use of selective compression of the audio signal before transmission and expansion back to the original upon reception, similar to the noise reduction technology used in audiocassette recorders.

New Functions

The idea of using the television screen for text-based information services, known as videotex, emerged in the mid-1970s. In 1976, the BBC and ITV introduced the world's first teletext services, called Ceefax and Oracle, respectively. Teletext was mainly conceived as a medium for news and program information, although, in the case of ITV, it was also another means of advertising. It could only be received by British viewers who had teletext-enabled sets. By 1986, 1.8 million teletext televisions had been sold in Britain. An interactive form of videotex, known as viewdata, involved the telephone as well as the television and was offered as a subscription service. Viewdata services operate from a central computer. In Britain, the Prestel viewdata system, launched in 1979, was a commercial failure. The world's most popular viewdata service, Minitel, was introduced in France in 1981.

Cable and Satellite Television

In the United States, transmission of television broadcasts by cable was introduced in 1948 in order to improve reception in areas that were far away from the nearest transmitter. The number of U.S. cable television users increased from 1 million in 1965 to 11 million in 1977 and 33 million in 1984. Growth was assisted by the adoption in the 1970s of optical fiber cable technology, which increased the number of simultaneous transmissions that were possible. Cable services in the United States consist of several tiers—channels, subscription channels, pay-per-view broadcasts, and interactive broadcasts. In Britain, remoteness was less of a problem, although some cable links were used from 1951. By 1982, 2.5 million homes were cabled, but there were no commercial cable television services until the mid-1980s. The 1983 Cable Act, based on the report of a government-appointed panel, set criteria for commercial cable television services. Eleven licenses were granted, but a number withdrew before getting started and the remainder had very limited success. Deregulation of telephone services in 1984 created the opportunity for companies to offer telephone and cable television using the same cable network, but this only came to fruition in the mid-1990s, largely as a result of developments in digital technologies.

Satellite television broadcasting began soon after the launch of Telstar, the world's first commercial communications satellite, in 1962. However, for twenty years, communications satellites were only used for indirect satellite broadcasts, where a television signal was transmitted from one television station to another by satellite and then to viewers in the normal way. Direct satellite broadcasting (DBS) is the transmission of programs via satellite to a home receiving dish, the equivalent of an aerial. DBS was introduced more or less simultaneously in the United States and Europe. In the United States, the FCC approved DBS in 1982, on the grounds that it would serve the 35 percent of the population that lived in remote rural areas not served by cable. Eight licenses were granted and in 1983, United Satellite Communications Incorporated (USCI) introduced a DBS service offering five channels of programs. When USCI only attracted 1,000 subscribers in its first four months, other licensees backed off. In Britain, a public-private consortium, including the BBC, was authorized to proceed with DBS in 1982, but withdrew in 1985 in the face of mounting costs and uncertain demand. Sky, the satellite television network owned by the Australian media magnate Rupert Murdoch, launched its service in mainland Europe in 1982 and in Britain in 1984. Since renamed BSkyB (British Sky Broadcasting), it remains the major DBS provider. In 1994, the RCA Digital Satellite System introduced digital satellite television broadcasting in the United States, with the programming provided by the recently formed DirecTV company.

High-Definition and Digital Television

Digital technology has also stimulated the development of high-definition television (HDTV). Concepts of HDTV have been redefined as scanning technology has improved, and by the mid-1980s, the television industry was considering HDTV formats that would require about five times more bandwidth than the existing formats, if broadcast by analog transmission methods. Because frequency bands in many countries were already congested, this was a significant issue. Digital compression techniques permit the transmitted digital signal to be reduced to as little as a fortieth of the original digital signal. In 1986, the International Radio Consultative Committee discussed the introduction of a global HDTV standard. Japan and the United States were agreed on a standard with three components—wide-screen (16:9 ratio, rather than 4:3), 1,125-line picture resolution, and a 60-Hz. frame rate. This was rejected by the European countries, which were working on a rival 1,250-line format called HD-MAC (high definition multiplexed analog components). Because the HD-MAC format was partly analog and partly digital, it soon became evident that it was far from satisfactory. In contrast, the Japanese-American standard was developed with digital transmission in mind.

Japan introduced its HiVision HDTV system in 1989 with the backing of the NHK (the Japanese Broadcasting Corporation) and a consortium of electronics companies. In 1992, a fully digital HDTV system was demonstrated in the United States. The Digital HDTV Grand Alliance, an international body formed in 1993 to agree to global standards, adopted the MPEG-2 method of data compression. In late 1996, the FCC allocated channels for HDTV broadcasting. DirecTV began transmitting digital HDTV broadcasts by satellite in early 1998. In Britain and the rest of Europe, the emphasis of digital TV broadcasting was on increasing the number of channels by making better use of frequency. HDTV was a secondary consideration. The world's first commercial digital terrestrial TV service, ONdigital, was launched in Britain in late 1998. It used a transmission system called OFDM (orthogonal frequency division multiplex), which is resistant to interference. Digital terrestrial broadcasting began almost simultaneously in the United States. As with previous changes in television

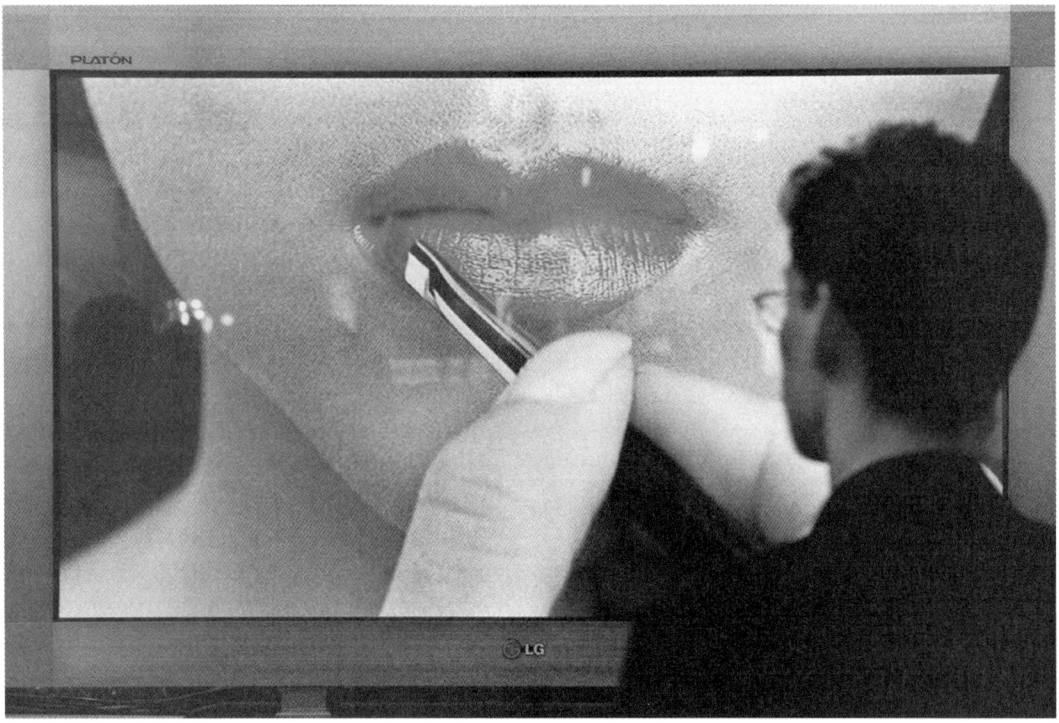

A man at the 1999 Korea Electronics Show in Seoul, South Korea, watches a model on a Flatron PD-60X1, a new slim and flat digital television monitor (AFP/Corbis)

broadcasting formats, regulators have stipulated that there must be a gradual transition. Broadly, the assumption is that ten years is a reasonable lifetime for a television set and that analog transmissions should therefore continue for a similar period. In order to receive digital broadcasts on analog television sets, viewers need to have a digital decoder. The decoder is encrypted to allow viewers access to only those services to which they have subscribed. The vision for the future is that as television becomes fully digital, it will be easier to integrate other digital technologies, such as Internet access, digital control systems for other household equipment, and automated payment for services by use of "smart cards," which are used like a credit or debit card.

The Impact of Television

Television broadcasting has undoubtedly changed the world. It has enabled people all over the world to watch major international events, such as the Olympic games, live in their homes. Television is a powerful educational medium in the broadest sense. It has transformed news coverage and, in so doing, broadened and deepened our understanding of current events. For example, television transmission of President John Fitzgerald Kennedy's assassination, the Vietnam War, and the Ethiopian famine had a profound effect on viewers. Today, there are fewer homes in Britain and United States without a television than there are homes with more than one television. However, some people choose not to own a television because they feel that it is a low-grade form of entertainment and has a bad effect on home life. Broadcasters respond to such accusations by stating that programming is influenced by audience numbers for particular types of programs and that they are simply giving viewers what they they want. Nevertheless, in the United States, where commercial pressures have been greatest, there was sufficient concern about program quality for the national Public Broadcasting System to be set up in the 1960s in order to increase educational content. The characterization of television viewers as "couch potatoes" and "square eyes" is a reflection of the compulsive nature of television. To some extent, concerns about television addiction have more recently been transferred to computers and the Internet.

See also Baird Television Company; Bang & Olufsen; Camcorders; Electrical and Musical Industries; Matsushita; RCA; Remote Controls; Sony Corporation; Videotape Recorders

References:

Bennett-Levy, Michael. *Historic Televisions and Video Recorders.* Midlothian, UK: MBL Publications, 1994.

Forester, Tom. *High-Tech Society: The Story of the Information Technology Revolution.* Cambridge, MA.: MIT Press, 1987.

Goodman, Robert. *How Electronic Things Work . . . and What to Do When They Don't.* New York: McGraw-Hill, 1999.

Scannell, Paddy. "Mass Media, Mass Democracy" in *The Making of Britain: Echos of Greatness,* edited by Lesley M. Smith. Basingstoke, UK, and London: Macmillan/LWT and Channel 4, 1988.

Sinclair, Ian. *Birth of the Box: The Story of Television.* Wilmslow, UK: Sigma Press, 1995.

Winn, Marie. *The Plug-In Drug: Television, Children, and the Family.* New York: Viking Penguin, 1977.

Texas Instruments

The U.S. company Texas Instruments (TI) is a major semiconductor manufacturer, with seventeen plants across the world. J. Clarence Karcher and Eugene McDermott founded the company in 1930 as Geophysical Service, which produced seismographic equipment and provided specialist seismological exploration services to oil companies. Its first contracts were in Texas, but within ten years it had undertaken many contracts outside the United States, in countries including Mexico, Canada, Venezuela, Colombia, Saudi Arabia, and India. The company was incorporated in Delaware in 1938 and a year later changed to its name to the Coronado Corporation, with Geophysical Service Inc. (GSI) as a subsidiary.

In 1941, McDermott and three partners acquired GSI and began to diversify its manufacturing operations. Immediately, the company picked up a contract to produce submarine-detection equipment for the U.S. Navy. Electronics work for the Navy and the Army Signal Corps continued throughout World War II. In 1948, the company received its first contract for airborne radar systems. The name Texas Instruments was adopted in 1951. A year later, TI entered the semiconductor business when it bought a license to manufacture transistors. In 1953, a restructuring resulted from the merger with the Intercontinental Rubber Company and the acquisition of Houston Technical Laboratories. The company's first commercial transistors were introduced in 1954, when it also provided the circuitry design for the IDEA Corporation's Regency radio, the first transistor radio.

The transistor boom led TI to set up its first overseas subsidiaries in Britain (1956) and Italy (1957) and to build a new semiconductor plant in Dallas (1958). One of the company's greatest achievements came in 1958, when one of its engineers, Jack Kilby, invented the integrated circuit. The first TI product based on integrated circuit technology, a computer for the U.S. Air Force, was completed in 1961, and the first consumer product, a hearing aid, was launched in 1964. In 1960, following the previous year's merger with the Metals and Controls Corporation, TI had nearly 17,000 employees and annual revenue exceeding $200 million. It continued to develop new products for both its specialist and consumer markets, including infrared imagers, solid-state thermal printers, and solid-state radar. In 1967, TI introduced the first electronic hand-held calculator. A consumer version of this calculator appeared in 1972. The company remained at the forefront of semiconductor technology, developing a microprocessor chip in 1971, the same year that Intel launched the first commercial microprocessor. The TI microprocessor was patented in 1973, when the company also introduced a 4K-bit DRAM (dynamic random access memory) chip. A new type of microchip, the first speech-synthesizer chip, was the basis for the TI's Speak and Spell learning aid, which came out in 1978.

By 1988, semiconductor production was such a dominant area of business for TI that a streamlining of operations was logical. Accordingly, TI sold a majority interest in its original business, Geophysical Service, to Halliburton Company. In the same year, the company joined forces with the Japanese company Hitachi to produce 16-Mb DRAM chips and demonstrated the world's first quantum effect transistor. In 1990, with annual revenues of $6.6 billion and 70,000 employees, TI embarked on a strategic program for the last decade of the century based on innovation and business rationalization. Innovative developments of the 1990s included the ThunderSwitch chip, an Ethernet switch on a single chip, and Timeline Technology, a technique for fitting 125 million transistors on one chip. By the end of the decade, TI was focused on semiconductor and digital technologies, having sold off its military and scientific product divisions. In 1992, TI sold its multi-user computer systems and service business to Hewlett-Packard. The major restructuring occurred in 1997 when TI made nine divestitures and three acquisitions. The opening of a $150 million research and development center, the Kilby Center, in the same year signaled TI's continuing commitment to advanced technology.

See also Calculators; Computer Printers; Computers; Radio

References:

Braun, Ernest, and Stuart Macdonald. *Revolution in Miniature.* New York: Cambridge University Press, 1982.

www.ti.com/corp/docs/company/history/

Toasters

Traditionally toast was made from slices of bread held before an open fire or grate on either a hand-held or tripod toasting fork. Once brought to the table toast needs to be eaten quickly so that it does not become cold and either hard or soggy, depending on the type of bread. Fashionable Edwardian tables had toast Crispers—covered toast racks heated by a spirit burner—to keep the toast warm. Stove-top toasters were also developed; they remain popular in France. They are made of pressed, tinned sheet iron and are either circular or rectangular in shape. The base is dimpled to distribute the heat to a grille above. They toast quickly, but can do only one side of a slice of bread at a time.

In Britain toast was universally popular for breakfast and afternoon tea. The toaster was one of the first domestic utensils to take advantage of the emergence of electricity. Together with an electric kettle, the toaster could provide the emerging "servantless" household with a quick meal. Crompton & Company of Chelmsford, England, manufactured the first electric toaster in 1893. It used bare "resistance wire" wound around flat mica sheets that heated up very quickly. Most early models were open, with a central double-sided heating element. A slice of bread was placed on either side; the bread was therefore toasted on only one side at a time. Once one side was "done," the bread had to be turned by hand. It was separated from the heat by wire grilles and kept in place by hinged "gates."

In the United States around 1900, the Simplex Electric Heating Company produced an electric griddle that was marketed to commercial kitchens. It was heated by an embedded element and claimed to be suitable as a toaster.

The key development for electric toasters was the invention of Nichrome by the young American engineer Albert Marsh in March 1905. An alloy of nickel and chromium, it was tough and sufficiently ductile to be drawn into resistance wire—the ideal heating element for toasters. Two months later George Schneider of the American Electric Heater Company of Detroit applied for a patent for an enclosed electric toaster using "suitable resistance wire," but the toaster was never produced.

There are two contestants for the first commercially successful electric toaster in the United States. One is the D-12 of 1909, designed by Frank Shailor of General Electric. It had four conical heating elements encased in a wire cage on which the bread rested. The bases were white earthenware, some with gold lining or floral transfer decoration. The other is the Pacific Electric Heating Company's E1 Tosto, which may have been in production as early as 1905. Pacific Electric became Hotpoint and continued to manufacture toasters. Simplex produced the T-211 in the same year, and other manufacturers followed. Landers, Frary and Clark introduced their Thermax in 1915.

The toaster was designed and marketed as a piece of table equipment rather than kitchenware. By 1920 most models were either nickel- or chromium-plated and the "gates" had become "doors," often with pierced decoration. The General Electric Company's Magnet toaster had spring-loaded sides. When it was opened, the bread slid down and aligned itself with the untoasted side facing the heating element. Most early toasters were not earthed (grounded).

A photograph in the English *Daily Mail* of 1920, entitled "Electricity's Aids to the Tea-Table" featured a family taking tea. Their electrical aids were a kettle, a toaster, and a heating stand, and the wires for the toaster and kettle were fed from the light socket above.

Modifications and improvements continued. Some toasters were able to toast on both sides by adding elements to the "doors." Flat tops were introduced so that toast could be piled on top and kept warm by the elements working below. The Universal E9410 featured twin buttons that operated swinging grilles that enabled the bread to be turned while remaining inside the grille. The Toastrite Blue Willow, produced by the Pan Manufacturing Company of Cleveland, Ohio, featured a ceramic body with a blue willow pattern to match dinner services.

The next logical step in development was the automatic thermostatically controlled model that stopped when the toast was "done" and/or "popped" the toast up away from the elements. This was invented by Charles Strite, a mechanic from Stillwater, Minnesota, who had patented the first spring-loaded automatic toaster in 1919. The Waters Genter Company was formed in 1921 to manufacture these toasters, which were intended for the restaurant trade. They produced the 1-A-1, the first domestic automatic pop-up toaster, in 1926. Branded as the Toastmaster, it hit the market in 1926, retailing at $12.50—expensive for the time. McGraw Electric of Minneapolis purchased Waters Genter in 1929. The Sunbeam Toastmaster appeared in the same year. This automatically shut off when the toast was ready. Its advertising proclaimed "Pop! up comes the toast automatically when it's done and the current is automatically turned off. The toast is made in a jiffy because both sides are toasted at the same time. There is no guesswork." Sales rose in the United States with the arrival of commercially packaged sliced bread. The Toastmaster was introduced in Great Britain in 1928.

Automatic toasters, which utilized a thermostat controlling a clock mechanism, were initially more expensive than the simpler "open" type. By the end of World War II they had become cheaper and increasingly reliable. Nevertheless, the even more affordable manual "open" type sold well throughout the 1930s and 1940s. Both types featured art deco styling. Kenneth Wood's Kenwood A 100 of 1947 was not unlike American models of the 1920s. The United States remained the technological leader. There were some interesting attempts at alternative designs, both in Europe and the United States. The British Saluta Number 584 of 1930 was cylindrical with a central heating element. Four rotating wire bread slots surrounded it and were swiveled into position close to the heat. The Toast-o-Lator of 1935 featured a mini–conveyor belt that fed the bread from one end to the other. It even had an eyehole for the operator to see how things were progressing inside (an idea revived by the Sunbeam Toast-Logic in the 1990s). Sometimes it was a little fast and violently shot the toast out the other end!

Despite these diversions the shape of the toaster became the now-familiar bulbous curved rectangle with two slots and external controls. The double-sided toasting slot was introduced in the the United States in 1937. Most toasters were made from chromium-plated pressed steel. Like the "open" models of the 1930s they continued to have streamlined or rounded features. One of the most elegant was the Sunbeam T-9 Half-Round. A popular model was the Toastmaster 1B 14, which was typical of 1950s toasters. The first British company to produce a similar model was Morphy Richards. In 1956 it introduced a rounded two-slot pop-up model with a browning control and quick release button. The steel body was available in either chrome plate or yellow enamel, resting on black Bakelite feet. It quickly became a bestseller. Variants of this model remained on the market until the 1970s. Patterned blue and orange colors were introduced to complement contemporary kitchen styles. A similarly long-lived toaster was the Sunbeam T-20, the first fully automatic toaster. It automatically lowered the bread and when the toasting cycle was completed, it shut itself off and raised the toast. The T-20 was in production from the late 1940s to the mid-1990s.

Morphy Richards experimented with other styles. In 1961 it introduced a squarer model made from metal panels, the ends being made of Bakelite. It required less metal to produce and was therefore cheaper. The company also produced a "sideways" toaster in the 1970s in which the toast was "tipped" rather than "popped" onto a hinged gangplank at one end.

Throughout the 1960s and 1970s toasters remained rectangular metal boxes. They were decorated with varying printed designs and colors that followed trends in kitchen color schemes. Sunbeam produced a model with silk-screened patterns on glass sides. The most popular subjects were flowers and wheatsheaves. Companies such as Philips and Braun produced toasters as part of their growing portfolio of domestic appliances. They became a ubiquitous item within the modern kitchen. Most were fairly reliable, but the toast could still come out burnt or be crushed in cheaper models with less durable components.

Plastics were used more and more and eventually became the most popular envelope for the heating elements. This was the main innovation during the 1980s, alongside the "one slot" toaster. Initially produced by the French Tefal Company, this featured a long single slot that could accommodate two slices of bread in line. It also had spring-loaded sides that could allow for different thicknesses of bread. Plastic-bodied toasters increasingly featured "coolwalls," which have a gap between the inner metal and the outer walls, preventing the outside from getting hotter than 40°C (104°). The early 1990s saw the introduction of a "frozen bread" setting, which illustrated the manufacturers' response to greater freezer ownership. Tests by the British consumer magazine *Which?* showed that toasters without this option made equally acceptable toast!

By 1985 all toasters shared the same basic technology and were largely distinguished by how they looked—long and thin, short and square, plain or patterned—yet this was not enough for the style conscious. For them these toasters lacked a certain integrity. Toasters entered the debate on "good design" that raged in the media of the day. A Russell Hobbs toaster with an Honesty floral pattern was displayed in the Victoria & Albert Museum's Boilerhouse project "Taste" exhibition of 1983 as an example of bad design. The exhibition's organizer, Stephen Bayley, declared, "Your choice of the *Honesty* pattern toaster declares you to be the sort of person who will cheerfully admit: 'I love buying cynical junk.'"

This attitude, linked to the growing influence of Alessi products, led to the search for more authentic and often retrospective designs. Even the microchip Braun electronic Control-Sensor of the late 1980s looked a little dull. Still-functioning 1950s chrome toasters were much sought after in Europe and the United States. A clear winner was the Dualit Company of London. Max Gort-Barton, an aircraft engineer, founded the company, which began making Dual-Light electric heaters in 1947 and then moved into high-quality metal toasters, grills, and boilers for the catering industry. In the 1980s their stainless steel toaster was hailed as a classic by architects and designers. Designed by Gort-Barton in 1946, it has remained virtually unchanged. Available as two-, four-, and six-slice models, these large, expensive, and elegant toasters, finished in either stainless steel or white or black trim, featured in all the fashionable kitchens of *House & Garden* magazine. They used a simple clockwork timer, and the toast was pushed up with a lever when done. The Combi model had a wider slot and grille for toasted sandwiches.

The Dualit remains popular and was even slightly redesigned to look curvier in 1995. It now comes with pastel green, blue, cream, and yellow trim. The success of the Dualit led other companies to reexamine their designs. The U.S. Toastmaster produced a Cool Steel toaster in 1990. Morphy Richards reintroduced a rather poor imitation of their 1956 model, and Philips has linked with Alessi to produce the postmodern Workshop toaster, which has elements of 1950s styling. The Italian Guzzini Company has also produced a postmodern toaster with a white or green plastic body.

The late 1990s also saw a technological step forward with the use of a "Black Heat Element" by Dualit. The etched-foil circuits are laminated onto a ceramic substrate and protected by a high-performance mesh, which is in turn encased in a tough outer binder. This gives the same heat output, but saves up to 17 percent in electricity use and gives the element a longer life.

Originally designed for the dining table, the toaster developed throughout the twentieth century, reflecting numerous design movements along the way. It has entered the twenty-first century as an essential item for most kitchens, either as an almost invisible white box or as a distinctive style statement.

See also Alessi; Braun; General Electric; Kenwood; McGraw Electric; Sunbeam

References:

de Haan, David. *Antique Household Gadgets and Appliances, 1860–1930.* Poole, UK: Blandford Press, 1977.

Sparke, Penny. *Electrical Appliances.* London and Sydney: Unwin Hyman, 1987.

Which? November 1993.

www.toaster.org

Toilets

See Water Closets

Toothbrushes

Absurd as it may seem today, in 1900 the humble toothbrush had luxury status. First mentioned in England two and a half centuries earlier, the toothbrush had not become widely adopted in the intervening period for two reasons: the toothbrush was an expensive item, with an ivory or bone handle and natural bristle head, and the understanding of oral hygiene and dental care was still rudimentary and unscientific. Alternatives to the toothbrush were twigs (particularly hazel or willow), scraps of linen, and simplest of all, fingers. Soot and chalk-based tooth powders were used as cleaning agents, while mouthwashes contained flavorings and sweeteners such as cinnamon, cloves, honey, and orange peel. In the late nineteenth century, toothpaste appeared on the market. William Colgate of New York produced toothpaste in jars starting in 1873 and introduced the first collapsible toothpaste tubes containing Colgate Dental Cream in 1896. In Britain, the first toothpaste to appear on sale was Dr. Zierner's Alexandra Dentifrices in 1891.

After World War I, standards of general personal hygiene, including dental hygiene, began to improve noticeably. One sign of rising standards was the inclusion of separate bathrooms in ordinary homes built in the 1920s. New materials reduced the price of many items. The first use of nylon, when it went into commercial production in the United States in 1938, was for toothbrush bristles. Nylon bristles were marketed as "the miracle tuft" because they had several advantages over natural bristles. Aside from being cheaper, nylon had good durability and the filaments were of uniform diameter that could be varied according to purpose. By the early 1950s, the all-synthetic toothbrush with plastic handle had arrived. Since then, the manual toothbrush has changed only in minor ways, such as the angling of the handle to improve reach and the use of softer bristles. Few companies have attempted to market the toothbrush as a design object. However, the French designer Philippe Starck, who also glamorized the toilet brush and holder, produced the highly stylized Fluocaril toothbrush in 1989.

In 1955, the U.S. company Proctor & Gamble introduced Crest, the first fluoride toothpaste, which soon became the best-selling brand. This came about as a result of research that established that fluoride reduced tooth decay. It also became normal practice for water utilities to add fluoride salts, in very low concentration, to the water supply. Toothpaste manufacturers turned their attention to improving the whitening properties of toothpaste. Colgate Ultra Brite toothpaste, introduced in 1968, had a monofluorophosphate (MFP) formula. Modern advanced whitening toothpastes include compounds that have been found to be effective stain removers.

An electric toothbrush was patented in the United States as early as 1885, but it was only in the early 1960s that the electric toothbrush became a practical, commercial product. The Squibb Company of New York led the way in this product area. Electric toothbrushes were initially perceived by many people as a strictly nonessential gadget. Early models had pulsating electric heads that complemented the basic manual brushing action. Similar models are still available, but the best modern electric toothbrushes, described as "plaque removers," have small circular heads that rotate. This action is effective, but also kinder on gums and tooth enamel than manual brushing. Developments in rechargeable battery technology have made the modern electric toothbrush more energy-efficient than earlier models powered by disposable batteries. For the ultimate electric dental care experience, consumers can also opt for the electric interdental cleaner, an alternative to dental floss, and the electric tooth polisher.

See also Piped Water; Plastics

Trash Compactors

Trash compactors are mainly found in U.S. kitchens. They are domestic versions of commercial baling machines that ram and compact rubbish. Domestic models are designed to fit under kitchen countertops. They have always been a relatively expensive appliance ($400 to $800 in 2000) and essentially limit the number of trips to the trashcan. Although not essential for a traditional house with a yard, they can be more of a benefit in an apartment or flat.

Most have a drop-front door; the trash is placed in a bag within a container and the door closed. Five thousand pounds of pressure then reduces the volume of the trash to 25 percent of its original size. Compactors have capacities of around 1.5 cubic feet and can contain about a week's worth of rubbish. Trash kept for this length of time can begin to smell, and this has led to the introduction of charcoal filters.

Trash compactors were introduced in the 1960s and continue to be manufactured by companies such as KitchenAid, General Electric, and Whirlpool.

See also Waste Disposal Units

Trouser Presses

Woolen and flannel men's trousers grew in popularity throughout the nineteenth century and had become ubiquitous by 1900. Much of this was due to a rising urban population and a growth in relatively cleaner types of work in offices and workshops. Industrial tailoring, mass-produced clothes made in factories, was led by companies such as Burton's and John Collier in the United Kingdom. A more upmarket American equivalent was Brooks Brothers, established in 1818.

Wool trousers are usually cut and finished to look at their best with a vertical crease. This was normally maintained by ironing. Manual presses were used to maintain the crease if the trousers were left in overnight. Electric presses developed in the 1950s in the United States, the United Kingdom, and Italy. They were often installed in hotel rooms as vertical wall-mounted models. They featured thermostatically controlled heating pads and lever-operated presses. Most domestic models are freestanding with 15- to 45-minute timers. The primary manufacturers are Corby and Morphy Richards.

Tupperware

Tupperware, a type of polyethylene food container, is perhaps as well known for its style of marketing and sales as it is for its practicality. During the 1950s and 1960s, the "Tupperware party" was a regular fixture in the calendars of many women in the United States.

Tupperware was invented by Earl Silas Tupper (1907–1983), a former tree surgeon from Berlin, New Hampshire. He was driven to become a success as either an inventor or an entrepreneur, and his opportunity to do both was aided by experimenting with samples of polyethylene produced by Du Pont. He began producing cheap novelty goods, but in 1942, after manufacturing gas mask parts, he turned to the domestic market. Perhaps in response to the wartime need to save food, he invented a container with a flexible plastic lid that fitted with an air-expelling

Manual trouser presses, like this Watts model of c. 1900, were sold in tailors' shops. (The Museum of Science and Industry in Manchester)

"burp," resulting in an almost airtight seal that was ideal for keeping food leftovers. He formed the Tupper Plastic Company in 1942 and opened his first factory in South Grafton, Massachusetts.

The early wares were simple injection molded pieces in candy-like colors. Their Bauhaus simplicity led the Museum of Modern Art to declare them "marvelously free of that vulgarity which characterizes so much of household equipment." In 1947, the U.S. magazine *House Beautiful* featured them in an article entitled "Fine Art for 39 cents." Despite these cultural endorsements, Tupperware was marketed in a traditional manner and sold through hardware and department stores with limited success. This was partly due to the low public perception of plastics in the immediate postwar period. Tupperware's salvation lay in the form of a hard-up divorcee named Brownie Wise. She desperately needed money to pay for hospital care for her sick son and began to sell Tupperware products through door-to-door demonstrations. Her technique worked, and by 1948 she was selling up to $1,500 worth a week.

Wise displayed a charismatic confidence in the product, and by 1951 Tupper had appointed her as vice president and general manager of Tupperware home parties. Wise was able to build on three important elements, the modern growth of suburbia, the increasing affluence, and the traditional needs of women to socialize, especially in new neighborhoods. Suburban housewives were expected to create and maintain the "dream home," with its expanding range of appliances such as refrigerators. For many this was isolated work and the Tupperware party was a way of meeting other people. Wise combined commercialism with friendship and neighborliness. Women sold to other women whom they knew and trusted, and the items that they bought were practical and useful. Tupperware hostesses received commission on their sales, and as Wise had illustrated, money was there to be made. Tupperware parties, in fact, offered housewives and mothers part-time paid work, masquerading as a leisure activity. The idea was so successful that Tupperware was withdrawn from the shops and only available through the party system. The initial line of boxes was perfect for storing food in the refrigerator. The line expanded to include salad bowls, party snack containers, and picnic ware.

Tupperware continued to sell well throughout the 1950s, and Wise became the first woman to appear on the cover of *Business Week* magazine. The company headquarters moved to Orlando, Florida, in 1995. Tupper, who has been described as an eccentric, an oddball, a hermit, and a perfectionist, cannot have been easy to work with. Despite owing his commercial success to Wise, they began to quarrel and he ousted her from the board in 1958. He sold the company to the Rexall Drug Corporation in the same year and retired to Costa Rica, where he died of a heart attack in 1983.

The product continued to bear his name and went on to be successful in Europe and Japan. It conquered Great Britain in the 1960s, although the U.S. method of selling through parties was initially suspect. The *Daily Mail* headlined a story on the Tupperware party with "Soft Sell Steals into Suburbia." British women, like their U.S. sisters, knew that they attended the parties and purchased the goods out of loyalty to their friends, but they also realized that Tupperware was a good product and performed well.

Tupperware continues to sell throughout the world and is now appreciated as a form of pure plastics design. The company has continued to develop the product line with Rock 'n' Serve, a line of virtually unbreakable containers for freezers and microwaves, and OvenWorks plastic cookware for microwaves and conventional ovens. A recent addition is FridgeSmart, with patented vents to produce an atmosphere-controlled environment to keep food fresher.

The Tupperware burp continues to sound in kitchens and at picnics around the world.

In 1998, it was estimated that a Tupperware party started somewhere in the world every 2.2 seconds, annually selling goods worth over $1.1 billion to 105 million people through 950,000 salespeople.

See also Plastics; Refrigerators

References:

Clarke, Alison J. *Tupperware: The Promise of Plastic in 1950s America.* Washington, DC, and London: Smithsonian Institution Press: 1999.

Katz, Sylvia, and Jeremy Myerson. *Kitchenware.* London: Conran Octopus, 1990.

Mossman, Susan, ed. *Early Plastics: Perspectives, 1850–1950.* London and Washington, DC: Science Museum/Leicester University Press, 1997.

www.tupperware.com

Typewriters

A typewriter is a machine that prints characters onto paper as the user strikes keys that operate movable type. The mechanization of the printing of documents for mass consumption, such as books and newspapers, began in the fifteenth century. However, it was not until the late nineteenth century that a successful design for a machine to replace the handwriting of documents such as letters and bills was devised. By the beginning of the twentieth century, the typewriter was well-established as an office machine, but little attempt was made to market the typewriter to domestic consumers until the 1930s.

The first patent for a mechanical writing machine was obtained in Britain by Henry Mill in 1714, but this machine was never put into production. In 1830, William Austin Burt of Detroit, Michigan, was granted a patent for his typographer, a machine with a type head in the form of type characters on circular band of metal. This was the first design to use a changeable font, a feature that only became common in typewriters in the 1960s. The modern typewriter owes its basic design to Christopher Latham Sholes, a printer in Milwaukee, Wisconsin. In 1867, with assistance from Carlos Glidden and Samuel Soulé, Sholes developed his first typewriter, which was patented in 1870. This

Poster with writing instruments surrounding an Olivetti portable typewriter, Milan, 1953 (Library of Congress)

typewriter consisted of a series of levers, called type bars, for individual characters. The type bars were placed radially so that they struck the center of an inked ribbon when the keys were depressed. The paper was held on a cylindrical platen, which moved after each keystroke, as did the ribbon. Sholes continued to refine the design, for example by replacing the original alphabetical layout of the keyboard with today's QWERTY layout. The QWERTY layout, still used in today's computer keyboards, was intended to reduce the problem of type bars jamming due to rapid consecutive striking of adjacent keys.

In 1873, Sholes sold the manufacturing rights to E. Remington & Sons, gunmakers in New York, and mass production began in 1874. The Remington No. 1 had a number of design flaws: it only printed capital letters (upper case), a foot pedal was used to move paper up, and the lines of type were pro-

"Perfect Day!" A 1950 advertisement for the IBM typewriter. (Library of Congress)

duced on the underside of the platen, so they were hidden from the typist. The Remington No. 2 of 1878 incorporated a shift key, which activated a set of lower case characters. Other U.S. makers, such as the Royal Typewriter Company, used the double keyboard with separate keys for upper and lower case characters. The double keyboard was abandoned when touch-typing became more common in the early twentieth century.

By 1900, at least forty companies were making typewriters for office use. These typewriters were extremely functional in appearance, typically with black metal frames that exposed the mechanism. Typing quickly became a predominantly female occupation. The 1881 census records only 7,000 women clerks in England and Wales, whereas in 1911, the number had risen to 146,000.

The standard office typewriters were too heavy and bulky to have much appeal for household use. Like sewing machines before them and computers afterward, typewriters needed to become more portable. Although earlier portable models existed, it was only in the 1920s and 1930s that makers began to mass-market portables. Makers realized that these machines needed to be attractive as well as comparatively cheap. The Italian company Olivetti was particularly active in setting design trends. Olivetti portable typewriters of the 1930s featured all-concealing slim steel cases in lighter colors. This was very much in keeping with the contemporary fashion for streamlining larger appliances such as refrigerators. The same design principles were not widely applied to office typewriters until the 1940s. In 1945, Olivetti employed Marcello Nizzoli to design its Lexicon 80 office typewriter. Nizzoli also designed the 1950s line of Olivetti Lettera portable typewriters. Olivetti continued to produce innovative designs, including the Valentine portable typewriter of 1969, designed by the Austrian-born designer Ettore Sottsass and the British designer Perry King. The brightly colored, molded-plastic typewriter body fitted snugly within the shell of its carrying case.

Manual typewriters required the keys to be struck cleanly and firmly in order to print cleanly. In 1917, a U.S. company, the Noiseless Typewriter Company, introduced a typewriter that was designed to respond to lighter pressure, thus making less noise, but it was electric—and later electronic—typewriters that ultimately solved the noise problem. The idea of the electrically powered typewriter was mooted in the nineteenth century, but the first electric models did not appear until the 1920s. The first commercially successful one was launched in the United States by IBM in 1935. Again, good design was regarded as crucial to marketing, as the electric models were more expensive than manual ones. The IBM Executive AA model of 1946 was designed by Eliot Noyes, who later became design director of IBM. Noyes was an advocate of sculptural, organic forms, rather than severe streamlining. The Executive had rounded lines and two-tone coloring. On the technical front, the Executive featured proportional character spacing. Electric portable typewriters became available from 1956. In the 1950s and 1960s, the Royal Typewriter Company also emphasized design and used colors, such as deep pink, to soften the appearance of their typewriters. By the late 1970s, makers had reverted to designs that expressed the technological character of the typewriter, in line with the prevailing trend for sophisticated monochromatic minimalism. Black became the dominant color, and plastic typewriter casings tended to be angular in shape.

The next major advance was the introduction by IBM in 1961 of the golf ball, a multifaceted type head that was easily removable for changing fonts. Although the golf ball was not the first changeable font device since Burt's pioneering typographer, it was the first to become widely adopted. The golf ball was an advance on its predecessors in that it was not fixed but movable. Instead of the paper having to move past a fixed type position, the paper remained in place and the golf ball traveled across it. This was less noisy than the moving platen carriage. Eliot Noyes conveyed this technical change in his compact and seamless design for the 1961 IBM Selectric golf-ball typewriter. The golf-ball typewriter also introduced carbon ribbon, which produced a crisper print. The other key advantage of carbon ribbon was that the carbon characters could be removed from the paper by sticky tape, rendering corrections almost invisible. In the 1970s, the golf ball was superseded by the daisy-wheel, a slim disk with the type characters around the edge. In the 1980s, the typewriter, unrivalled for its purpose for 100 years, had to compete with the new generation of word-processors and personal computers. The daisy-wheel typewriter acquired limited electronic functionality in the form of memory and small display screens, which allowed line-by-line previewing of type prior to printing. However, the end was in sight, and as the price of computers fell, typewriters were relegated to an occasional fall-back role.

See also Computer Printers; Computers; IBM

References:

Beeching, Wilfred A. *A Century of the Typewriter.* London: Heinemann, 1974.

Lupton, Ellen. *Mechanical Brides: Women and Machines from Home to Office.* New York: Cooper-Hewitt National Museum of Design, Smithsonian Institution, and Princeton Architectural Press, 1993.

V

Vacuum Cleaners

Before the invention of the electric vacuum cleaner in 1901, the cleaning of floors, particularly carpeted floors, was an inefficient and time-consuming job. The basic tool was the broom or brush, which produced reasonable results on smooth surfaces that could be finished by scrubbing, wiping, or polishing. Carpets were loosely laid so that they could be periodically removed and hung outside for dust to be knocked out with a carpet beater or sent away for cleaning. In poorer homes, rag rugs took the place of carpets. Compact carpet sweepers, which became widely available in the late nineteenth century, were easy to use and improved light cleaning of carpets.

Bellows Vacuum Cleaners

Patents for bellows vacuum cleaners, operated by hand or foot, first appeared in about 1860. Closing the bellows causes air to be driven out, thereby drawing air into the nozzle to restore equal air pressure. Foot-operated versions allowed the user to have both hands free for directing the cleaner head, whereas the heavier hand-operated versions required two users, one to work the bellows and one to direct the cleaner head. Models based on the bellows principle became widely available from the 1890s, but because the bellows provided only weak suction, they performed very moderately. British models included the rather unwieldy Baby Daisy and the lightweight, hand-held Star, which was very similar to the Success model produced in the United States.

Fan Vacuum Cleaners

The first floor cleaner to incorporate a fan was patented in the United States in 1859. It was a wheeled cleaner with a fan driven by the wheel axle, but the fan blew dust away rather than sucking it up. Ives M. McGaffey of Chicago devised the first vacuum cleaner, patented in 1869. McGaffey's Whirlwind was a wheeled cleaner producing suction via a hand-cranked fan. A continuously operating fan is a more effective mechanism than bellows, but user-powered cleaners could not sustain sufficient suction. In 1901, Hubert Cecil Booth, a British civil engineer, patented the first electric vacuum cleaner. Having seen a demonstration of a U.S. system involving blowing compressed air at carpets or upholstery to drive dust out, he concluded that suction cleaning would be more effective at collecting the dust. Booth's first model, called Puffing Billy, consisted of a large electric motor (or petrol [gas] engine), fan, and dust-collecting box on a pushcart. The cart stayed

Model 35 vacuum cleaner made by the Air-Way Electrical Appliance Company of Toledo, Ohio, c. 1925, featuring its patented Sanitary System (The Museum of Science and Industry in Manchester)

outside the building to be cleaned and long hoses were fed through doors or windows to extract the dust. Booth set up the British Vacuum Cleaner Company, which became Goblin (BVC) in 1924, and offered a commercial service from 1902, with his vacuum cleaners mounted on horse-drawn carts for ease of transport. Similar vacuum cleaners by other makers were available in France in 1903. In 1904, Booth developed a smaller version, the Trolley Vac, intended for home ownership, but it was still very heavy and expensive. A U.S. patent similar to the Booth design was issued to William Noe in 1905. In a more compact form, the Booth-type vacuum cleaner became a practical domestic appliance, now described as the cylinder vacuum cleaner.

Central Vacuum Cleaning Systems

In the United States, vacuum cleaning developed in a different direction with David T. Kenney's patents of 1903 and 1905 for a central vacuum-cleaning system. This was conceived primarily for use in large buildings such as hotels and office blocks. The communal machinery—the electric motor, pump, and dust collector—was a fixed installation, typically in the cellar, and was linked to all floors and rooms by a network of ducting with inlets in the walls for the attachment of hoses. The first of Kenney's systems was installed in a building in New York in 1902. Central heating manufacturers saw the complementarity of the central vacuum system and entered the market. However, the expense of installation was a major obstacle for all but wealthier households. In Britain, a central vacuum system was installed in the House of Commons, but the system never reached more than a small market.

Upright and Cylinder Vacuum Cleaners

The first truly practical domestic vacuum cleaner was patented in the United States by James Murray Spangler in 1908. Spangler called his invention the electric suction sweeper. His prototype consisted of a small electric motor and fan mounted on a wheeled base with rotating brushes below, and a pivoted broom handle with an attached pillowcase to act as the dust bag. Rotating brushes were a feature derived from the manual carpet sweeper. At 40 lb (18 kg), it was less than half the weight of the Trolley Vac. The patent rights were acquired by William Hoover, who began mass production of the first upright vacuum cleaners in 1908. The Hoover vacuum cleaner was introduced in Britain in 1912. The weight of vacuum cleaners was soon greatly reduced by the use of the small, high-speed, fractional horsepower universal motor developed by the U.S. mechanic Chester Beach in 1905. Spangler went on to

Mother and daughter with vacuum cleaners, 1939 (Urban Archives, Temple University)

develop an improved patent in 1915, which established a more compact design.

By 1920, lightweight versions of both the upright and the cylinder (or canister) vacuum cleaner were available. Manufacturers promoted vacuum cleaners as not just labor-saving aids but also instruments of sanitization, protecting family health and well-being by eradicating dust and mites. The upright cleaner produced from 1920 by the Air-Way Corporation of Toledo, Ohio, was called the Sanitizor. Air-Way's major contribution to vacuum cleaner design was the disposable "filter fiber" dust bag. The Swedish company Electrolux led the way in improvements to the cylinder cleaner. Their models had light metal bodies covered with leatherette, toboggan-style runners for ease of movement, a characteristic pistol grip, and a flexible hose with a wide range of attachments to suit different floor surfaces, upholstery, and draperies. After Electrolux models appeared on the U.S. market in 1924, U.S. manufacturers developed similar sets of attachments for upright cleaners in order to compete in terms of versatility. For example, Hoover patented a side-mounted hose attachment in 1936. Eureka, based in Detroit, Michigan, had provided attachments, including hairdryers, for its upright vacuum cleaners since 1913. Eureka's factory had a production capacity of 2,000 cleaners a day, and by 1927 it was responsible for a third of U.S. vacuum cleaner production.

Hoover's major innovation of the 1920s was the addition of a beater bar to assist the rotating brushes. This led to the launch of the Hoover Junior model in 1926 with the slogan "It Beats As It Sweeps As It Cleans." Hoover was not only technically innovative, but also commercially innovative. In order to encourage sales, it engaged in direct selling and promoted methods of payment on credit. From 1917, Hoover vacuum cleaners were sold by mail order in the United States. In both the United States and Britain, Hoover employed teams of door-to-door salespeople offering products on easy terms.

The Bagless Vacuum Cleaner

An original concept in vacuum cleaner technology was introduced in the United States by Rexair in 1936. It began life in the 1920s as John W. Newcombe's Separator, a bagless cleaner designed for use in industrial contexts. Leslie H. Green spotted the potential of adapting the Separator for household use. In 1929, Green set up the Rexair company in Michigan. The final element in the design, a water reservoir to collect the dust, was incorporated in the 1936 model. The Rexair vacuum cleaner, known as the Rainbow since 1955, was promoted with the slogan "Wet dust can't fly."

Design Issues

By the mid-1930s, the price of vacuum cleaners had dropped substantially. Makers began to pay more attention to the design of appliances, in line with the general trend in appliances toward a more streamlined appearance. The U.S. designer Lurelle Guild designed the Electrolux Model 30 in 1937. The use of chrome plating, polished and enameled steel, and cast aluminum was reminiscent of finishes used on automobiles. Hoover employed the U.S. designer Henry Dreyfuss to design the Model 150 of 1936. It was the first domestic appliance to use lightweight magnesium alloy castings (only a third of the weight of aluminum); the use of Bakelite to encase the fan and motor produced smoother curves. However, neither of these models appeared as futuristic as the 1920s Scandinavian Tellus vacuum cleaner. The Tellus broke with the traditional elongated form of a cylinder in favor of a squat all-metallic canister. It was also designed to spray paint and dry hair or clothes. In the 1940s, more efficient motors led to improved suction. The efficiency of the vacuum cleaner continued to be expressed through streamlining. In the mid-1940s, the Swedish designer Sixten Sason remodeled the Electrolux cylinder vacuum cleaner to resemble a torpedo.

Growth of Ownership

Small, cheap hand-held vacuum cleaners opened up the market to new purchasers. The Royal Prince was launched by P. A. Geier Company (later the Royal Appliance Manufacturing Company) of Cleveland, Ohio, in 1937. Consisting of a short cylinder containing the mechanism and an attached dust bag, this had the advantage of being easier to use in awkward spaces. Royal Appliance has continued to be a leader in the hand-held vacuum cleaner market. Hoover developed a model called the Dustette. By World War II, the vacuum cleaner was well established as a genuinely beneficial appliance. In 1941, home ownership of vacuum cleaners in the United States had reached 47 percent. In Britain, the growth in ownership was slower, partly because few working-class homes had carpet in any rooms. Indeed, the existence of the vacuum cleaner helped to encourage the use of carpeting. By 1948, ownership in Britain had risen to 40 percent.

From the 1950s to the late 1970s, there were no major advances in vacuum cleaner technology, but sales grew steadily as more people aspired to and achieved the home with fitted carpets. Plastic shells were introduced for cylinder models, bringing lighter and brighter colors. One novelty was the Hoover Constellation cylinder vacuum cleaner, available from 1956, which had a spherical body and hover motion, with the fan providing a cushion of air. It was advertised as the first self-propelled vacuum cleaner.

The Wet and Dry Vacuum Cleaner

Apart from minor innovations such as flex winders and disposable dust bags, the next significant development was the wet and dry vacuum cleaner, patented by the U.S. inventor Dan Brazier and launched as the Vax in 1977. Described as a tank shape, the Vax had a large chamber accessed from the top and was designed to suck up liquid as well as dust. Other manufacturers brought out similar machines. The wet and dry vacuum cleaner became the 3-in-1 cleaner when carpet cleaning was added to its functions.

Bagless and Cordless Vacuum Cleaners

The biggest innovation since the first of Spangler's patents was the invention of the cyclonic bagless vacuum cleaner by Britain's James Dyson. In 1979, based on his conclusions that a clogged dust bag only allowed 70 percent suction, Dyson produced an industrial scale version of his cyclone cleaner. He spent five years developing his first domestic bagless cleaner, the G-Force, a name referring to the centrifugal suction that keeps the airstream clear. In spite of the disinterest of existing manufacturers, Dyson persevered with his idea. His first mass-produced model, the Dual Cyclone upright cleaner, was launched in 1993. The Dual Cyclone range became the fastest selling vacuum cleaner in Britain, in spite of being considerably above the average price. Dyson vacuum cleaners continue to embody new ideas, such the use of recycled plastic waste in the manufacture of the Recyclone. Major manufacturers, including Hoover and Electrolux, have since launched cyclonic vacuum cleaners.

Another development that took place in 1979 was the launch of the Dustbuster by the U.S. company Black & Decker. The Dustbuster was the first cordless vacuum cleaner, based on technology developed by Black & Decker for the Apollo moon landing. From 1979 through 1997 the Dustbuster has reached sales of 100 million. In 1996, Black & Decker produced the cordless FloorBuster upright model. Health concerns have again become a marketing theme used to promote microfilter vacuum cleaners and reflected in a revived interest in central vacuum-cleaning systems. Microfilter models typically claim to remove 99.97 percent of dust and allergens. Novelty gadgets—such as the computer keyboard vacuum cleaner and the attachment for grooming pets—continue to emerge where there is niche marketing potential. The next major step could be the commercial development of the robot vacuum cleaner, which is becoming increasingly feasible in the age of smart microchips.

See also Carpet Sweepers; Dyson Appliances; Electrolux; The Hoover Company

References:

Giedion, Siegfried. *Mechanization Takes Command.* New York: Oxford University Press, 1948.

Hardyment, Christina. *From Mangle to Microwave: The Mechanisation of Household Work.* Cambridge, UK: Polity Press, 1988.

Matranga, Victoria Kasuba, and Karen Kohn. *America at Home: A Celebration of Twentieth-Century Housewares.* Rosemont, IL: National Housewares Manufacturers Association, 1997.

VCRs

See Videotape Recorders

Videodisc Players

The videodisc player is a device for playing prerecorded moving images and sound, which it transmits through a television set.

The videodisc was the first of the family of optical discs, which includes the compact disc (CD), the CD-ROM (compact disc with read-only memory), the CD-I (or CDi, interactive compact disc), and DVD (digital versatile disk). The term "optical disc" can be applied to any disc on which the recording of sound, images, or data is made and read by a laser. The first working laser was produced in the United States in 1960, followed three years later by the videodisc. Philips Electronics of the Netherlands demonstrated a videodisc system in 1972, but the commercial development of the videodisc only began in the mid-1970s in the wake of the videocassette.

The development of the videodisc player followed a similar pattern to that of the videocassette recorder, with the active companies each using a different format. In this case, there were three main rivals: Philips with its LaserVision system, RCA of the United States with its SelectaVision system, and JVC (Japanese Victor Company) of Japan with its VHD system. Only LaserVision used optical disc technology; the other two formats were more akin to gramophone disc technology, using a diamond stylus rather than a laser to pick up the recorded signals. The Japanese company Pioneer developed a variant of LaserVision named LaserDisc. Videodisc players finally hit the market in 1978. None of the makers achieved their early sales targets. The sales levels of prerecorded videocassettes had encouraged the companies to believe that play-only videodisc systems would be attractive to consumers. RCA's SelectaVision players were the cheapest—indeed, cheaper than Sony's Betamax videocassette recorder. However, most consumers failed to see the merit of a technology that was less functional than its recordable tape-based counterpart.

In 1984, RCA withdrew from production after sustaining losses of over $500 million on its videodisc operations. Philips, with the most sophisticated of the videodisc formats, concentrated on the educational and business markets. Both Philips and JVC began to explore the potential of interactive videodiscs. Philips brought out a CD-I player for attaching to a television set in 1991. The CD-I is essentially a hybrid of the videodisc and the audio disc, providing music and related video clips. It has only achieved a minority market.

The latest videodisc technology is DVD, which emerged as a single standard for compact videodiscs in 1995. A repeat of the earlier videocassette and videodisc format wars was forestalled in 1993 when the two consortia developing rival compact videodiscs—the Japanese–American partnership, Toshiba/Time Warner, and the Japanese–Dutch partnership, Sony/Philips—agreed to combine. The DVD is a high-density optical disc, like the audio CD, 12 cm (4.7 in) in diameter compared to the 30 cm (12 in) diameter of first generation videodiscs. A double-layer DVD can store up to four hours of video. In 1996, Toshiba introduced both DVD-ROM drives, for playing DVDs through a computer, and DVD video players, for use with television sets. The initial launch was in Japan. In both Britain and the United States, DVD players achieved unprecedented market impact, beating the equivalent early sales rate of audio CD players. Movie companies played a big role in stimulating demand by packaging films with interviews and behind-the-scenes footage for early release on DVD. More than 900,000 DVD players and 7 million DVDs were sold in the United States in 1998, and 250,000 DVD players were sold in Britain in 1999. The next step is expected to be the introduction of DVD recorders.

See also Compact Disc Players; Computers; Matsushita; Philips Electronics; RCA; Sony Corporation

References:

Forester, Tom. *High-Tech Society: The Story of the Information Technology Revolution.* Cambridge, MA: MIT Press, 1987.

Videophones

A videophone consists of a composite telephone and video transceiver (camera and monitor) to enable the simultaneous transmission and reception of sound and pictures. A standard telephone may be converted into a videophone by attaching a telephone video transceiver. The same effect may also be achieved by connecting together a telephone, video telephone camera, and television.

The videophone was already an image of futuristic communication before it became a reality when AT&T (American Telephone and Telegraph) showed its prototype Picturephone at the 1964 New York World's Fair. AT&T had been developing the videophone since 1956. However, the bulkiness and small screen of the 1964 model limited its appeal. AT&T persevered with the Picturephone and introduced a commercial service in Pittsburgh in 1970. Meanwhile, in the same year, Toshiba of Japan developed the world's first color videophone. In terms of commercial potential, video telephony was perceived more as a business tool than a social tool. Telephone companies reasoned that because of the cost group-to-group video communication, known as video conferencing, would be more attractive to the business community than one-to-one communication. The delay between the reception of sound and picture, however, discouraged would-be users. The Confravision service introduced by the General Post Office in Britain in the early 1970s was a commercial failure as was AT&T's Picturephone Meeting Service, launched in 1982.

The upgrading of telephone networks from analog to digital technology in the 1980s, by the laying of optical-fiber cable trunks and installation of exchanges with microelectronic switching, has improved the performance of videophones. Digital signals are much more compressed than their analog equivalents, so transmission is faster and cheaper. This has led to the production of videophone equipment designed for the home, albeit still at luxury prices. The standard format is a table phone with a small, hinged pop-up unit incorporating the screen and camera. However, it is the integration of the videophone concept with the computer that appears to offer the greatest potential. AT&T introduced its Model 70 computer videophone in 1993. Since then, the technology has become both more sophisticated, with the development of special consoles that operate panning cameras for true group video-conferencing, and more simple, with the availability of cheap digital cameras, the improvement in modem performance, and the development of mass-market software, such as Microsoft's NetMeeting.

See also Camcorders; Telephones; Videotape Recorders

References:

Forester, Tom. *High-Tech Society: The Story of the Information Technology Revolution.* Cambridge, MA: MIT Press, 1987.

www.att.com/technology/history/chronology/

Videotape Recorders

A videotape recorder is a device that records and plays back moving images and sound. It records sound in the same way as an audiotape recorder, except that the sound track only runs along one edge, leaving most of the width of the tape free for the video track. The recording of the moving image is more complex and was the focus of considerable research and development efforts during the 1950s, when television companies became interested in videotape technology. Since consumer videocassette recorders (VCRs) first appeared on the market in the mid-1970s, they have become a commonplace household item.

Early Developments

Early television experiments in the 1920s awakened interest in methods for recording pictures as well as sound. John Logie Baird, the British inventor of a mechanical television system, carried out many experiments in the attempt to make his system technically and commercially viable. One of his ideas, put into practice in 1927, was to record sound and vision on a gramophone record so that it could be played back through a record player connected to his televisor. However, his Phonovi-

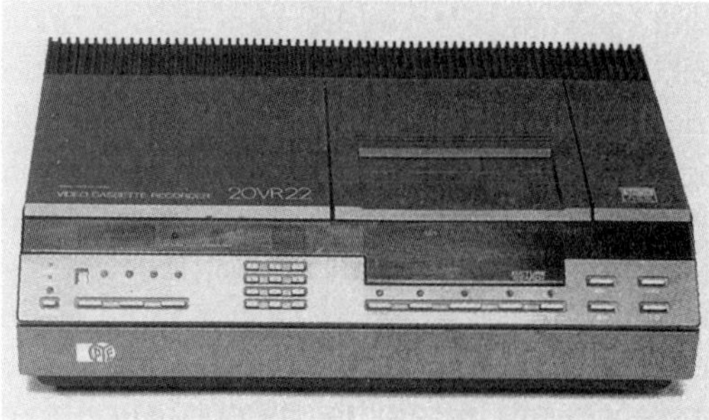

This 1980 Pye model 20VR22 VCR was made to the Philips Video 2000 specification. (The Museum of Science and Industry in Manchester)

sion discs failed to deliver adequate picture quality. The potential of magnetic tape for recording pictures as well as sound was also recognized in the late 1920s. Boris Ritcheouloff took out a British patent for a picture recorder based on the wire sound recorder developed in 1898 by the Danish engineer Valdemar Poulsen. As there was no outlet for such recordings, the idea never became a reality. Germany was well in advance of other countries in the development of audiotape recording, and a number of 1930s concepts relating to the design of recording heads were later adopted for videotape technology.

The Development of Professional Videotape Recording

After World War II, U.S. radio networks adopted audiotape technology, in place of sound recording on disc, for prerecorded broadcasts. Because the same networks were active in television broadcasting, the idea of using magnetic tape for prerecording television broadcasts was a natural afterthought. Prerecording television broadcasts on film was unsatisfactory in two respects: the film speed of 24 frames per second did not match the television picture rate of 30 frames per second; time had to be allowed for the film to be processed before broadcast. In 1950, Jack Mullin, the chief engineer at Crosby Enterprises, began to develop a prototype videotape recorder using tape made by the Minnesota Mining and Manufacturing Company (3M). During the next few years, a number of videotape systems went into development. In the United States, the main contenders were the Ampex Corporation, the leading manufacturer of professional audio recorders, and the Radio Corporation of America (RCA). In Britain, the British Broadcasting Corporation (BBC) was similarly engaged, and in Japan, Toshiba was just one of the companies at work in this field. RCA was the first company to make the transition from private demonstrations of the technology to experimental use in public broadcasting, albeit very briefly, in 1955.

As video occupies a much greater frequency bandwidth than sound—at least 2.5 MHz, compared to 20 kHz—the main problem was how to fit the video signal on the tape. The audiotape recorder had stationary heads that laid down sound as longitudinal tracks. When the same technique was used for video, a tape reel about 2 ft (600-cm) in diameter provided just 15 minutes of playing time. Ampex was the company that came up with a commercially viable solution. After initially experimenting with an arcuate scanning system, Ampex engineers devised a transverse scanning method, whereby the tape was wound transversely across a rotating drum holding the four recording heads. Thus, the video tracks were laid diagonally across the 2-in tape. This became known as the quadruplex video system. An Italian patent covering rotary heads for sound recording had been granted just before World War II. The Ampex system received a rapturous response when it was demonstrated at the 1956 National Association of Radio and Television Broadcasters convention in Chicago. It went into production as the VR-1000 model and achieved sales of 600 units by 1960. RCA introduced a quadruplex model, the TRT-1A, in 1957. While these first-generation videotape systems recorded black-and-white pictures, Ampex and RCA collaborated from 1958 to develop color video recording.

Meanwhile, Japanese companies, including Toshiba, JVC, and Sony, were developing an alternative scanning method, known as helical scanning. The principle of helical scanning had been expounded by a director of the German company Telefunken in the early 1930s, and Telefunken took out a patent for a two-head helical recorder in the mid-1950s. The helical system only needs two recording heads because it has a sloping video drum. As the sloping drum spins, the heads alternately traverse the tape obliquely. Japanese helical-scan videotape recorders were marketed from the early 1960s. Helical scanning was the method adopted for all consumer videotape recorders.

Compact Videotape Recorders

Operated by electronic valve technology, the professional videotape recorders of the late 1950s weighed almost half a ton. Thus, RCA's TR-22 videotape recorder, the first solid-state (or transistorized) model represented a major breakthrough when it was introduced in 1962. A year later, Sony launched the first compact solid-state videotape recorder, the PV-100 model, which used 1/2-in tape. A model intended for home use, the CV-2000, followed in 1965. The first compact videotape recorder to go on sale in Britain was the Telcan of 1963, manufactured by the Nottingham Electronic Valve Company. The Telcan was based on the VERA (Vision Electronic Recording Apparatus) videotape format developed by the BBC in the 1950s. However, videotape recorders were not only expensive luxuries, but also inconvenient for casual use. As had been the case with the audiotape recorder, the reel-to-reel format, which involved fiddly threading of tape reels, put off many consumers.

Early Videocassette Recorders

Taking inspiration from the success of the audiocassette format introduced by the Dutch company Philips in 1963, a number of companies began to develop videocassette formats. In 1969, Sony demonstrated the first videocassette format. Its 3/4-in U-Matic videocassette and recorders, introduced commercially in 1971, provided one hour of running time. U-Matic videocassette recorders were successful in attracting educational and business users, but not general consumers. A major disadvantage of the U-Matic format in terms of home use was that recordings could not be played back through a television set, but required a special monitor. Other early videocassette formats included Instavision, jointly developed by Ampex and Toshiba, and CartriVision, produced by the U.S. company Avco. The most long-lived of these early videocassette formats was the Philips version. The first Philips VCR, the N-1500 model, was launched in Europe in 1972.

Popular Videocassette Recorders

For the VCR to become a popular home appliance, it needed to be affordable and easy to use. Sony and JVC (Japanese Victor Company), an affiliate of the biggest Japanese electronics company, Matsushita, began to develop rival consumer VCR formats in the early 1970s. Sony's new Betamax format, which was basically the U-Matic system scaled down to 1/2-in tape, was launched in 1975. Priced at about £1,000 or $2,300, the SL-7300 Betamax VCR still fell into the luxury bracket. A year later, JVC/Matsushita launched its VHS (Video Home System) 1/2-in tape format. VHS had two key advantages over Betamax: VHS VCRs were little more than a third of the price of Betamax VCRs, and VHS cassettes provided two hours of running time. This was achieved by compressing the video signal to take up less bandwidth, but still provide good enough picture quality in the context of home use. While Sony concentrated on the quality end of the market, other manufacturers, such as Sanyo, managed to bring Betamax VCR prices down to VHS level.

Meanwhile, Philips continued to evolve its system through the N-1700 and VCR-LP formats to the final version, Video 2000, launched in 1980. Video 2000 was innovative in that, as with audiocassettes, the videocassette could be turned over, doubling the running time. It offered eight hours running time, compared to VHS's three hours, but was far more expensive. In 1979, the German company Grundig introduced a variant of the Philips format, SVR, that offered both high picture quality and a long running time (four hours). Inevitably, the proliferation of formats created some confusion amongst consumers and an initial reluctance to purchase, in case of making the wrong choice. As a result, for many, VCR rental was a preferable option. Videocassettes were also comparatively expensive at first, so overnight rental of prerecorded videos became popular.

The video invasion was resisted by the motion-picture industry, which was concerned about illegal copying of copyright material and a potential decline in cinema attendance. Illegal recording fell into two categories: home recording of material broadcast on television and commercial piracy. These issues had also arisen when the audiocassette recorder entered the home. While commercial piracy was clearly a breach of copyright and one that could reasonably be enforced, the issue of home recording was more contentious. In 1976, the Disney Corporation and Universal Studios sued Sony in the U.S. courts for providing the means for their copyrights to be infringed. A final judicial ruling took eight years to arrive, because the original decision in favor of Sony was successfully appealed and then reversed by the Supreme Court in 1984. In Britain, the 1988 Copyright, Designs, and Patents Act made home video recording legal if recordings were only kept for four weeks. By enabling the recording of television programs for delayed viewing, a practice referred to as time-shifting, home VCRs have transformed viewing habits.

By the mid-1980s, VHS had achieved an unassailable degree of dominance in the home VCR market. Philips abandoned Video 2000 in 1985, and Sony conceded to consumer demand by making VHS VCRs from 1988. However, Betamax's successor formats, Betacam, Betacam SP, and DigiBeta, captured the professional market, as they offered broadcast-quality recording. Standardization of the VHS format and the development of a long-play (LP) model, which doubled recording time, resulted in dramatic sales growth. In 1982, 19 percent of British homes and 6 percent of U.S. homes had VCRs. The lower ownership level in the United States may have reflected the wider existing viewing choice, owing to the greater number of television channels. Four years later, home ownership of VCRs had swelled to 46 percent in Britain and 33 percent in the United States, and Japan was well ahead with 60 percent household penetration. Since 1986 basic VCR technology has remained the same, but manufacturers have introduced a range of features to increase convenience and functionality. These include stereo sound and various forms of advanced programming, including bar-code reading. Another innovation, designed to counteract the problem of broadcast delays or changes when using timer recording, enables VCRs to detect and respond to a program delivery signal, giving precise start and end times.

See also Camcorders; Matsushita; Philips Electronics; RCA; Sony Corporation; Tape Recorders; Television; Videodisc Players

References:

Goodman, Robert. *How Electronic Things Work . . . and What to Do When They Don't.* New York: McGraw-Hill, 1999.

Rushin, Don. "The History of Magnetic Tape." On www.technicalpress.com/Articles/History/History_tape.htm.

Schoenherr, Steve. "Recording Technology History." On http://ac.acusd.edu/history/recording/notes.html.

www.popadom.demon.co.uk/vidhist/

Virtual Pets

A variant of the hand-held electronic game, the Tamagotchi virtual pet was launched in Japan in late 1996. The concept is simple: the virtual pet needs feeding, cleaning, training, and attention—and if the owner fails to provide these with specified regularity, the pet will die. Virtual pets have since been joined by other virtual characters, including virtual mermaids, virtual warriors, and virtual androids.

See also Electronic Games

Vodafone

In less than twenty years, the British mobile phone operator Vodafone has enjoyed spectacular financial success. The merger of the Vodafone Group with the American company AirTouch Communications in 1999 created the largest international mobile telecommunications company with a capital value of about $135 million. Vodafone AirTouch is ranked second on the British FTSE (Financial Times Stock Exchange) index, third in Europe and among the world's top 25 companies. It has 35.5 million customers worldwide and operates in twenty-four countries.

Vodafone began life as a result of the British government's decision in 1982 to grant two nationwide licenses for new cellular phone services. One license was awarded to the Cellnet consortium led by British Telecom (BT), the state telephone operator, and the other went to a consortium led by Racal Electronics. The Racal-Millicom consortium set up the Racal Telecommunications Division to develop and implement the system and chose Vodafone, to express the union of voice and data by phone, as the name for the cellular network. A wholly owned subsidiary, Vodac, was set up in 1984 to act as the Vodafone service provider in order to comply with the license terms, which forbade the network operator from selling direct to the public. The Vodafone network came on stream in 1985, with its first exchange in Birmingham, followed by the London exchange and then five more regional exchanges. By the end of 1986, it was serving 63,000 customers.

In 1988, Vodafone was listed on the London and New York stock exchanges for the first time after the flotation of 20 percent of its shares. In the same year, following the finalization of the GSM (Groupe Spécial Mobile) digital cellular standard, Vodafone placed a major order with Ericsson for GSM equipment. The digital cellular service, the first in Britain, began operating in 1991, the year that Vodafone became independent from Racal in Britain's largest-ever corporate demerger. At the end of 1991, the company had nearly 700,000 customers. Prior to the Vodafone–AirTouch merger in June 1999, the two companies were linked by their role as founding members and investors in the multinational satellite-based mobile telecommunications system, Globalstar. In the six months after the merger, Vodafone-AirTouch revealed ambitions to strengthen its dominant position in the major cellular network markets. In the United States, the company reached an agreement with Bell Atlantic to merge their cellular networks to create a nationwide service. Vodafone AirTouch then launched a successful bid for the German telecommunications group, Mannesmann, thus ending the year and the century on a high note.

See also British Telecom; Ericsson; Mobile Phones

References:

Forester, Tom. *High-Tech Society: The Story of the Information Technology Revolution.* Cambridge, MA: MIT Press, 1987.

www.vodafone.co.uk

www.vodafone-airtouch-plc.com

W

Waffle Irons

Waffles are hot moist cakes made from a batter of eggs, flour, butter, and cream and were traditionally made in cast-iron waffle irons. These are thought to have originated in Scandinavia. The irons were circular double-sided pans with raised grids, hinged together. They had long arms for use on open fires. The bottom pan was lightly buttered and heated, the top pan was greased, the batter was then poured on, the pans closed, the handles clipped together, and the iron returned to the fire. When the underside had browned the iron was turned over to cook the other side. The resulting waffle was then covered with melted butter, treacle, or maple syrup.

Waffles were a popular dish in the nineteenth century, and such irons were common in Europe and, especially, the United States. As a result it is unsurprising that electric models were produced. One of the first was produced by Landers, Frary and Clark in 1918, followed by models by McGraw, Hotpoint, and Westinghouse. The earliest models had circular pans and were usually chrome-plated and had Bakelite handles. By the 1940s, waffle irons had become larger and rectangular with thermostatic control. Dual models appeared; the Universal Cook-a-matic could be either a waffle maker or a grill by a change of grids.

As with other cooking appliances, the principles have remained constant but the materials have changed. Cases are now in plastics and the irons are Teflon. Current models include the Chef's Choice WafflePro line that allows the user to select the texture and color of the waffle, which is baked in 90 seconds.

Wallpaper Stripper

Stripping off old wallpaper prior to redecoration is a time-consuming job if done by hand with water and a metal scraper. The paper has to be well soaked for the job to go smoothly. Professional decorators use electrically heated, high-wattage strippers that consist of a heated tank of water connected to a rectangular panel with perforated pipes to allow steam to escape. The panel (or steam-plate) is placed against the wall so that condensation builds up and the steam soaks into the paper. These machines can be rented from tool rental shops on a daily basis, but smaller domestic models are also available.

See also Do-It-Yourself

Wal-Mart

Founded in 1962, the American discount retailing company Wal-Mart took less than thirty years to establish itself as the world's leading retailer, overtaking long-established companies such as Sears, Roebuck, Kmart, F. W. Woolworth, and Montgomery Ward. By 1997, when it became the first retail company to achieve annual sales of $100 billion, it was the biggest employer in the United States, with a workforce of 680,000.

In 1962, Sam Walton was a small-scale retailer whose experience of operating variety stores in Missouri and Arkansas convinced him that there was a growing market for discount stores. The retail map of the United States was shifting in response to increasing car ownership. Consumers were showing an increasing preference for shopping at suburban retail centers, which were able to provide ample parking space. Convenient parking meant that the cost-cutting self-service principle of supermarkets could be extended to stores selling bulkier goods, such as electrical equipment. Such stores became known as discount stores or big box stores. The first discount store, the Ann and Hope Mill Outlet in Rhode Island, opened in 1953. Sam Walton opened his first Wal-Mart discount store in Rogers, Arkansas, in 1962.

In 1969, the company was incorporated as Wal-Mart Stores, Inc., and five years later, it had thirty-eight stores. Booming business meant that in 1970 it was ready to centralize administrative services and opened a distribution center and offices in Bentonville, Arkansas. In the late 1970s, Wal-Mart began to accelerate its expansion by acquiring other retail chains, such as Mohr-Value and the Hutcheson Shoe Company. It also expanded its range of retail services to include pharmacy, jewelry, and car servicing. In the 1980s, Wal-Mart continued to develop new variants on its successful formula. In 1983, it launched Woolco Stores and the first of the SAM's Club line of membership warehouse clubs in Midwest City, Oklahoma. It also introduced a one-hour photo lab at its store in Tulsa, Oklahoma. Its first Supercenter, selling groceries as well as housewares and electric appliances, opened in Washington, Missouri, in 1988. The Supersaver pharmacy and toiletries chain was a major acquisition in 1989.

In the 1990s, the company turned its attention to expansion through acquisition and new overseas ventures. Wal-Mart acquired Western Merchandisers, Inc., of Amarillo Texas, and the McLane Company of Temple, Texas, in 1990. Its first overseas store opened in Mexico City in 1991, followed by the start of business in Puerto Rico in 1992. The company went on to develop new markets in Canada (through the acquisition of 122 Woolco stores in 1994), Argentina, Korea, China, and Germany. When Sam Walton died in 1992, Wal-Mart was thriving as the leading American retailer. In 1993, the company hit the phenomenal landmark of $1 billion sales in a single week.

Wal-Mart developed a distinctive image derived from its roots in small-town Middle America; from the beginning, it placed great emphasis on hometown identity and friendly service. In 1983, it introduced a new form of personal service by employing People Greeters to assist customers on arrival. Wal-Mart's environmental initiatives have included the employment of Green Coordinators and the designation of selected stores as Environmental Demonstration Stores. Each Wal-Mart store provides college scholarships and charitable donations to the local community. The company has also operated a vigorous Buy American program. In 1999, Wal-Mart gained a foothold in the United Kingdom market when it purchased the ASDA group, a chain of 229 supermarkets.

See also Kmart; Sears, Roebuck and Company

References:
www.walmart.com/estore/pages/about1.jsp

Washing Machines

A washing machine, in its most basic form, is a washtub with a built-in mechanism for loosening dirt from fabrics through suction, friction, or agitation. As one of the most laborious household tasks, washing clothes was an obvious candidate for mechanical im-

"Home Washing Machine and Wringer"; lithograph, 1869 (Library of Congress)

provement. The earliest manual washing machines were made in Britain in the late seventeenth century, but few households had washing machines until the invention and development of the electric washing machine in the twentieth century.

Electric washing machines were not more mechanically efficient than their steam-driven or hand-operated predecessors. They featured the same washing mechanisms devised in the nineteenth century. The innovation lay purely in the substitution of electricity as the power source. Steam power was impractical in smaller houses, and manual washing machines were more expensive but scarcely less labor-consuming than the traditional washtub and separate washboard, dolly, or posser.

Alva John Fisher, an American engineer, is credited as the inventor of the first electric washing machine. His design of 1907, patented in 1909, attached the electric motor to the dolly-type agitator with a long belt. The agitator was fixed to the lid rather than the base. The drawbacks of his design were that it was easy for the motor to get wet and for users to get electric shocks from the exposed belt. The Hurley Machine Company of Chicago produced the first commercial washing machines based on Fisher's patent under the Thor brand name.

By 1914, U.S. and Canadian companies had developed a range of designs. Hurley produced a washing machine with a mechanism similar to James King's U.S. patent of 1851. The tub was a trough containing a perforated cylinder that rotated horizontally, with the motor placed below the tub. The Thor washing machine also had a powered wringer. It was marketed as being operable from a light socket. Maytag of Iowa and the Canadian company, Beatty Brothers, produced washing machines with vertical agitators fitted in the base of a cylindrical tub and driven by a motor below the tub. Both types of machine were top-loaders. Washing with these machines was still quite laborious because water had to be put in and removed by hand.

These two mechanisms—the rotating drum, producing friction and suction, and the vertical agitator, producing agitation and friction—proved to be the basis for all commercially successful washing machines through to the present day. The top-loader with vertical agitator and power wringer was the most common type of washing machine, with changes in materials and styling, until the late 1950s. The wooden tubs of early washers were superseded by tubs made of lighter metals, a legacy of World War I production processes. Water heaters and emptying pumps were fitted. Maytag's Model 80 of 1922 featured a shallow, square, cast-aluminum tub, pointing the way to the more standardized box-like washing machines of the future. A very similar British model, the Hotpoint "gyrator" washer, was chosen for the King George V Jubilee House of 1935. A British advertisement of the 1930s entitled "Washing Day, Washing Play" encouraged British housewives to imitate their U.S. counterparts by hastening to acquire a washing machine. It stated that thousands of women across the United States were playing tennis on washing day owing to their newfound leisure time!

Other washing machine variants of the 1920s and 1930s, such as the washer with an oscillating tank and the gas-heated washer, had little impact. There was also a short-lived spate of multipurpose washing machines that were designed to take attachments such as mincers, food mixers, and knife cleaners. A more limited and logical space-saving idea was the combined washing machine and dishwasher, which merely required interchangeable mechanisms to be fitted in the tub. This idea was revived in the 1940s, but by then washing machines were developing in other directions.

In the 1930s, the fashion for streamlining appliances in line with the vision of a more scientifically managed and efficient kitchen included the washing machine. This had a more immediate impact in the United States than in Britain, where streamlining was largely a postwar phenomenon. The first changes were the extension of the tub casing to hide the motor below and the encasing of the wringer. An early example was Kenmore Toperator of 1933, a cylindrical model designed for the U.S. retail company Sears, Roebuck by the American designer Henry Dreyfuss. The next stage was the promotion of the rectangular shape combined with the shift from a mottled enamel finish to a white porcelain-enamel finish. Maytag was a leader in this further streamlining of the washing machine, commissioning the U.S. design studio Harold van Doren Associates to produce a line that included the Master model of 1939 and the Commander model of 1940. These masculine model names emphasized the values implicit in the design: the washing machine as a tool that empowered the housewife to conquer the drudgery of washing. Whiteness was both a symbol of the ultrahygienic standards of the washing process and a motivator to keep the appliance as clean as the clothes it washed. Streamlining did not reach its ultimate form until the late 1940s when legs were completely banished in favor of casing from top to bottom.

Combining washing and spin-drying functions in a single machine required a means of reconciling the low-speed rotation of the washing action with the high-speed rotation of the spinning action. The solution was the two-speed motor with automatic speed-changer. In the 1920s, the development of motors of this kind enabled manufacturers to

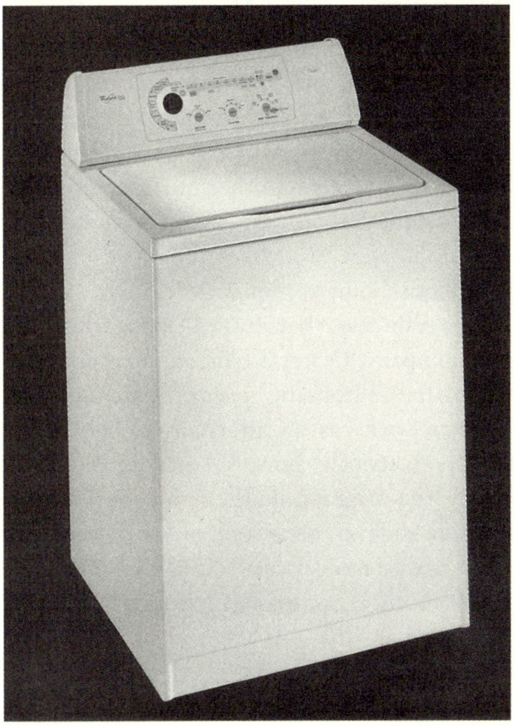

The Whirlpool Catalyst TM washer, 2000 (Courtesy, Whirlpool)

produce the first combined washing machine–spin dryers, which were top-loaders. These machines were too expensive to find a market. The first attempt to mass-market the front-loading automatic washing machine came in 1938, when Bendix launched the automatic washer in both the United States and Britain. World War II and its consequent restrictions on nonessential production intervened before the automatic washer could make any impact, but this type of washing machine was launched again in the United States after World War II. It is based on the rotating drum principle, with clothes placed inside a perforated drum and revealed by a porthole window in the door. The Westinghouse Laundromat model of 1948 featured a handy drop-down front panel for resting clothes on removal.

By the mid-1950s, the wringer was rendered obsolete by the spin dryer, either integrated with the washer or as a separate appliance. Although twin tubs (washers and dryers housed side-by-side in a single cabinet or joined on a base frame and sharing a power cable) were less compact and required more effort from the user than automatics, they were more efficient spinners. In Britain, Wilkins Servis was a popular brand of twin tubs. The automation of the washing process was completed when tumble drying was incorporated into automatic washers to produce the composite washer-dryer. The leader in the washer-dryer market was Bendix, an early proponent of the automatic washing machine. The first washer-dryer, the Bendix Dumatic, was launched in the United States in 1953 and arrived in Britain in 1957.

The dominance of white was challenged in the 1950s and 1960s by the availability of washing machines in colors that matched the current fashions in kitchen design. This attraction faded as people realized that the washing machine would outlive the decor. By 1970, with the washing machine established as a staple domestic appliance, manufacturers needed to find new ways to cultivate consumers. Washability had become a factor influencing selection of fabrics by consumers, and textile technologists continued to develop new synthetic fabrics and to modify natural fabrics. Consumers also became increasingly concerned about environmental issues, encouraging manufacturers to reduce the water, power, and detergent consumption of washing machines. Manufacturers were able to supply conventional solutions to these concerns by developing washing machines that offered a wider range of wash programs and could operate efficiently with less water at lower temperatures, controlled by standard thermostats and gauges.

The 1986 Jetsystem washing machine, produced by the Italian company Zanussi, introduced the innovative spray washing technique, which economized on water and therefore also electricity and detergent. A more innovative approach was to harness the abilities of new "smart" microchips. These chips, developed from the mid-1980s, embody "fuzzy logic," a set of instructions that allow for best-fit solutions rather than just exact solutions. Consequently, washing machines with advanced fuzzy logic can adjust water and detergent intake according to load. Electronic chips can also detect load distribution and, with well-balanced loads, permit automatic washing machines to spin at higher speeds than previously possible. Manufacturers such as the German companies, AEG (Allgemeine Elektrizitats Gesellschaft) and Bosch, have focused on the market for more sophisticated, and therefore more expensive, washing machines. The future of environmentally friendly washing machines could lie in a different direction: ultrasonic waves are capable of releasing dirt without the agency of water or detergent. Although this technology was available in the 1950s, under the brand name Vibrasonic, it had no market impact so its reliability and commercial viability is untested.

Today, the vast majority of homes in both the United States, Canada, and Europe have automatic washing machines, but the patterns of growth of ownership were very different. Early standardization of electricity in the United States encouraged faster growth of domestic electrification and ownership of appliances than in Europe. In Europe, British homes were slower to acquire washing machines than German ones, but faster than French. Thus, while just over half of U.S. homes had washing machines in 1941, only 4 percent of British homes had washing machines in 1948, rising to 29 percent in 1958. It was only in 1963 that half of British homes had washing machines.

See also Clothes Dryers; Detergents; Dishwashers; Maytag Corporation; Synthetic Fabrics; Whirlpool Corporation; Zanussi

References:

Giedion, Siegfried. *Mechanization Takes Command.* New York: Oxford University Press, 1948.

Hardyment, Christina. *From Mangle to Microwave: The Mechanisation of Household Work.* Cambridge, UK: Polity Press, 1988.

Lupton, Ellen. *Mechanical Brides: Women and Machines from Home to Office.* New York: Cooper-Hewitt National Museum of Design, Smithsonian Institute, and Princeton Architectural Press, 1993.

Waste Disposal Units

The twentieth century saw more and more people living in cities and suburbs, accompanied by rising levels of domestic waste that needed to be disposed off. Previously most household waste was either burnt or used for composting. Urban sanitary reforms had led city authorities to collect domestic garbage from households at regular intervals.

Nevertheless, there was a cost to this, and one company, General Electric, began to develop the automatic garbage disposer or "electric sink" as an alternative. The result was an electrically powered grinder mounted below the strainer in the drain outlet of a kitchen sink. It consisted of a control element, waste-receiving chamber, grinding plate, strainer, pump, and motor. Essentially, it ground up the garbage and flushed it directly into the sewer system.

General Electric began to design their Disposall in 1929, and it went on sale in 1935. Although influenced by larger grinding machines used at sewage plants, it was specifically designed for the domestic market. It was developed at the General Electric laboratories in Schenectady, Pennsylvania, and Fort Wayne, Indiana. Morris M. Cohn, the Schenectady sanitary engineer, was a consultant and active promoter of such garbage "dual disposal" systems, which handled both domestic and human waste. Opponents felt that it was too expensive and impractical in large cities with already strained sewerage systems. General Electric was the sole manufacturer and had only sold 175,000 units prior to World War II. They were expensive and noisy.

The war stopped production of the domestic model, but General Electric introduced a larger type for the armed forces. In doing so, they refined the home version, using fewer parts and introducing a "water flow interlock." This was linked to the sink faucet or tap by an electrical contact and allowed water to flow through the unit before it operated, thus cutting down on the buildup of grease.

Postwar prosperity helped increase sales. By 1948 there were seventeen companies manufacturing disposers, and around 200,000 were being installed annually. Consumers were able to add this appliance to the increasingly electrified U.S. kitchen in the 1950s, provided they were permitted to install them. Disposers were forbidden in Boston and New York City. A few smaller towns such as Jasper, Indiana, opted completely for the dual disposal system, installing disposers in every home in 1950.

Prices were between $75 and $135 for a unit; installation was between $20 and $150 in an old house and, significantly, only $10 in a new one. The manufacturers saw that the product was becoming more acceptable and began to target property developers in the rapidly expanding suburbs, where they could offer enticing discounts. By 1958 disposer sales topped $50 million and in 1959 figures indicated that more than four million households had them installed. Manufacturers promoted disposers as being safe, healthy, and clean, aided by the fact that most cities did not prohibit them. By the 1960s, many required that all new houses have them. Safety was a potential problem as the opening at the top had to be wide enough to be filled (60 to 80 mm); therefore careless fingers could come into contact with the cutting blades.

The most popular disposal unit is the cheaper and simpler "continuous-feed" type in use in most U.S. homes. Less popular is the "batch-feed" type that does not start until the sink plug is put in.

The garbage disposer has not traveled so well outside the United States. Although sometimes a feature of new houses and luxury flats, it is far from ubiquitous in Europe. Sales in the United Kingdom in 1990 were only around 100,000.

References:

Giedion, Siegfried. *Mechanization Takes Command.* New York: Oxford University Press, 1948.

Hoy, Suellen. "The Garbage Disposer, the Public Health, and the Good Life." In *Technology and Choice: Readings from Technology and Culture,* edited by Marcel C. LaFollette and Jeffrey K. Strine. Chicago and London: University of Chicago Press, 1991.

Water Closets

By the end of the twentieth century, the indoor water closet (or toilet) had become one of life's essentials in the developed world. Yet, only 50 years earlier, many homes still had outdoor water closets, and a century earlier, many homes still had earth closets. The late arrival of the water closet as a home fixture is all the more surprising given that its ancestry goes back several thousand years.

Water-borne sewerage systems existed in late prehistoric times in places including Mesopotamia and the Indus Valley. Water engineering was one of many technologies in which the ancient Romans excelled. Roman technology spread throughout much of western Europe in the wake of the Roman army's successful campaigns. Roman forts and fortresses featured communal latrines, which were cleaned out by water flowing through channels beneath the rows of toilet seats. This kind of toilet system was revived in medieval monasteries, but the chamber pot was the dominant toilet receptacle until the nineteenth century. Closestools concealed the chamber pot in a wooden box, often with a padded lid. The eighteenth century fashion for polished wooden cabinets produced a more stylish variant, known as the night table or commode.

The first modern water closet was invented in 1594 by the English courtier and writer Sir John Harington, who installed one in his own house near Bath. In 1596, he published a description of his invention and its historical context. A year later, he made one for Queen Elizabeth I, which was installed in Richmond Palace. Harington's water closet consisted of a lead-lined wooden pan with sloping sides, which was filled by a water cistern. A plug in the bottom held the water in place and kept smells from wafting back from the sewer. After use, it was emptied by pulling a rod to lift the plug so that the con-

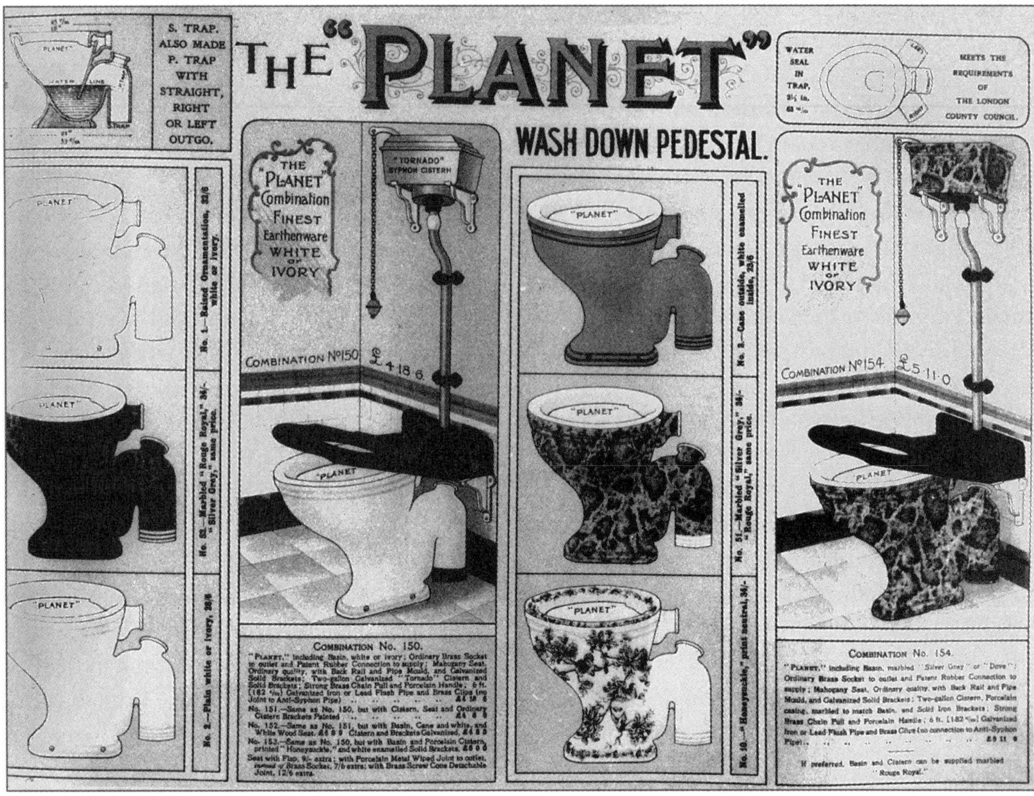

The Planet line of wash-down water closets from the 1901 Twyfords' catalog (The Museum of Science and Industry in Manchester)

tents drained away. However, there was little incentive for the wealthy householders who could afford such a system to install one because the unpleasant task of emptying chamber pots outside the home fell to servants. In the late eighteenth century, the plug mechanism was replaced by a valve, introducing the valve closet. The English watchmaker Alexander Cumming developed a design with a sliding valve in 1775, and the English inventor Joseph Bramah patented a design involving two hinged valves in 1778. Valve closets were encased in polished wooden cabinets. Metal closet pans began to be superseded by ceramic pans in the early nineteenth century.

During the nineteenth century, valve closets became widespread in wealthy and middle-class households, which were able to afford piped water. Even in these homes, water supplies were restricted and there was a need to conserve water by fitting cisterns with balls and stopcocks, which prevented overflows. A cheaper water closet, the hopper closet, was developed as a working-class equivalent of the valve closet in the early nineteenth century, but was rarely adopted. It consisted of a ceramic conical pan, flushed by jets of water spiralling from the rim, and a separate S-trap. To discourage the fouling of streets and rivers through the disposal of chamber-pot contents, communal privy middens, also called privies, were provided in working-class areas. These were nothing more than cesspits with seats constructed above. If these pits were badly lined or allowed to overflow, the contents fouled the groundwater. From about 1860, privies began to be replaced by earth closets, also known as dry closets or pail closets. The placing of a pail below the seat allowed easy removal and disposal of waste, which was neutralized by the addition of dry earth or ashes from a hopper after each use.

The modern valveless water closet began to evolve in the mid-nineteenth century. It was distinguished from the valve water closet by its siphonic flushing and its containment within an earthenware pedestal that removed the need for a supporting and concealing wooden cabinet. The wash-out closet, such as that patented in 1852 by the British sanitary engineer Josiah George Jennings had a shallow pan with a concave tray that retained a seal of water above the level of the outflow trap. The shallow design with a large contact area meant that the flushing action was rather gentle. A concurrent development was the wash-down closet, which was essentially an improvement of the hopper closet. The key improvement lay in the one-piece design, whereby the pan and S-trap were formed as a single molding, preventing leaks. The wash-out closet was also one piece, but the wash-down closet had other advantages in its deeper, narrower pan with a higher water seal and the more efficient siphonic action achieved by the tighter configuration of the S-trap. Accordingly, the wash-down closet became the standard type. From 1870, many designs of wash-down closet evolved and were proudly exhibited by specialist sanitary ware manufacturers. The grander designs had highly decorative molded or transfer-printed sides and almost equally grand names, such as Jennings' Pedestal Vase and Twyfords' Unitas.

By the late nineteenth century, most urban areas in Britain and the United States had solved their water supply problems by building reservoirs and through better-engineered systems of distribution pipes. The fear of creating water shortages had been a justification for not piping water to all homes and, consequently, not providing sewerage systems to allow the use of water closets universally. From about 1880, this began to change and water-closet sewerage systems were gradually introduced throughout urban areas. The water closets designed for working-class homes were made of cheaper fireclay in simple shapes and with plain glazes, often buff-colored. In some parts of Britain, local authorities were still sufficiently concerned about potential water shortages to adopt a type of water closet known as a slop-water, tipper, or tippler closet. The pedestals of these water closets were placed over wide pipes leading to a shallow pan holding a tipper, at a depth level with the kitchen wastewater sewer. The tipper collected the waste and was periodically flushed by the flow of kitchen wastewater, which caused it to tip and the contents to be washed out. Hence, it needed no cistern and used no extra water. Better-designed cisterns also helped to prevent water wastage. Victorian cisterns were lead- or zinc-lined wooden chambers, placed high above the water closet to increase the force of water and operated by pulling a long chain with a ceramic handle. The two broad types were the valve type and the siphonic type. The valve type released the contents of the cistern to the flush pipe when a lever opened the valve. Siphonic cisterns fell into two main subtypes, the piston type and the inverted-cup type. The piston type used a lever to lift a piston that drew water through an s-shaped siphon and down the flush pipe, while the inverted-cup or bell type used a lever to raise and then lower the inverted cup, thereby drawing the water into the exposed flush pipe.

During the twentieth century, the technology of the water closet has evolved very little, but the style and materials have changed in line with prevailing tastes and developments. While many homes still had outside toilets, the appearance of the water closet was of little importance. The interwar period brought the modernization of water-closet design, along with the concept of the matching bathroom suite. One example of this is the bathroom suite patented in 1936 by the American designer Henry Dreyfuss and manufactured by the Crane Company. The water closet has a slim, low-level cistern that sits on a streamlined box, which exposes only the water closet basin, concealing the trap and pipes. Because low-level cisterns did not require a long chain pull, a compact lever or push-button sufficed. This style of water closet and matching fixtures only found its way into middle-class homes at that time, but the basic design concept became more widespread in the 1950s.

Vitreous china, which is less permeable, less susceptible to staining, and more durable than earthenware, also came into use in the 1950s. While, like earthenware, it was originally press-molded, the current technique of slip casting liquid clay means that it can be made into almost any shape. This has made it possible for the cistern and pedestal to be formed as one piece, the ultimate in streamlining and ease of cleaning. Alternatively, plastic has become a cheap, durable, and lightweight alternative for cisterns and toilet seats. White, symbolic of hygiene, has been the standard color for water closets for much of the century. The fashion for colored bathroom suites has fluctuated, peaking in the 1970s, when avocado became particularly popular. Since then, white (which does not date) has enjoyed a revival in popularity, as it was more in keeping with the minimalist, high-tech aesthetic of the 1980s and 1990s. Late-twentieth-century retro styling revived decorative Victorian and Edwardian water closet designs. However, a growing architectural salvage industry has made the genuine period article more widely available and often a more attractive option for owners of period houses seeking authentic fixtures.

See also Piped Water

References:

Muir, Frank. *A Book at Bathtime.* London: Heinemann, 1982.

Wohl, Anthony S. *Endangered Lives: Public Health in Victorian Britain.* London: Methuen, 1984.

Wright, Lawrence. *Clean and Decent: The Fascinating History of the Bathroom and the Water Closet.* London: Routledge, 1980.

Water Heaters

Before World War I, heating large volumes of water for washing clothes and bathing was a laborious and inefficient process in most homes. In 1900, many households were still without piped cold water and only a small minority of households had piped hot water. Using large pots and pans to heat water on the kitchen range or above open fires was time-consuming and inconvenient. Large receptacles called coppers, fixed above fire grates in brick enclosures, were more convenient, but still had to be filled by hand. Originally made of copper, which was a good conductor of heat, they were more commonly made of cheaper cast iron. They were usually located in basements or outhouses to fit in with laundry needs. A similar idea for heating bath water was the gas bath, which had gas burners below the bathtub for heating the filled bath.

In the first half of the nineteenth century, a piped hot water supply involved fitting a boiler at the back or side of a kitchen range or installing a freestanding boiler. In both cases, the stove burned solid fuel, such as coal, coke, or wood, and the hot water was stored in a cylindrical tank, often placed in a linen cupboard. The disadvantage of such boilers was the length of time that it took to heat a full tank of water and the inability to vary the volume of water heated according to particular needs. The first water heater to address these deficiencies was the gas geyser, invented by Benjamin Waddy Maughan in 1868. The term "geyser" was taken from the Icelandic word "geysir" for a natural hot spring that gushes intermittently. Water was drawn into the geyser through a spiral coil of pipe, which was heated by gas burners. This meant that hot water was available instantaneously on demand, with no heat lost through storage. Maughan's geyser was designed to be used in the bathroom and took the form of a gleaming brass and copper cylinder on elegant cabriole legs. There was no means of supplying hot water from the bathroom geyser to the kitchen, scullery, or laundry room, creating an opening for other manufacturers. For example, the British gas appliance company Sugg developed a geyser called the Boiling Steam Therma that was de-

A 1950s Santon looped-element immersion heater (The Museum of Science and Industry in Manchester)

signed to be wall-mounted above the kitchen sink.

A major shortcoming of these early gas geysers was their lack of basic safety features, which made them prone to accidents through misuse. For instance, they had no pilot jets, which meant that there might be a build-up of gas before the burners could be lit, causing a dramatic and off-putting bang. Equally dangerous, the water and gas controls were completely separate, so there was nothing to prevent users from lighting the gas burners with no water flowing, which would melt the pipe solder and create smoke and fumes. Built-in safety measures improved the appeal of gas geysers, although some consumers continued to be apprehensive. In 1899, the British company Ewart introduced the Califont geyser, which was an advance in technical design as it could withstand a pressurized water supply. This meant that the geyser was not confined to producing hot water locally through its own outlet, but could be activated by turning on any tap in the home. The change in water pressure caused by turning on the tap opened the valve that controlled the gas supply to the water heater and the pilot jet lit the burners.

The first electric water heaters, which appeared in the early 1890s, were copper urns with elements fastened to the underside. They were inefficient and expensive to use because much of the heat was wasted on heating the surrounding air and the urn itself. Electric geysers became available just before World War I. While they were compact and much more efficient than electric urns, they demanded a 10-kilowatt load, which imposed a severe strain on not just the domestic circuit, but also the local mains supply. Hot water remained a luxury until small gas water heaters became widely available in the 1920s and 1930s. These were more affordable for the average householder and were adequate for the needs of smaller homes. Gas water heaters became more efficient as a result of the development in Germany of the aerated gas burner. This was introduced in Britain by Bernard Friedman, an immigrant German engineer, who launched the Ascot water heater in 1932. Even by the end of the twentieth century, the gas instantaneous multipoint water heater remained one of the most convenient and efficient water heaters.

The load demands of instantaneous water heating meant that electricity was unable to compete with gas in this market. Therefore, by the 1920s, electrical appliance manufacturers were concentrating on improving water storage heaters. One advantage was that some electric heaters could be fitted to existing hot tanks, reducing installation costs. For example, electric belts were designed to fit around cylindrical tanks. However, as with the earlier electric urn, these external heaters wasted energy. The introduction of the immersion heater, which was placed inside the tank, made electric water heating more efficient. Immersion heaters came in various shapes, of which the most common was the rod type. The development process was similar to that of the electric kettle, with the important distinctions that the kettle did not need to store hot water and the water heater did not need to boil water. Storage efficiency could be improved by surrounding the tank with insulation material, while water temperature could be regulated by thermostatic controls. There were also improvements to the management of water intake and outlet. In early immersion tanks, as hot water was drawn off, there was a corresponding intake of cold water, maintaining the tank at full capacity. Thus if the user did not use the entire tank, the remaining hot water would be diluted with cold water and reheated. In the late 1930s, manufacturers developed another option, which involved fitting a ball-cock or float to the outlet pipe (rather than the intake pipe, as in the case of a water closet cistern) to prevent the intake of cold water until the tank was drained.

Self-contained gas or electric water heaters were the norm in less wealthy households in both Britain and the United States up until World War II. After the war, central heating, which combines space heating and water heating, soon became more widespread in the United States as it began to be adopted as a standard feature in new housing. In Britain, central heating was still regarded as a luxury until the late 1960s and 1970s, and new homes built in the 1950s were more likely to have modest new features, such as small electric water heaters specifically designed to fit discreetly inside kitchen sink units. In 1981, almost two-thirds of homes in Britain relied on immersion heaters for bulk water heating.

See also Aga Cookers; Kettles; Heating; Piped Water; Washing Machines

References:

Williams, Trevor I. *A History of Technology, Vols. 6 and 7: The Twentieth Century.* Oxford, UK: Clarendon Press, 1978.

Wright, Lawrence. *Clean and Decent: The Fascinating History of the Bathroom and the Water Closet.* London: Routledge, 1980.

West Bend

The West Bend Company is a major U.S. manufacturer of a variety of small appliances and takes its name from the town of West Bend, Wisconsin. It was founded by Bernhardt C. Ziegler, a young entrepreneur who wished to establish work for local people following a disastrous fire that closed down the main local employer, a pocketbook factory, in 1911. He and six others put down $1,000 each and founded the West Bend Aluminum Company in the same year. They rented an old button factory and began production of saucepans, first exhibiting their products at a hardware association meeting in Milwaukee in 1913. Like the fledgling Whirlpool Company, West Bend produced wares for Sears, Roebuck. It began to introduce new lines in the 1920s and 1930s with the successful Waterless cooker and the Flavo-Seal range. It also produced the Flavo-Drip (1922) and Flavo-Perk coffeemakers, a portable coffee urn, the Speedmaster electric kettle, and the Cadet, a water heater for cars.

The company emerged from wartime production with new products such as popcorn poppers and colored anodized aluminum cookware. These were available in a range of colors. The Trig singing kettle came in cherry red, delphinium blue, and sunset gold. West Bend continued to expand the line in the 1960s, introducing Teflon-coated products and buffet appliances such as the Automatic Buffet Chef and the Smokeless Broiler-Rotisserie.

West Bend remained privately owned until 1968, when it was acquired by the Rexall Drug and Chemical Company, which merged with Kraft in 1980. During the 1970s, the company branched out into more specialized appliances such as electric pizza bakers, yogurt makers, and woks. It began to concentrate on electrical appliances, dropping its retail cookware products in 1982. It continued to manufacture for distributors, however. It also diversified into home exercise systems and bathroom products. Its best-selling product of the early 1980s was a cordless iron, followed by the first, and very successful, U.S.-manufactured breadmaker of 1993.

West Bend is still based in the town whose name it took. It is now a division of Premark International.

See also Breadmakers; Coffeemakers; Cookware; Kettles; Popcorn Poppers

References:

www.westbend.com

Westinghouse Electric

The electrical engineering company founded by the American inventor and entrepreneur George Westinghouse in 1886 was a dominant force in the U.S. electricity industry for many years. It also made a significant contribution to the development of radio and television broadcasting. Today, the former Westinghouse Electric has been split into three independent companies, linked through interlocking ownership. These are the Westinghouse Electric Company, Westinghouse Government Services Company, and Westinghouse Government Environmental Services Company.

George Westinghouse was born in Central Bridge, New York, in 1846. At the age of ten, he moved with his family to Schenectady, New York, where his father opened a shop that sold agricultural machinery and small steam engines. There, Westinghouse became familiar with and fascinated by machinery. He returned to this interest after fighting in the American Civil War and filed the first of his 400 patents in 1865. Westinghouse invented a compressed-air brake for railway locomotives in 1868 and set up the Westinghouse Air Brake Company to manufacture his invention in Pittsburgh in 1869. Continuing his railway interests, he set up the Union Switch and Signal Company in 1881 to produce signaling and switching equipment. After developing a reduction valve to control the distribution of natural gas, he began to apply the same line of thinking to electricity and developed an improved transformer. Convinced of the superior efficiency of alternating current (AC) over the direct current system pioneered by Thomas Edison, he founded Westinghouse Electric in 1886 and equipped the first commercial U.S. AC generating station in Buffalo, New York.

In 1888, Westinghouse gained control of the key AC technology by acquiring the rights to commercially exploit the patents of the Croatian engineer Nikola Tesla. In 1884, Tesla had immigrated to the United States and begun to develop AC generating and transmission equipment, including the first practical AC induction motor. In 1889, Westinghouse Electric manufactured an electric fan, the first appliance based on the compact AC motor. In the same year, Westinghouse set up a model town for his workforce in Turtle Creek Valley, near Pittsburgh. In 1890, Westinghouse undertook his first long-distance AC commission when he equipped a power station at the Willamette Falls, Oregon, which transmitted electricity 14 miles to Portland. The 1893 Columbian Exposition in Chicago provided an opportunity for Westinghouse to mount a huge public display of electric lighting. Industrial use of alternating current was facilitated by the introduction in 1894 of the first polyphase induction motors, suitable for driving machinery.

Westinghouse's greatest electrical engineering achievement to date was to manufacture the generators for the new AC power plant at Niagara Falls in 1895. The Niagara Falls hydroelectric project was a publicity triumph for Westinghouse, culminating in the 1901 Pan-American Exposition in Buffalo, which had electricity as its central theme.

Edison finally accepted that his direct current system was doomed and, in 1895, General Electric and Westinghouse agreed to pool their patents through a cross-licensing arrangement. This gave the two companies a virtual monopoly of heavy electrical engineering and benefited users by standardizing electricity supply. In 1900, Westinghouse built the first U.S. steam turbine-generator, which was installed at the Hartford Electric Light Company in 1901.

In spite of these technological successes, George Westinghouse lost control of Westinghouse Electric in 1908 after a series of financial crises. Nevertheless, Westinghouse Electric continued to be at the forefront of U.S. electrical engineering. In 1917, it produced the first automatic substation and the first fully automatic electric range. Before World War I, Westinghouse and General Electric had left consumer electrical appliances, other than lighting, to smaller companies because the profit potential of the small domestic market was limited. This changed in the 1920s as more and more homes were electrified. In 1924, Westinghouse introduced the first commercial model of the automatic electric iron. Ten years later, it created an all-electric home in Mansfield, Ohio. The domestic appliance division was acquired by White Consolidated in 1975 and lives on as the White-Westinghouse line, still a major U.S. brand.

Radio communications became a new business area for Westinghouse when the U.S. government negotiated the acquisition of Marconi's U.S. subsidiary by a consortium of U.S. companies, including Westinghouse, in 1919. Under this arrangement, the resulting Radio Corporation of America marketed radios made by Westinghouse and General Electric under the Radiola brand name. In 1920, Westinghouse launched regular U.S. radio broadcasting with the transmission of the presidential election results from its KDKA radio station in East Pittsburgh, followed by a program of daily broadcasts. In 1921, Westinghouse introduced its first radio receivers for home use. Two years later, Westinghouse made the first international radio broadcast by short wave. In 1926, Westinghouse, RCA, and General Electric formed the National Broadcasting Corporation (NBC), which soon operated a coast-to-coast network of 25 radio stations.

While radio broadcasting was still in its infancy, Westinghouse Electric became involved in the new field of television. A Westinghouse research engineer, Vladimir Zworykin, who had immigrated to the United States from Russia in 1919, began to develop an electronic television system. He demonstrated his television camera, known as the iconoscope, in 1923, and developed the picture-tube receiver, known as kinescope, between 1924 and 1928. In 1929, Westinghouse ceded its television research and development to RCA, after lobbying by David Sarnoff, later president of RCA.

In the late 1930s, Westinghouse moved into the fields of nuclear and defense engineering, new interests that were strengthened during World War II. Among its military production achievements were the first long-range warning ground radar and the first gyroscopic gun control for tanks. In 1941, Westinghouse developed the first U.S. jet engine. Wartime production of pure uranium for nuclear weapons was followed by the production of zirconium for nuclear reactors in 1952. Westinghouse continued to be involved in both military and civilian applications of nuclear energy, developing the atomic engine for the Nautilus submarine in 1953–1954 and equipping the first U.S. nuclear power plant at Shippington, Pennsylvania, in 1957. In the 1960s, its radar equipment was tailored to meet the new demands of the U.S. space program, while Westinghouse television cameras filmed the 1969 moon landing. The company continued to develop its nuclear and defense technologies during the 1970s, alongside its production of steam turbines and generators.

In 1999, CBS Corporation, which had been acquired by Westinghouse in 1995 and then emerged as the parent company, decided to sell off Westinghouse as three businesses. Westinghouse Electric, the nuclear electrical engineering business, was acquired by British Nuclear Fuels Limited (BNFL). The Westinghouse Government Services Company, the defense products business, was acquired by the U.S. company Morrison Knudsen. The new ownership of the third business, the Westinghouse Government Environmental Services Company, provided the link between the other two, as Westinghouse Government Services took a 60 percent interest and BNFL the remaining 40 percent.

See also Electricity; General Electric; General Electric Company; Radio; RCA; Television

References:

Nye, David E. *Electrifying America: Social Meanings of a New Technology.* Cambridge, MA: MIT Press, 1990.

www.westinghouse.com/corp/

Whirlpool Corporation

Whirlpool is one of the largest appliance manufacturers in the world. Although not a major innovator, it is good example of the globalization of products and brands. Whirlpool traces its origins to the Upton Machine Company, which was founded in St. Joseph, Michigan, in 1911 to manufacture motor-driven washing machines and wringers. It received an order to manufacture for Sears, Roebuck in 1916.

The Upton Machine Company merged with the Nineteen Hundred Corporation of Binghamton, New York, in 1929, operating from both sites until 1939. In 1949, the company produced its first automatic washing machine, which was marketed under the Sears brand. It introduced its own brand, Whirlpool, a year later and changed its name to the Whirlpool Corporation in 1950. As a producer of larger appliances such as refrigerators and washing machines, it was well placed to take advantage of the rising prosperity of the 1950s and 1960s. The line was able to expand to include dishwashers and electric dryers.

Whirlpool expanded during the 1950s, acquiring new manufacturing sites and other companies, including the Seeger Refrigerator Company of St. Paul, Minnesota, and the Estate brand and air-conditioning division of RCA. It was beginning to look to foreign markets in the late 1950s, purchasing an interest in a Brazilian company in 1958. This trend continued during the 1960s and 1970s, with expansion into the Canadian market and further Brazilian purchases. It acquired an interest in the Canadian Inglis Company in 1969, taking sole ownership in 1990.

In 1986, Whirlpool acquired the KitchenAid division of the Hobart Corporation, as well as developing interests in Italy and India. In 1989, it entered into an arrangement with Philips to manufacture and market appliances in Europe. The Roper brand name was also acquired. Whirlpool became the sole owner of this enterprise in 1991. With the collapse of the Soviet Union, the company looked to Eastern Europe, establishing interests in Hungary and Slovakia in 1992, Poland and the Czech Republic in 1993, and Russia in 1995. It also strengthened its presence in China and India. Apart from the potential of new markets, manufacturing could be relocated to take advantage of cheaper labor costs. Whirlpool closed factories in the United States and Canada in 1994. By 1996 it employed over 9,000 people in Asia.

Whirlpool remains one of the world's leading manufacturers of appliances, operating in over 170 countries. Its main products are refrigerators, freezers, cookers, microwaves, dishwashers, dryers, and air conditioners. It remains strong in its original line of laundry appliances, continues to manufacture for Sears, Roebuck's Kenmore brand, and has a number of its own brands: in the United States, Whirlpool is the main line, with Whirlpool Gold as the premium line; KitchenAid is the upscale brand, and Roper the basic, budget line. Whirlpool is also marketed as a global line.

See also KitchenAid; Philips Electronics; Sears, Roebuck and Company; Washing Machines

References:

www.whirlpool.com

Woks

See Cookware

Women and Housework

Cast in the role of housewife by the shift of production from the home to the factory, women initially looked on the new household technologies of the twentieth century as a means of liberation from household drudgery. It was in the interests of utilities and manufacturers to nurture this illusion. So, in the 1920s and 1930s, electricity was presented to the newly enfranchised middle-class woman as a household panacea. Analysis of household tasks suggested that the use of electric appliances could reduce time spent on housework by as much as three-quarters. However, this assumed the range of tasks and the expected standards of performance would remain the same. This was not the case. Society imposed new standards and expectations on women in their role as guardians of home and family. Feminist historians have argued that this was a natural consequence of patriarchal capitalism, wherein the woman's role was to look after the male breadwinner and the family, and the household's role was to consume goods. Just as the politicization of women's issues had little effect on the gender division of household tasks, the availability of labor-saving technologies did not greatly reduce the burden of housework in terms of time input. Today, for many women, the potential of labor-saving technology as a provider of expanded leisure time and opportunities has not been realized.

The Birth of the Housewife

Paid and unpaid work coexisted in the preindustrial home, and both types of work were shared by male and female family members, primarily according to physical suitability. The traditional gender division of labor became more polarized when the Industrial Revolution established the factory as the ideal production unit. Women continued to undertake paid work both inside and outside the home, but the home became characterized as the female sphere or domain, while the workplace was the male sphere or domain. This distinction was reflected by the adoption of the term "housework" from the mid-nineteenth century to describe women's unpaid work in the home. Laborious and time-consuming as unmechanized household tasks might be, in most homes they were unpaid and therefore not equivalent to work outside the home. Even as paid employment in wealthier households, housework was regarded as an inferior type of work. By 1900, domestic service was for many a last resort, particularly in the United States, as it held vestiges of feudal dependence and deference.

For many housewives, continued reliance on solid fuels for heating and cooking and on communal pumps or standpipes for water for washing made these tasks more onerous. In the pre-industrial home, men had been at hand to take on the heavy chores of fetching and carrying water and fuels. Cleaning was a thankless task in homes where solid fuels and piped gas produced dirty residues, and materials such as the cast iron used for grates and ranges were hard to clean. Industrialization did potentially lighten the load of housework in some ways, for example, by introducing commercial bottled and canned foods, but these were luxury goods. Similarly, cotton clothes were easier to wash than their woolen predecessors were, but as standards of personal hygiene rose, clothes were washed more frequently. The home became romanticized and expectations rose.

Female Emancipation

Gaining the right to vote became symbolic of women's aspirations to achieve a wider role in society. Women's suffrage societies emerged in Britain and the United States in the mid-nineteenth century. Britain's National Society for Women's Suffrage was founded in 1867, and the U.S. equivalent, the National Woman Suffrage Association, was founded in 1869. In Britain, women ratepayers (taxpayers) gained the right to vote in

A woman scrubbing the floor, 1961 (Hulton-Deutsch Collection/Corbis)

local elections in 1869, the same year that the territory of Wyoming granted women the right to vote in local elections. The state of Colorado followed suit in 1893. The National Union of Women's Suffrage Societies, founded by Millicent Fawcett in 1887, became the largest British women's suffrage organization. In 1903, dissatisfied with the failure of a number of bills promoting female suffrage to gain full parliamentary approval, Emmeline Pankhurst formed the Women's Social and Political Union, which began to mount an increasingly militant campaign. While the militant approach succeeded in capturing public attention, it was less successful in engendering wider sympathy and support. In the United States, the National Woman Suffrage Association tended to frown on militancy. One of its cofounders, Susan Brownell Anthony, formed the International Woman Suffrage Alliance in 1904 as a means of encouraging international solidarity. However, the activities of the Women's Social and Political Union did stimulate the development of new women's organizations in the United States, most notably the Equality League of Self-Supporting Women, founded in 1907, and the National Woman's Party, founded in 1916.

Ultimately, the enfranchisement of women owed as much to the social and political impact of World War I as to the tireless campaigning of women suffragists. Prior to the war, the most significant change in the range of paid work available to women had arisen as a result of the adoption of the typewriter and telephone as standard items of business equipment. Women entered office work, which had previously been almost exclusively the province of men, as typists, switchboard operators, and receptionists. The war itself opened the doors for women to other male work strongholds, such as engineering, where they were needed to fill the vacancies created by military conscription. As many factories were by then electrically powered, women war workers had the opportunity to experience at first hand the new form of power that could also transform their homes. After the war, governments in Britain and the United States seemed alert to their obligations to women. In Britain, women householders over the age of 30 gained the right to vote in 1918. It was another ten years before women were given equal voting rights to men by the extension of the franchise to all women over the age of 21. In the United States, after ratification by the requisite number of states, the constitution was amended in 1920 to give women the same voting rights as men.

The Rational Household and Servant Substitution

In the early twentieth century, as women progressed toward equal political rights and began to find new employment opportunities, they faced a simultaneous barrage of propaganda encouraging them to become better household managers and guardians of family welfare. The male view of the woman's role was concisely expressed by the German emperor Kaiser Wilhelm II, who advocated the doctrine of Kinder, Küche, Kirche (children, kitchen, church), a slogan later adopted by the Nazi party. Female household economists cast a similar doctrine in a more progressive light by espousing the application of scientific management theories to housework. The concept of household management was directed primarily toward middle-class housewives, who had been the main audience for household manuals popularized by writers such as the United States' Catherine Beecher and Britain's Mrs. (Isabella Mary Mayson) Beeton in the mid-nineteenth century. In the wealthiest homes, household management was the job of the paid housekeeper under the direction of the mistress of the home, while in poorer homes, the housewife had no one to manage but herself and limited technological resources.

The pioneering theorist of scientific management of manufacturing industry was the U.S. engineer and management consultant Frederick Winslow Taylor. His ideas were translated to the household context by American household economists, notably Christine Frederick and Mary Pattison, who both produced Taylorist household manuals (*Household Engineering* and *The Business of Home Management*) in 1915. Taylor had used time and motion studies to analyze workflow and formulate the theoretical basis for assembly line production. The same approach was applied to analyzing household tasks and formulating efficient kitchen layouts. This approach gained favor in middle-class homes in the 1920s, as a solution to the problem of servant shortages. After World War I, the availability of better paid unskilled work in factories was, for many, a more attractive option than domestic service. Statistics on domestic servants suggest that the 1920s saw a contraction in employment in multiple-servant households and a trend in middle-class households away from live-in servants toward "daily helps." Aspiring middle-class households, which grew in number after the war, regarded employment of a servant as a badge of status.

The Myth of the Self-sustaining Electrical Home

The electricity industry strove to convince the aspiring middle-class housewife that appliances could fill the void created by the servant shortage. Indeed, the silent electrical servant was described as superior to the human servant—always available, inexhaustible, and infinitely dependable. The industry promoted the myth of the self-sustaining electrical home wherein the housewife merely needed to manage the electric appliances, with no physical effort required. While it was, for example, undoubtedly less time-consuming and less tiring to wash clothes in an electric washing machine than by hand, the washing machine did not load or unload the clothes. Nevertheless, many women chose to be optimistic about the labor-saving and liber-

From a 1938 British Electrical Development Association booklet, "How to Protect Your Family from Food Danger!" (The Museum of Science and Industry in Manchester)

ating potential of electricity, and embraced the new technology with evangelical zeal.

In Britain, the body that spearheaded women's involvement in the promotion of electricity as a liberating force was the Electrical Association for Women (EAW). It was launched at a meeting of the Women's Engineering Society in 1924 and adopted the broadly educational mission of informing women about "electrical apparatus" and providing a platform for women's views on all matters relating to electricity. While it was constitutionally independent of the electricity industry, it received funding from the industry, which sometimes tried to use its funding leverage to influence the EAW's program. The EAW was in touch with the circumstances and aspirations of the kind of middle-class women who formed its core membership. For example, its 1930 Bachelor Girls Exhibition in London was built entirely by women and was directed at young, single working women who probably worked in offices. However, it tended to adopt a well-meaning, but potentially offensive brand of patronage toward working-class women. The nearest U.S. equivalent to the EAW was the Women's Committee of the National Electric Light Association (NELA). The NELA also conferred with other women's organizations, such as the Federation of Women's Clubs, and invited their representatives to address NELA conventions.

Feminist campaigners took a rather different view of the solution to household drudgery. In 1922, the British journalist Leonora Eyles proposed a more cooperative approach to housework and childcare through the setting up of communal kitchens and nurseries. In the United States too, in spite of the higher rates of ownership of domestic appliances, there were women who rejected the model of empowerment through individual consumption. For example, the American Federation of Labor funded cooperative laundries and the West Seattle Co-operative Club set up its own day-care center.

"Housewifery Expands to Fill the Time Available"

As did World War I, World War II brought women into the labor force to fill the jobs vacated by men on military service. After the war, women were unceremoniously banished to the home as governments wished to avoid a return to the unemployment levels of the 1930s that the war had reversed. In order to keep married women in the home, the role of housewife was elevated and glorified. The ideas of rational household management, ergonomics, and household technological expertise were marshaled and packaged to present housework and family care as the equivalent of an outside job, albeit unpaid. New appliances—freezers, dishwashers, tumble dryers—were dangled as incentives to stay at home. Rising wages and the growing availability of rent-purchase and other credit schemes meant that these potential acquisitions were realistically attainable. Increased spending on household consumer goods boosted production and hence employment, while saddling women with an attendant definition of household perfection that was forever rising beyond their reach.

Resentment of this kind of manipulation was one of the motivating factors behind the rise of the women's movement in the 1960s. In the United States, the publication of Betty Friedan's *The Feminine Mystique* in 1963 exposed the reality behind the myth of the "happy housewife heroine." Betty Friedan found that many housewives suffered from "the problem that has no name," a state of boredom and lack of fulfillment, and masked their depression by taking prescription tranquilizers. She concluded that full-time housework was unnecessary and unrewarding, and that in practice "housewifery expands to fill the time available."

Women's Liberation

A new, high-profile wave of feminism was set in motion by the publication in 1949 of *The Second Sex,* by the French socialist and feminist writer Simone de Beauvoir. De Beauvoir asserted that the subordinate role of women was the result of calculated repression by a male-dominated society. This was borne out by Betty Friedan's research. By 1960, it was in the interests of the male-dominated power groups in Britain and the United States to address some of these issues, as employment was still buoyant and women were once more welcome to join the workforce. In 1964, the U.S. Civil Rights Act introduced a government agency, the Equal Employment Opportunities Commission, to police discriminatory employment practices. However, the proposed Equal Rights Amendment that would have outlawed sex discrimination, although passed by Congress in 1972, failed by a short margin to achieve ratification by the necessary three-quarters of the fifty states by 1982. In Britain, the 1970 Equal Pay Act and the 1975 Sex Discrimination Act addressed some of the demands of the women's movement, but proved largely ineffective as even as late as 1993, when women made up 49 percent of the workforce, the average woman's wage was 40 percent lower than the average man's wage. This was partly because many more women than men were in part-time work, but also because women were underrepresented in better paid jobs, such as management, and in certain professions, including the law and politics.

However, women activists had a more extensive agenda than just equal employment opportunities. In 1966, Betty Friedan cofounded the National Organization for Women (NOW) and became its first president. NOW's agenda covered issues such as portrayal of women by the media and women's reproductive rights. By this time, as a result of the introduction of an improved intrauterine device (IUD, or the coil) and successful trials of a combined estrogen-progestin contraceptive pill, women had the choice of more effective contraceptives. This made it possible for women to plan a family to suit their career needs, rather than the reverse. In Britain, the introduction of the 1967 Abortion Act, which made legal terminations easier to obtain, further increased women's control of their reproductive rights. American women gained more limited abortion rights as a result of the Supreme Court ruling in the test *Roe v. Wade* case in 1973. Campaigns in support of wages for housework and childcare predictably failed to make any headway, leading many feminists to conclude that the inequality of the power structure would perpetuate the oppression of women. While women have the vote on equal terms, only about a tenth of elected members in both the British and U.S. legislatures are women.

Who Does the Housework Now?

At the end of the twentieth century, almost as many women as men in Britain and the United States were in paid employment. Yet, studies consistently showed that in the majority of cases where both female and male household partners worked, women still shouldered most of the household chores. Multiple household appliances may make those chores less strenuous, but housework still consumes "free" time that men can devote to leisure pursuits. Thus, the growth of women's employment has made little difference to women's role in the home. Because, in spite of equal opportunities legislation, women are significantly less well rewarded in their paid work, the addition of an unequal share of unpaid housework further diminishes their economic status.

See also Electrical Association for Women; Electricity; Servants

References:

Arnold, Erik, and Lesley Burr. "Housework and the Appliance of Science." In *Smothered by Invention: Technology in Women's Lives,* edited by Wendy Faulkner and Erik Arnold. London: Pluto Press, 1985.

Bose, Christine E., Philip L. Bereano, and Mary Malloy. "Household Technology and the Social Construction of Housework." In *Technology and Choice: Readings from Technology and Culture,* edited by Marcel C. LaFollett and Jeffrey K. Stine. Chicago and London: University of Chicago Press, 1991.

Cowan, Ruth Schwartz. *More Work for Mother: The Ironies of Household Technology from the Open Hearth to the Microwave.* London: Free Association Books, 1989.

Davidson, Caroline. *A Woman's Work Is Never Done: A History of Housework in the British Isles, 1650–1950.* London: Chatto & Windus, 1982.

Day, Tanis. "Capital-Labor Substitution in the Home." In *Technology and Choice: Readings from Technology and Culture,* edited by Marcel C. LaFollett and Jeffrey K. Stine. Chicago and London: University of Chicago Press, 1991.

Friedan, Betty. *The Feminine Mystique.* Harmondsworth, UK: Penguin Books, 1965.

Lupton, Ellen. *Mechanical Brides: Women and Machines from Home to Office.* New York: Cooper-Hewitt National Museum of Design, Smithsonian Institute, and Princeton Architectural Press, 1993.

Oakley, Ann. *Housewife.* Harmondsworth, UK: Penguin Books, 1976.

———. *The Sociology of Housework.* London: Martin Robertson, 1974.

Rothschild, Joan. *Machina ex Dea: Feminist Perspectives on Technology.* New York: Pergamon Press, 1983.

Rowbotham, Sheila. *A Century of Women: The History of Women in Britain and the United States.* Harmondsworth, UK: Penguin Books, 1999.

Thane, Pat. "A Woman's Place." In *The Making of Britain: Echoes of Greatness,* edited by Lesley M. Smith. Basingstoke, UK, and London: Macmillan/LWT and Channel 4, 1988.

Wajcman, Judy. *Feminism Confronts Technology.* Cambridge, UK: Polity Press, 1991.

Workmates

The Workmate was patented in 1968 by an Englishman, Ron Hickman. It was originally called the Minibench and has become one of the most popular accessories for the do-it-yourself enthusiast.

Hickman had left Lotus Cars, a manufacturer of performance sports cars in the United Kingdom, where he had been a project engineer, to concentrate on developing his own products. His inspiration for the Workmate came from a dilemma familiar to many people forced to work without the proper equipment. He had been sawing some wood using a good chair as a bench and had cut through the wood into the chair! His solution was initially for his own use, but Hickman soon realized that he had a marketable product that would appeal to the growing thousands of handy men and women living in apartments or small modern houses that could not accommodate a bench for such tasks.

The Workmate uses two wooden beams mounted on a light alloy frame designed to fold away for easy storage. One of the beams moves like a vice, being capable of holding wood, pipes, and tiles. It could remain stable and carry loads up to 350 lb (160 kg).

Hickman had been to see the Black & Decker Company with an early design in 1967, but they had rejected the idea, as had other companies. Convinced of its promise, he established his own Mate-Tool company. Sales reached 1,500 in the first year and doubled every year up to 1972. This success convinced Black & Decker, which came back and commissioned him to design the Mark Two model that was launched in 1972, followed by a U.S. launch in 1975.

Since then, over 20 million Workmates have been sold worldwide. They are used both by amateurs within the home and by professionals working on site. Although the essentials of the design remain unchanged, it has been steadily improved with the introduction of dual working heights and increased vice jaw lengths. A smaller, portable version has also been introduced. The global success of the Workmate owes much to the power of a large multinational company and successful marketing, but it is equally true that its success lay in it being a simple, but innovative solution to a problem that had arisen out of changing social and economic patterns.

See also Black & Decker; Do-It-Yourself

References:

Hillman, David, and David Gibbs. *Century Makers.* London: Seven Dials, 1999.

Y

Yogurt Makers

Yogurt is a sour, semisolid food made from fermented milk that originated in Turkey. It was commercially produced and marketed in the 1960s, being promoted as a healthy low-fat food, although the early brands were sweetened and fruit flavored. The French company Danone was a leader in Europe. The company was originally founded in Barcelona as a family yogurt-making business, but the owner's two sons left Spain during the civil war. One went to France and the other to the United States. Both began yogurt businesses in their new homes. Danone is now one of the largest food companies in Europe. The 1970s and 1980s saw a reduction in the amount of sugar in the yogurt, as manufacturers responded to Western media-fueled obsessions with a healthy looking, slim figure.

Domestic yogurt makers appeared in the 1970s. The milk is placed into pots, which are placed in a rectangular, thermostatically controlled heater that raises the milk to the correct temperature and keeps it there for the fermenting period, which is around four hours.

Z

Zanussi

Founded by Antonio Zanussi in northeast Italy in 1916, Zanussi is one of the companies that helped to establish Italy's reputation for innovative product design. Established as a leading European manufacturer of white goods (stoves, washing machines, and refrigerators), the company was acquired in 1984 by the Swedish Electrolux group, but has retained its distinctive identity. The takeover provided new investment to maintain Zanussi's technological edge.

Zanussi began as a manufacturer of wood-burning stoves, from 1933 marketed under the Rex trademark. The company began to expand and diversify after Lino Zanussi, son of the founder, took charge in 1946. Initially, Zanussi stayed within the confines of the cooker market, where it extended its line to include gas, electric, and dual-fuel cookers. In 1954, it began to branch out into new product lines with the launch of its first refrigerator and to cultivate a strong research and development strategy through its Study Center. Zanussi's first washing machine, the Rex 201 twin tub (a washer and dryer in one unit) of 1958, showed an attention to space-saving efficiency that has become a characteristic trait. By the end of the 1950s, Zanussi's commitment to innovation was undeniable. In 1959, it set up a design and research center and launched the Rex Supermarket 280 SM2, the first European fridge-freezer to feature automatic defrosting.

In 1962, Zanussi received the first of its Compasso d'Oro awards, the most prestigious Italian design accolade, for its Rex 700 gas cooker. The company was also enjoying considerable commercial success, exporting its products to seventy countries. In the late 1960s, Zanussi launched a line of coordinated kitchen appliances, followed by a line of built-in appliances. By the mid-1970s, it held a 25 percent share of the Italian market and 10 percent of the European market. Zanussi was also one of the first manufacturers to recognize the potential of incorporating electronic functions in kitchen appliances. Its slogan "the appliance of science" was not just a marketing catchphrase but a genuine description of its product philosophy. The 1986 Jetsystem washing machine introduced the innovative spray washing technique, which economized on water and, therefore, also electricity and detergent.

The essence of Zanussi's strongest products is their seamless fusion of high performance and aesthetic expression. Ironically, one of Zanussi's most high-profile designs, the 1987 Wizard refrigerator, was one of its least commercially successful products. Designed by Roberto Pezzetta, Zanussi's head of industrial design, the Wizard challenged the discreet, streamlined image of the classic white refrigerator. With its pyramid top, protruding red base, and availability in stark black, the Wizard brought postmodernism to refrigerator design, but at the cost of space efficiency. In the early 1990s, Zanussi returned to its tried and tested consumer-oriented product values. Pezzetta then began to develop a series of designs that were both innovative and sympathetic to mass consumer tastes. "Soft Tech" styling gave rounded lines to the BST 6 cooker and the Rondo fridge-freezer. Appliances were made available in a range of finishes from colored glass in yellow and blue to polished steel. Pezzetta's most dramatic design since the Wizard was the 1998 OZ fridge-freezer, with its deep door providing larder storage and undulating lines.

Zanussi has also maintained its quest for improved product efficiency. The TL15 Input washing machine of 1997 uses fuzzy logic to optimize performance—the user only has to select the setting by fabric type and the fuzzy logic chip determines the best wash program. Like its parent company Electrolux, Zanussi has been developing a system whereby the microprocessor has a more fully integrated role in the kitchen. Its Live-In system features a computerized control unit with a folding LCD screen. The controller operates appliances, taking into account off-peak electricity savings, and acts as an entertainment and communication device.

See also Cookers; Electrolux; Refrigerators; Washing Machines

References:

www.zanussi.com/storia/

BIBLIOGRAPHY

Books and Articles

Adamson, Ian, and Richard Kennedy. *Sinclair and the Sunrise Technology*. Harmondsworth, UK: Penguin Books, 1986.

Alessi, Alberto. *The Dream Factory: Alessi since 1921*. Cologne, Germany: Konemann, 1998.

Attfield, Judy, and P. Kirkham, eds. *A View from the Interior: Feminism, Women, and Design*. London: Women's Press, 1989.

Baker, Eric. *Great Inventions, Good Intentions*. San Francisco: Chronicle Books, 1990.

Baker, John C., Stuart Evans, and Cindy Snow. *Pyrex: 60 Years of Design*. 1983.

Banham, Reyner. *Theory and Design in the First Machine Age*. Oxford, UK: Butterworth-Heinemann, 1972.

Barrett, Helena, and John Phillips. *Suburban Style: The British Home, 1840–1960*. London: Macdonald Orbis, 1987.

Bayley, Stephen. *In Good Shape: Style in Industrial Products, 1900–1960*. London: Design Council, 1979.

Bayley, Stephen, ed. *The Conran Dictionary of Design*. London: Octopus Conran, 1985.

Beeching, Wilfred A. *A Century of the Typewriter*. London: Heinemann, 1974.

Belling Company. *The Story of Belling*. 1962.

Bennett-Levy, Michael. *Historic Televisions and Video Recorders*. Midlothian, UK: MBL Publications, 1994.

Bertram, Antony. *Design*. London: Pelican, 1938.

Bijker, Wiebe E., and John Law. *Shaping Technology/Building Society*. Cambridge, MA, and London: MIT Press, 1992.

Blake, Peter. *The Master Builders*. New York: Alfred A. Knopf, 1960.

Boxshall, Jan. *Every Home Should Have One: Seventy Five Years of Change in the Home*. London: Ebury Press/Good Housekeeping, 1997.

Braun, Ernest, and Stuart Macdonald. *Revolution in Miniature*. New York: Cambridge University Press, 1982.

Brears, Peter. *The Kitchen Catalogue*. York, UK: York Castle Museum, 1979.

———. *Traditional Food in Yorkshire*. Edinburgh, UK: John Donald, 1987.

Brears, Peter, and Stephen Harrison. *The Dairy Catalogue*. York, UK: York Castle Museum, 1979.

Brook, James, and Iain A. Beal, eds. *Resisting the Virtual Life: The Culture and Politics of Information*. San Francisco: City Lights Books, 1995.

Bruce, Robert V. *Alexander Graham Bell and the Conquest of Solitude*. Ithaca, NY: Cornell University Press, 1990.

Brummer, Alex, and Roger Crowe. *Weinstock: The Life and Times of Britain's Premier Industrialist*. London: HarperCollins, 1998.

Burnett, John. *Plenty and Want: A Social History of Diet in England from 1815 to the Present Day*. London and New York: Methuen, 1983.

Butcher, Lee. *Accidental Millionaire: The Rise and Fall of Steve Jobs at Apple Computer*. New York: Paragon House, 1988.

Burney, Jan. *Ettore Sottsass*. London: HarperCollins, 1991.

Byers, Anthony. *Centenary of Service: A History of Electricity in the Home*. London: Electricity Council, 1981.

Campbell, Susan. *The Cook's Companion*. London: Chancellor Press, 1980.

Chant, Colin, ed. *Science, Technology, and Everyday Life, 1870–1950.* London: Open University/Routledge, 1990.

Chesney, Kellow. *The Victorian Underworld.* London: Pelican, 1976.

Clarke, Alison J. *Tupperware: The Promise of Plastic in 1950s America.* Washington, DC, and London: Smithsonian Institution Press, 1999.

Coe, Brian. *Cameras from Daguerreotypes to Instant Pictures.* London: Marshall Cavendish, 1978.

Collins, Michael. *Towards Post-Modernism: Design since 1851.* London: British Museum, 1987.

Consumers Association. *Thirty Years of Which? 1957–1987.* London: Consumers Association, 1987.

Consumers Union. *Consumer Reports Best Buys for Your Home, 2000.* New York: Consumers Union, 1999.

Cowan, Ruth Schwartz. "The Industrial Revolution in the Home—Household Technology and Social Change in the Twentieth Century." In *Material Culture Studies in America,* edited by Thomas J. Schlereth. Nashville, TN: American Association for State and Local History, 1984.

———. *More Work for Mother: The Ironies of Household Technology from the Open Hearth to the Microwave.* London: Free Association Books, 1989.

Davidson, Caroline. *A Woman's Work Is Never Done: A History of Housework in the British Isles, 1650–1950.* London: Chatto & Windus, 1982.

Davidson, H. C., ed. *The Book of the Home.* London: Gresham Publishing, 1904.

Davis, Mike. *City of Quartz: Excavating the Future in Los Angeles.* London: Pimlico, 1998.

de Haan, David. *Antique Household Gadgets and Appliances, 1860–1930.* Poole, UK: Blandford Press, 1977.

Dormer, Peter. *Design since 1945.* London: Thames and Hudson, 1998.

———. *The Meanings of Modern Design: Towards the Twenty-first Century.* London: Thames and Hudson, 1991.

Droste, Magdalena. *Bauhaus, 1919–1933.* Cologne, Germany: Benedikt Taschen, 1990.

Druitt, Sylvia. *Antique Personal Possessions.* Poole, UK: Blandford Press, 1980.

du Gay, Paul, Stuart Hall, Linda Janes, Hugh Mackay, and Keith Negus. *Doing Cultural Studies: The Story of the Sony Walkman.* London: Sage Publications, 1997.

Electricity Supply in the United Kingdom: A Chronology. London: Electricity Council, 1987.

Elena, Pierre. "The Rise of the Fitted Kitchen." *Social History Curators Group Journal* 18 (1990–1991).

Faulkner, Wendy, and Erik Arnold, eds. *Smothered by Invention: Technology in Women's Lives.* London: Pluto Press, 1985.

Feild, Rachael. *Irons in the Fire: A History of Cooking Equipment.* Marlborough, UK: Crowood Press, 1984.

Fiell, Charlotte, and Peter Fiell. *Design of the Twentieth Century.* Cologne, Germany: Benedikt Taschen, 1999.

Fischer, Claude S. *America Calling: A Social History of the Telephone to 1940.* Chicago and London: University of Chicago Press, 1994.

Forester, Tom. *High-Tech Society: The Story of the Information Technology Revolution.* Cambridge, MA: MIT Press, 1987.

Forty, Adrian. *Objects of Desire: Design and Society, 1750–1980.* London: Thames and Hudson, 1986.

Friedan, Betty. *The Feminine Mystique.* Harmondsworth, UK: Penguin Books, 1965.

Fumagalli, Ambrogio. *Coffeemakers.* San Francisco: Chronicle Books, 1995.

Furnival, Jane. *Suck, Don't Blow.* London: Michael O'Mare, 1998.

Gardner, Robert, and Dennis Shortelle. *From Talking Drums to the Internet: An Encyclopedia of Communications Technology.* Santa Barbara, CA: ABC-CLIO, 1997.

Gathorne-Hardy, Jonathan. *The Rise and Fall of the British Nanny.* London: Arrow Books, 1974.

Giedion, Siegfried. *Mechanization Takes Command.* New York: Oxford University Press, 1948.

Godfrey, Frank P. *An International History of the Sewing Machine.* London: Robert Hale, 1982.

Going, Tom. "A Resounding Ekco." *Plastiquarian* 6 (1990).

Goldberg, Michael J. *Groovy Kitchen Designs for Collectors, 1935–1965.* Atglen, PA: Schiffer Publishing, 1996.

Goodman, Robert. *How Electronic Things Work . . . and What to Do When They Don't.* New York: McGraw-Hill, 1999.

Haddock, Thomas F. *A Collector's Guide to Personal Computers and Pocket Calculators.* Florence, AL: Books Americana, 1993.

Hardyment, Christina. *From Mangle to Microwave: The Mechanisation of Household Work.* Cambridge, UK: Polity Press, 1988.

Hannah, Leslie. *Electricity before Nationalisation: A Study of the Development of the Electricity Supply*

Industry in Britain to 1948. London: Macmillan, 1979.

———. *Engineers, Managers and Politicians.* London: Macmillan, 1982.

Harris, Jennifer, Sarah Hyde, and Greg Smith. *1966 and All That: Design and the Consumer in Britain, 1960–1969.* London: Trefoil Books, 1986.

Herz, J. C. *Joystick Nation.* London: Abacus, 1997.

Heskett, John. *Industrial Design.* New York: Oxford University Press, 1980.

Hick, Roger. *A History of the 35 mm Still Camera.* London and Boston, MA: Focal Press, 1984.

Hill, Jonathan. *The Cat's Whisker: Fifty Years of Wireless Design.* London: Oresko, 1978.

———. *Radio! Radio!* Bampton, UK: Sunrise Press, 1986.

Hillman, David, and David Gibbs. *Century Makers.* London: Seven Dials, 1999.

Hine, Thomas. *Populuxe.* New York: Alfred A. Knopf, 1986.

Hogben, Carol. *The Wireless Show!* London: Victoria and Albert Museum, 1977.

Hoggart, Richard. *The Uses of Literacy.* London: Chatto & Windus, 1957.

Holdsworth, Angela. *Out of the Doll's House: Women in the 20th Century.* London: BBC, 1988.

Hoover's 500: Profiles of America's Largest Business Enterprises. Austin, TX: Hoover's Business Press, 1996.

Horn, Pamela. *The Rise and Fall of the Victorian Servant.* Gloucester, UK: Alan Sutton, 1986.

Horn, Richard. *Fifties Style: Then and Now.* London: Columbus Books, 1985.

Hudson, Kenneth. *The Archaeology of the Consumer Society.* London: Heinemann, 1983.

Jackson, Lesley. *The New Look: Design in the Fifties.* London: Thames and Hudson, 1991.

Jewell, F. Brian. *Veteran Sewing Machines: A Collectors' Guide.* Newton Abbot, UK: David & Charles, 1975.

Jones, R., and O. Marriot. *Anatomy of a Merger: A History of GEC, AEI, and English Electric.* London: Jonathan Cape, 1970.

Julier, Guy. *Dictionary of Twentieth-Century Design and Designers.* London: Thames and Hudson, 1993.

Katz, Sylvia. *Classic Plastics.* London: Thames and Hudson, 1984.

Katz, Sylvia, and Jeremy Myerson. *Kitchenware.* London: Conran Octopus, 1990.

Kenna, Rudolph, and William Grandison. *Somethin' Else: 50s Life and Style.* Glasgow, UK: Richard Drew, 1989.

Kochan, Nicholas, ed. *The World's Greatest Brands.* London: Macmillan/Interbrand, 1996.

LaFollette, Marcel C., and Jeffrey K. Stine, eds. *Technology and Choice: Readings from Technology and Culture.* Chicago and London: University of Chicago Press, 1991.

Lancaster, Maud. *Electrical Cooking, Heating, Cleaning, etc.: Being a Manual of Electricity in the Service of the Home.* London: Constable, 1914.

Lupton, Ellen. *Mechanical Brides: Women and Machines from Home to Office.* New York: Cooper-Hewitt National Museum of Design, Smithsonian Institution, and Princeton Architectural Press, 1993.

Lyons, Nick. *The Sony Vision.* New York: Crown Publishers, 1976.

McDermott, Catherine. *The Design Museum Book of Twentieth-Century Design.* London: Carlton Books, 1998.

MacDonald, Sally, and Julia Porter. *Putting on the Style: Setting up Home in the 1950s.* London: Geffrye Museum, 1990.

McLaren, Graham. *Ceramics of the 1950s.* Princes Risborough, UK: Shire Publications, 1997.

MacKenzie, Donald, and Judy Wajcman, eds. *The Social Shaping of Technology: How the Refrigerator Got Its Hum.* Milton Keynes, UK: Open University Press, 1985.

Malos, Ellen. *The Politics of Housework.* London and New York: Allison & Busby, 1982.

Martland, Peter. *Since Records Began: EMI First One Hundred Years.* London: Batsford, 1997.

Marwick, Arthur. *British Society since 1945.* Harmondsworth, UK: Penguin Books, 1982.

Massey, Anne. *Interior Design of the Twentieth Century.* London: Thames and Hudson, 1996.

Matranga, Victoria Kasuba, and Karen Kohn. *America at Home: A Celebration of Twentieth-Century Housewares.* Rosemont, IL: National Housewares Manufacturers Association, 1997.

Meadows, Daniel. *Nattering in Paradise: Suburbia in the 1980s.* London: Simon & Schuster, 1988.

Meikle, Jeffrey L. *Twentieth Century Limited: Industrial Design in America, 1925–1939.* Philadelphia: Temple University Press, 1979.

Morita, Akio, Edwin M. Reingold, and Mitsuko Shimomura. *Made in Japan: Akio Morita and Sony.* New York: Dutton, 1986.

Mossman, Susan, ed. *Early Plastics: Perspectives, 1850–1950.* London and Washington, DC: Science Museum/Leicester University Press, 1997.

Muir, Frank. *A Book at Bathtime*. London: Heinemann, 1982.

Muthesius, Stefan. *The English Terraced House*. New Haven, CT, and London: Yale University Press, 1982.

Myers, Kathy. *Understains: The Sense and Seduction of Advertising*. London: Comedia, 1986.

Norcross, Eric. "Toastmaster—A History with Dates." On www.toaster.org.

Nye, David E. *Electrifying America: Social Meanings of a New Technology*. Cambridge, MA: MIT Press, 1990.

Oakley, Ann. *Housewife*. Harmondsworth, UK: Penguin Books, 1976.

———. *The Sociology of Housework*. London: Martin Robertson, 1974.

O'Brien, Robert. *Machines*. Amsterdam: Time-Life International, 1972.

O'Dea, William. *The Social History of Lighting*. London: Routledge & Kegan Paul, 1958.

The Origins of Everyday Things. London: Reader's Digest Association, 1998.

Packard, Vance. *The Hidden Persuaders*. Harmondsworth, UK: Penguin Books, 1957.

———. *The Waste Makers*. Harmondsworth, UK: Penguin Books, 1961.

Pandit, S. A. *From Making to Music: The History of Thorn-EMI*. London: Hodder & Stoughton, 1996.

Pearce, Christopher. *Fifties Source Book*. London: Quarto, 1990.

Peto, James, and Donna Loveday, eds. *Modern Britain, 1929–1939*. London: Design Museum, 1999.

Phillips, Barty. *Conran and the Habitat Story*. London: Weidenfeld and Nicolson, 1984.

Pugh, Emerson W. *Building IBM: Shaping an Industry and Its Technology*. Cambridge, MA: MIT Press, 1995.

Randell, Wilfred. *Electricity and Woman*. London: Electrical Association for Women, 1945.

Robertson, J. *The Story of the Telephone: A History of the Telecommunications Industry of Britain*. London: Pitman, 1947.

Robertson, Patrick. *The Shell Book of Firsts*. London: Ebury Press and Michael Joseph, 1983.

Rogers, Everett M., and Judith K. Larsen. *Silicon Valley Fever*. New York: Basic Books, 1984.

Rothschild, Joan. *Machina Ex Dea: Feminist Perspectives on Technology*. New York: Pergamon Press, 1983.

Rowbotham, Sheila. *A Century of Women: The History of Women in Britain and the United States*. Harmondsworth, UK: Penguin Books, 1999.

Rushin, Don. "The History of Magnetic Tape." On www.technicalpress.com/Articles/History/History_Tape.htm.

Ryan, Deborah S. *The Ideal Home through the Twentieth Century*. London: Hazar Publishing, 1997.

Saxby Stephen. *The Age of Information*. London: Macmillan, 1990.

Schoenherr, Steve. "Recording Technology History." On http://ac.acusd.edu/history/recording/notes.html.

Scott, Peggy. *An Electrical Adventure*. London: Electrical Association for Women, 1934.

Shanks, Michael. *The Innovators: The Economics of Technology*. Harmondsworth, UK: Penguin Books, 1967.

Sinclair, Ian. *Birth of the Box: The Story of Television*. Wilmslow, UK: Sigma Press, 1995.

Smith, Lesley M., ed. *The Making of Britain: Echoes of Greatness*. Basingstoke, UK, and London: Macmillan/LWT and Channel 4, 1988.

Sparke, Penny. *A Century of Design: Design Pioneers of the Twentieth Century*. London: Mitchell Beazley, 1998.

———. *Design in Context*. London: Bloomsbury Publishing, 1991.

———. *Electrical Appliances*. London and Sydney: Unwin Hyman, 1987.

Sparke, Penny, ed. *Did Britain Make It? British Design in Context, 1946–1986*. London: Design Council, 1986.

———. *The Plastics Age: From Bakelite to Beanbags and Beyond*. Woodstock, NY: Overlook Press, 1993.

Stevenson, John. *British Society, 1914–45*. Harmondsworth, UK: Penguin Books, 1984.

Stewart, Richard. *Design and British Industry*. London: John Murray, 1987.

Sudjic, Deyan. *Cult Objects*. London: Paladin/Granada, 1985.

Sweet, Fay. *Alessi: Art and Poetry*. London: Thames and Hudson, 1998.

Tambini, Michael. *The Look of the Century*. London: Dorling Kindersley, 1996.

Trager, James, ed. *The People's Chronology: A Year-by-Year Record of Human Events from Prehistory to the Present*. London: Heinemann, 1985.

Tunstall, Jeremy. *Communications Deregulation: The Unleashing of America's Communications Industry.* Oxford, UK, and New York: Basil Blackwell, 1986.

Turner, Mark, ed. *Little Palaces: The Suburban House in North London, 1919–1939.* London: Middlesex Polytechnic, 1987.

Tweedale, Geoffrey. *Calculating Machines and Computers.* Princes Risborough, UK: Shire Publications, 1990.

Wajcman, Judy. *Feminism Confronts Technology.* Cambridge, UK: Polity Press, 1991.

Ward, Mary, and Neville Ward. *Home in the Twenties and Thirties.* London: Ian Allan, 1978.

Whitford, Frank. *Bauhaus.* London: Thames and Hudson, 1984.

Williams, Trevor I. *A History of Technology, Vols. 6 and 7: The Twentieth Century*. Oxford, UK: Clarendon Press, 1978.

Winn, Marie. *The Plug-in Drug: Television, Children, and the Family.* New York: Viking Penguin, 1977.

Wohl, Anthony S. *Endangered Lives: Public Health in Victorian Britain.* London: Methuen, 1984.

Woodham, John. *Kettle: An Appreciation.* London: Arum Press, 1997.

Worthy, James C. *Shaping an American Institution.* Urbana: University of Illinois Press, 1984.

Wright, Lawrence. *Clean and Decent: The Fascinating History of the Bathroom and the Water Closet.* London: Routledge, 1980.

———. *Home Fires Burning: The History of Domestic Heating and Cooking.* London: Routledge & Kegan Paul, 1964.

Yarsley, V. E., and E. G. Couzens. *Plastics.* London: Pelican, 1941.

York Oral History Society. *York Memories at Home: Personal Accounts of Domestic Life in York, 1900–1960.* York, UK: York Oral History Society/York Castle Museum, 1987.

Zerzan, John, and Alice Carnes. *Questioning Technology: A Critical Anthology.* Philadelphia: New Society Publishers, 1991.

Web Sites

Adventures in Cybersound, fax machine history: www.hffax.de

American Frozen Food Institute: www.affi.com/facts/decades/

Apple Computer, Inc.: www.apple.com; www.apple-history.com/history.html

Asahi Optical Corporation: www.pentax.com

AT&T: www.att.com; www.research.att.com

Atari: www.hasbrointeractive.com/atari/

Atari Historical Society: www.atari-history.com

Atari Time Machine: homepage.tinet.ie/~morrikar/

Bang & Olufsen: www.bang-olufsen.com

Birds Eye: www.birdseye.com

Black & Decker: www.blackanddecker.com

Bosch: www.boschappliances.com

Braun: www.braun.com

Breadman: www.breadman.com

British Telecom: www.bt.com

Brother: www.brother.com

Carrier Company: www.carrier.com

Crate & Barrel: www.crateandbarrel.com

Cuisinart: www.cuisinart.com

DeLonghi: www.delonghi.com

Design Council: www.design-council.org.uk

www.dolcevita.com

Duracell: www.duracell.com

Eastman Kodak: www.kodak.com

Electrical Musical Industries: www.emi.com

Electrolux: www.electrolux.com

Ericsson: www.ericsonus.com/us/consumer; www.ericsson.co.uk/UK/corporate/history.shtml

Farberware: www.farberware.com

General Electric (U.S.): www.ge.com

General Electric Company (UK): www.gec.com; www.marconi.com/about_marconi/heritage/

Glen Dimplex: www.glendimplex.co.uk

Goblin: www.glendimplex.co.uk

The Hoover Company: www.hoover.com

Hewlett-Packard: www.hp.com

HMV Media Group: www.hmv.co.uk/hmv/Customer_Service/about_hmv.html

IBM: www.ibm.com

Ideafinder: www.ideafinder.com

Ikea: www.ikea.com

Inventors and Inventions: http://inventors.about.com

Jacuzzi: www.jacuzzi.com

Kenmore: www.sears.com

KitchenAid: www.kitchenaid.com

Kmart: www.Kmart.com/corp.

Kraft Foods: www.kraft.com/corporate/about/

Magic Chef: www.maytag.com

Marconi: www.marconi.com/about_marconi/heritage/

Matsushita: www.matsushita.com; www.mei.co.jp/corp/hist-e.html
Maytag Corporation: www.maytagcorp.com
Microsoft Corporation: www.microsoft.com/MSCorp/Museum/
Morphy Richards: www.morphyrichards.co.uk
Motorola: www.motorola.com/General/Timeline/
Nabisco: www.nabisco.com
Nestlé: www.nestle.com/all_about/history/
Nintendo: www.nintendo.com/corp/history.html
OSRAM (lamp manufacturers): www.osram.com
Oster: www.sunbeam.com
Panasonic: www.panasonic.com
Pentax: www.pentax.com
Philips Electronics: www.philips.com
Procter & Gamble: www.proctergamble.com
Proctor-Silex: www.hambeach.com
RCA: www.rca.com; www.thomson-multimedia.com
Sears, Roebuck: www.sears.com
Sharp: www.sharp-usa.com
Siemens: www.siemens.com; www.siemens.de
Sinclair: www.nvg.ntnu.no/sinclair/
Singer: www.singerco.com
Sony Corporation: www.world.sony.com
Sunbeam: www.sunbeam.com
Sylvania: www.sylvania.com
Technical Press: www.technicalpress.com
Texas Instruments: www.ti.com/corp/docs/company/history/
Toasters: www.toaster.org
Toastmaster: www.toastmaster.com
Total rewind: www.popadom.demon.co.uk/vidhist/
Tupperware: www.tupperware.com
University of San Diego: ac.acusd.edu/history/recording/notes.html
U.S. National Inventors Hall of Fame: www.invent.org
Vodafone: www.vodafone.co.uk; www.vodaphone-airtouch-plc.com
WalMart: www.walmart.com
Waring: www.waringproducts.com
Waveguide: www.wave-guide.org
West Bend: www.westbend.com
Westinghouse: www.westinghouse.com/corp/
Whirlpool: www.whirlpool.com
White-Westinghouse: www.frigidaire.com
Williams-Sonoma: www.williams-sonoma.com
World Wide Web Consortium: www.w3.org
Zanussi: www.zanussi.com
Zenith Electronics: www.zenith.com

DESIGNERS

Many domestic appliances are the result of anonymous designers or in-house design teams. A wide range of companies employed free-lance designers, including some of the most influential industrial designers and architects of the twentieth century. Although not intended to be completely comprehensive, this section lists some of the key designers who have shaped the development of twentieth-century domestic appliances. Some of these have worked directly on appliance design; others have influenced the look of such products through their writing or approach to design. It is illuminating to note that the majority of these "name" designers are men. Throughout most of the twentieth century men ran the manufacturing companies, although their target market was predominantly female.

Egmont Arens (1888–1966)
American designer and advocate of streamlining. Arens was well known for package design. Arens restyled the KitchenAid food mixer and also designed spun aluminum saucepan sets.

Christian Barman (1898–1980)
Architect, designer, and writer. Barman was editor of the UK *Architects' Journal* and *Architectural Review.* He designed an electric iron and an electric heater for the His Master's Voice (HMV) Company in the 1930s.

Peter Behrens (1869–1940)
Influential German designer. Behrens was a founder member of the Deutsche Werkbund and artistic director at AEG from 1907 to 1914. Behrens designed both products and buildings and was responsible for volume production and corporate identity. His domestic appliances included electric kettles, lamps, and fans. After leaving AEG, he concentrated on architectural work.

Norman Bel Geddes (1893–1958)
Influential American designer, writer, and advocate of streamlining. Many of Bel Geddes's designs were fanciful and only a few were produced. Bel Geddes designed radio cabinets for RCA, the SGE *Oriel* stove of 1932, and the Electrolux refrigerator of 1934.

Mario Bellini (b. 1935)
Italian industrial designer. Bellini has been a consultant to Olivetti since 1963, designing computers and calculators. He also designed the TC 800D cassette deck for Yamaha (1973). Bellini also designs furniture for Italian manufacturers.

Misha Black (1910–1977)
Russian-born designer and later professor of industrial design at the Royal College of Art (UK). Black was active in both the Britain Can Make It exhibition (1946) and the Festival of Britain (1951). He designed distinctive bakelite radio cabinets for Ekco.

Hugo Blomberg (b. 1897)
Swedish designer. Head of design at Ericsson, Blomberg designed the Ericofon one-piece telephone with **Ralph Lysell** and **Gösta Thames.**

Marianne Brandt (1893–1983)
German designer. Brandt trained at the Bauhaus and designed craft prototypes for industrial production. She was responsible for the Tandem night-light, a rare example of a Bauhaus design achieving mass production. She taught industrial design after 1945.

Marcel Breuer (1902–1981)
Hungarian-born designer and architect. Breuer studied at the Bauhaus. Best known for his tubular steel furniture, Breuer designed a fitted, galley-style kitchen for the Bauhaus Haus am Horn of 1923.

Achille Castiglioni (b. 1918)
Important Italian designer. Castiglioni was inspired by his two brothers, Pier Giacomo and Livio, who designed a plastic radio case in 1938. His main work has been in furniture, lighting, and electrical appliances.

George Carwardine (1887–1948)
English automobile engineer. Carwardine's most famous design is the Anglepoise light of 1934 for Terry & Sons. Based on the jointing principles of the human arm, it still remains in production.

Serge Chermayeff (1900–1996)
Modernist architect and designer. Chermayeff was born in Russia and moved to the UK in 1910. He was a member of the MARS (Modern Architecture Research Group) group with **Wells Coates.** He designed Bakelite radio cases for Ekco in the 1930s. Chermayeff immigrated to the United States in 1940 to teach in New York and later in Chicago and Harvard and Yale Universities.

Wells Coates (1895–1958)
Canadian architect and designer. Coates was probably the leading Modernist architect based in the UK. He was a founder of MARS (Modern Architecture Research Group) and the designer of the Isokon flats in London. He designed innovative bakelite radio cases and an electric fire for the Ekco Company in the 1930s and 1940s. In 1955 Coates left Britain to lecture at Harvard.

Gino Colombini (b. 1915)
Technical director of pioneering Italian plastics company Kartell (f. 1949) since 1951. Colombini was responsible for their stylish domestic plastics such as lemon squeezers and colanders in the 1950s and 1960s.

Cesare "Joe" Colombo (1930–1971)
Italian designer. Colombo trained as a painter and an architect. He was a key figure in Italian design during the 1950s and 1960s, designing furniture and interiors. His work was futuristic, intending to provide an "integral habitat." He designed the Mini-kitchen of 1963 as well as lamps and an air conditioner for Candy.

Terence Conran (b.1931)
Initially a furniture and textile designer, Conran's lasting influence has been in the style of retailing through the Habitat group of stores established in London in 1964. Conran's foundation established the Boilerhouse Project design exhibitions at the Victoria and Albert Museum (1980) and the Design Museum (1989).

Henry Dreyfuss (1903–1972)
Highly influential American industrial designer. Dreyfuss opened his design office in 1929 and established his reputation in 1930s with rational and functional designs. His first job was reworking of kitchen storage jars. Dreyfuss worked in many areas and designed Bell telephones, Hoover vacuum cleaners, an RCA television, and a General Electric refrigerator. He was a pioneer of ergonomic design.

Fritz Eichler (b. 1911)
A theater designer from 1945 to 1963. Eichler worked with Braun in 1954 and coordinated its design policies, promoting the commission of designs from the Hoschule fur Gestaltung (College of Design) in Ulm (the successor to the Bauhaus, it closed in 1968).

Frogdesign
German design company. It was founded in 1969 by Harmut Esslinger (b. 1945); the name is an acronym of the Federal Republic of Germany. Frogdesign combines the functionalism of Braun products with a dash of postmodernism. It has worked for Wega Radio, AEG, Apple Computers, and AT&T. Frog designed the casing for the Apple IIc computer of 1984.

Abram Games (b. 1914)
British graphic designer famous for posters for Shell, BP, London Transport, and Guinness. Games restyled the Cona coffee percolator in 1951.

Bruno Gecchelin (b. 1939)
Italian architect and designer. Gecchelin has worked for various companies, designing kitchen utensils for Guzzini and refrigerators and gas cookers for Indesit.

Kenneth Grange (b. 1929)
Leading UK "name" product designer. Grange was influenced by the Braun products of the 1950s. He founded the Pentagram consultancy in 1971. Grange's most famous appliance design

was the highly successful Kenwood Chef. He also designed the Kodak Instamatic camera and the Wilkinson Sword Protector razor of 1992.

Michael Graves (b. 1934)
American postmodernist architect. Graves has also designed furniture for Memphis and kettles and tableware for Alessi since 1985.

Hans Gugelot (1920–1965)
Dutch-Swiss functionalist architect and designer. Gugelot worked for Braun and also taught **Dieter Rams**. He designed the Kodak Carousel slide projector (1963) and worked with Rams on the Braun Phonosuper of 1956.

Lurelle Van Arsdale Guild (1898–1986)
American designer. Guild was originally a theatre designer. His best-known appliance design is a vacuum cleaner for Electrolux in 1937. Guild also designed metal work and the Wear-Ever coffee pot of 1934.

Jean Heiberg (1884–1976)
Norwegian painter who studied in Munich and Paris. Heiberg was invited by Ericsson to assist in the design of their bakelite case for the Ericsson DBH 1001 telephone of 1932. The result was one of the most popular telephones ever produced. Many were manufactured in the UK.

Poul Henningsen (1895–1967)
Danish architect and designer. Henningsen's most famous design is the PH lamp of 1924/5, which is still in production. Henningsen's PH Artichoke light fitting of 1958 was also influential.

Jakob Jensen (b. 1926)
Danish chief designer at Bang & Olufsen. Jensen was responsible for the simple, understated style of the B&O product range and the innovative Beogram 4000 record deck of 1972 that featured a tangential pick-up arm.

Ivar Jepson (1903–1965)
Swedish born industrial designer who studied at the University of Berlin. Jepson moved to Chicago and by 1925 was working as a draftsman at the Chicago Flexible Shaft Company (Sunbeam). He redesigned the Mixmaster, the electric frying pan, and many other Sunbeam products.

René Lalique (1860–1945)
French designer. Lalique is best known for his jewelry and glassware. He established Cristal Lalique in 1909, which produced a variety of upscale products, including lighting.

Le Corbusier (1887–1965)
Swiss architect Charles-Edouard Jeanneret. A key figure of the Modern Movement, Le Corbusier initially worked in **Peter Behrens**'s office. His main work was in architecture and furniture. He coined the often misquoted phrase "The house is a machine for living in."

Raymond Loewy (1893–1986)
Influential advocate of streamlining. Born in Paris, Loewy had studied engineering and came to the United States in 1919. Initially a fashion illustrator, he opened his studio in 1927 and became established as an industrial designer in the 1930s. Loewy was essentially a stylist and had great promotional flair. He designed the Sears Coldspot refrigerator lines and Electrolux products in the 1930s.

Ralph Lysell (b. 1907)
Swedish designer. Lysell worked with **Hugo Blomberg** and **Gösta Thames** on the 1949 Ericsson Ericofon telephone.

Gerd Alfred Müller (b. 1932)
German designer. Müller worked on kitchen appliances and shavers at Braun between 1955 and 1960 with **Dieter Rams** and **Hans Gugelot.**

Marcello Nizzoli (1887–1969)
Italian designer. Nizzoli was originally a painter and interior designer and only turned to industrial design in 1938 at the age of 51 after joining Olivetti. He designed the Olivetti Lettera 22 portable typewriter of 1950 and the Mirella sewing machine of 1956.

Isamu Noguchi (1904–1988)
Japanese-American designer. Noguchi originally trained as a sculptor. Much of his work is in furniture, but he is best known for his Akari paper lamp shades based on the traditional Japanese *chochin* lamp. These have been copied worldwide.

Eliot Noyes (1919–1977)
American architect and industrial designer. Influenced by Modernism, Noyes was briefly head of the Department of Industrial Design at the Museum of Modern Art, New York. He is best known for his highly successful corporate design and identity work for IBM in the 1950s and 1960s.

Noyes designed the Executive (1959) and Selectric or "golfball" (1961) typewriters. He was also a consultant to Mobil, Westinghouse, and Pan Am.

Roberto Pezzetta (b. 1946)
Italian designer. Pezzetta joined Zanussi in the mid-1970s and became head of the industrial design section in 1984. Pezetta designed the postmodern Wizard refrigerator of 1987.

Dieter Rams (b. 1932)
German designer best known for his work with Braun in the 1950s and 1960s. Rams initially studied and worked as an architect before joining Braun in 1955. He designed many products, including the KM321 Kitchen Machine of 1957 and the SK4 Phonosuper record player of 1956 (with **Hans Gugelot**). Rams was appointed professor at the Hochschule fur Bildende Kunste (University of Fine Arts) in Hamburg in 1981. He has also designed furniture for the Vitsoe Company.

Gordon Russell (1892–1980)
English Arts and Crafts–influenced furniture designer and director of the UK Council of Industrial Design. His company made "modern-style" wooden cabinets for Murphy radios in the 1930s. They were designed by his brother, Dick Russell.

Richard Sapper (b. 1932)
German-born designer who has worked for Italian companies. Sapper worked at Mercedes-Benz until 1957 and moved to Milan in 1958. He worked with **Marco Zanuso** on televisions and later designed the Bollitore kettle of 1983 for Alessi and the Tizio lamp of 1978 for Artemide. Sapper returned to Germany in 1972 and has been a consultant to IBM since 1981, designing a number of products, including the Leapfrog computer of 1992.

Sixten Sason (1912–1969)
Swedish designer. Sason designed cars for the SAAB motor company in the 1940s and 1950s. He received commissions from the Husqvarna Company for kitchen products, sewing machines, and irons. Sason was also a consultant to Electrolux and the Hasselblad camera company.

Douglas Scott (1913–1990)
English designer. Scott is best known for his Routemaster bus for London Transport (1953). He worked in the London office of **Raymond Loewy,** where he designed refrigerators and vacuum cleaners for Electrolux. He was professor of Industrial Design at the Autonomous National University of Mexico from 1976 to 1979.

Clive Sinclair (b. 1940)
English journalist and inventor. Sinclair established Sinclair Radionics in 1962. Sinclair produced the first pocket calculator in 1972 and a miniature television in 1977. He also pioneered home computing with the Sinclair ZX80 of 1980. Sinclair's move into transport with the C5 electric car of 1985 was unsuccessful.

Ettore Sottsass (b. 1917)
Influential Italian architect and designer. Sottsass made his name with Olivetti in the 1960s. He founded the Memphis group in 1981 and has designed for Alessi. The Memphis style of "anti-design" and use of color has influenced mainstream appliance design.

Philippe Starck (b. 1949)
French product designer. Starck originally trained as an architect. He established himself in the 1980s with witty designs for lamps, lemon squeezers, kettles, and domestic plastics for Alessi, as well as lamps and plastic furniture for Kartell. Starck has also produced stylish casings for televisions and computer disk drives.

Walter Dorwin Teague (1883–1960)
Influential American designer. Teague opened his consultancy in 1926. He established his reputation in the 1930s and had numerous corporate clients. His major work related to transport, with projects for Boeing airplanes, car bodies, and service stations. He also designed cameras for Kodak from 1928 until his death, as well as products for Pyrex.

Gösta Thames (b. 1916)
Swedish designer. Worked with **Hugo Blomberg** and **Ralph Lysell** on the 1949 Ericsson Ericofon telephone.

Louis Comfort Tiffany (1848–1933)
American decorative artist and interior designer. Tiffany established his glass company in New York in 1885. It produced Art Nouveau glassware and oil and electric table lamps with stained-glass shades. By 1905 the company employed over 200 people; its designs have been much copied.

Harold van Doren (1895–1957)
Initially an academic and assistant director of the Minneapolis Museum of Arts, van Doren entered

industrial design in the early 1930s. Van Doren designed plastic cases for Air-King radios and washing machines for Maytag.

Frank Lloyd Wright (1867–1949)
Highly influential American architect. Wright was influenced by Japanese architecture and the functionalism of the Chicago architect Louis Sullivan. He worked for private and corporate clients and developed what he termed "organic architecture"—buildings that grew naturally from the landscape with fixtures and fittings designed by him.

Marco Zanuso (b. 1916)
Italian designer. Zanuso has worked for Olivetti and Kartell. He designed the 1956 Borletti sewing machine and collaborated with **Richard Sapper** on the portable television Doney14 of 1962, the TS 502 radio, and the Black ST/201 television of 1969, all for BrionVega.

Piet Zwart (1885–1977)
Dutch designer. Zwart worked in architecture and graphic design and was a guest teacher at the Bauhaus. He developed the Bruynzeel kitchen in the late 1930s and 1940s.

TECHNICAL GLOSSARY

Accumulator
An electric battery that can be recharged by passing a reverse current through it. Also known as a storage battery or secondary cell.

Active matrix
Also known as TFT (thin film transistor), an active matrix LCD (liquid crystal display) provides high-quality graphics as each pixel has its own tiny transistor, which is suspended in a film between two transparent sheets. The transistors respond to voltage applied across the vertical and horizontal wires. A color display requires three transistors (for red, blue, and green) for each pixel. The active matrix LCD of a computer screen may contain a billion transistors.

Alternating current (AC)
An electric current that reverses its direction periodically to flow back and forth. Mains electricity is supplied as alternating current.

Alternator
A generator that produces alternating current.

Ampère (A or amp)
Unit of electric current, named after the French scientist André Ampère (1775–1836). It is variously defined as the current that can send one volt through one ohm and the current that can produce a specific magnetic effect between two long, straight, parallel conductors 1 meter (3.3 feet) apart in a vacuum.

Amplifier
A single electronic device, such as a triode valve or transistor, or an assembly of devices that increases the strength of an input signal to deliver a greater output signal.

Amplitude
The range or width of an electromagnetic wave.

Amplitude modulation (AM)
The variation of the amplitude of a carrier radio wave by superimposing sound waves of greater amplitude and lower frequency.

Anion
A negatively charged ion that moves toward the anode during the process of electrolysis.

Anode
The positive electrode which receives electrons from the heated cathode of a device such as an electric cell or thermionic valve.

Armature
The coil of the electromagnet in an electric motor or generator.

Battery
A combination of two or more electric cells placed in the same polar orientation so that opposite ends (positive and negative) meet. When connected to an electric circuit, the battery provides a supply of electric current.

Bit
A single binary digit, being either 1 or 0. A bit is the smallest unit of information stored by a computer. The bit rating of a microprocessor indicates the number of bits that it can process simultaneously.

Boolean logic
A set of algebraic rules widely used in computing whereby a problem is stated as a series of true-false propositions. The propositions are expressed through operators such as "NOT" and "OR."

Byte
A byte is the amount of computer memory or disk space required to store a single character and is expressed as a number of bits. The number of bits required depends on how many characters the system needs to recognize. Most computer systems use an 8-bit byte.

Capacitor
A device storing an electric charge. It consists of two conducting plates separated by a dielectric. When electric charge is applied to one plate, it induces opposite charge on the other plate. The charge is stored because it cannot be transmitted across the dielectric.

Cathode
The negative electrode which, when heated, releases electrons toward the anode(s) of a device such as an electric cell or thermionic valve.

Cathode ray tube
A glass vacuum tube with an electron gun at one end and a fluorescent screen, a coating of phosphors, at the other end. The electron gun, consisting of an electrically heated cathode and three anodes, fires a beam of high-energy electrons at the screen, where the phosphors glow in reaction to bombardment by electrons. When the electrons rebound against the sides of the tube, which are coated with conducting black graphite paint, the energy is absorbed as heat and fed back to the electron gun. The electron beam is focused by sets of vertical and horizontal plates between the electron gun and the screen.

Cation
A positively charged ion that moves toward the cathode during the process of electrolysis.

Charge-coupled device (CCD)
A metal-oxide semiconductor device that stores charges and releases electrons in response to exposure to light.

Circuit
An assemblage of electric and electronic components through which an electric current can flow. The components may be arranged in series so that the current passes from one to another or in parallel so that the current passes through side-by-side components simultaneously.

Commutator
An opposite pair of insulated conductors connected to an electric circuit by metal or carbon brushes. Commutators are used to reverse current in direct current motors and generators.

Compressor
A device that compresses a gas, forcing it to change to the liquid state and lose heat. In refrigerators, gas compression produces the required cooling effect.

Condenser
A device that converts gases or vapors into liquids; also, another name for a capacitor.

Conduction
The act or property of transmitting electricity or heat through solids by means of contact between molecules.

Conductor
A substance capable of transmitting electricity or heat.

Convection
The act of transmitting electricity or heat through liquids or gases by currents.

Cycle
One of a series of periodic changes from an original state back to that state.

Cyclonic
Pertaining to a spiral passage of air drawn toward an area of lower pressure.

Dielectric
A substance incapable of conducting electricity; an insulator.

Diode
A type of thermionic valve or transistor containing two electrodes, a cold anode and heated cathode. It allows current to flow in only one direction and is therefore commonly used as a rectifier.

Direct current
An electric current that flows in only one direction.

Discharge tube
A glass tube that has been evacuated, leaving only traces of gas, through which an electric current is passed between electrodes. The ionized gas atoms emit light as their energy levels change in response to the passage of electrons.

Dynamo
A simple electric generator consisting of a bar magnet placed between the poles of a coil (or armature).

Electric cell
Two electrodes separated by an electrolyte.

Electric motor
A device that converts electrical energy into mechanical energy, allowing electric current to power functions such as turning fans or driving pumps. Motors may be supplied by direct current, alternating current, or, in the case of universal motors, either. They may also be fixed or variable speed. Size and power output are tailored to the intended use.

Electrode
A conductor through which electric current enters or leaves an electric cell or electron tube. A pair of electrodes consists of an anode and cathode.

Electrolysis
The process of the chemical decomposition of a substance by the application of an electric current, causing the migration of ions to the electrodes.

Electrolyte
A liquid that can be changed chemically by the application of electric current though an electrode.

Electromagnet
The basis of the electric generator, motor, and transformer, it consists of a soft iron core (a bar magnet) and a coil.

Electromagnetic waves
Oscillating electric and magnetic waves within a specific spectrum of wavelengths and frequencies traveling together. Radio waves have long wavelength and low frequency, whereas ultraviolet radiation has short wavelength and high frequency.

Electromagnetism
Magnetism produced by applying electric current to a natural magnet, such as an iron bar or coil.

Electron
A negatively charged particle within an atom. Electrons orbit the nucleus of the atom. The number of electrons in an atom varies from element to element. The atomic number of an element is the number of electrons in a neutral atom.

Fiber optics
Technology based on the transmission of light through very thin strands of glass or plastic.

Filament
A thread, usually intricately coiled, of a substance such as carbon or tungsten that provides high resistance to the passage of an electric current in an incandescent lamp or thermionic valve.

Frequency
The speed of variation of electromagnetic waves or current in terms of the number of oscillations, vibrations, or waves per unit of time. The standard unit of measurement of frequency is the hertz (Hz), which represents one cycle per second.

Frequency modulation (FM)
The variation of the frequency of a carrier radio wave in accordance with the frequency and amplitude of the sound signal.

Fuzzy logic
A form of computer logic that, unlike Boolean logic, does not rely on a series of true-false propositions, but combines a number of propositions simultaneously to define a point within a spectrum of probability. Fuzzy logic is regarded as being closer to normal human problem-solving than Boolean logic.

Generator
A machine that converts mechanical energy to electrical energy. A generator may produce alternating or direct current. "Dynamo" and "alternator" are terms used for particular types of generator.

Hertz
The standard unit of measurement of frequency.

Incandescence
White heat and light produced when, for example, a filament resists the passage of electric current.

Induction
The production of an electric current by the forces of magnetic attraction; the production by one conductor of an opposite electric charge in another nearby conductor.

Induction coil
An electric device consisting of two coils, one coil induces elctromagnetism in the other.

Integrated circuit
A miniature electronic circuit contained on a single chip or wafer of a semiconducting substance, such as silicon or germanium. It may contain millions of electronic components.

Ion
Atom or atoms that have become positively or negatively charged through the loss or gain of electrons caused by certain chemical reactions such as electrolysis.

Laser
Acronym for light amplification by stimulated emission of radiation. A device that produces a narrower, focused, and high-energy beam of light. Pulses of light can represent sound, pictures, or data, depending on the laser application. The high speed of the laser is the key advantage.

Liquid crystal display (LCD)
A display screen, used in portable computers, consisting of a thin layer of semiliquid crystals between transparent sheets. The rod-like crystals naturally align in rows. The application of an electric field twists the molecules, bending light so that it is reflected or transmitted, producing an image.

Metal-oxide semiconductor (MOS)
A particular type of semiconductor used for computer chips. MOS chips consume little power, but are slower than the bipolar chips based on transitor-transitor logic (TTL).

Microphone
A device that converts sound waves into electrical signals. The sound waves produce mechanical energy, expressed as the vibration of a thin diaphragm. The vibrations are converted to electrical pulses by means of a moving coil or a capacitor.

Microprocessor
An integrated circuit that contains the entire central processing unit for a computer or similar appliance.

Modulation
The variation of the frequency or amplitude of a radio carrier wave according to the characteristics of the sound signal being transmitted.

Optical fibers
Thin strands of glass or plastic that can transmit electrical energy as light.

Passive matrix
A type of liquid crystal display that passes current between electrodes at each end of the grid of rows and columns that make up the screen. They produce low-quality images and are slow to respond.

Photoelectric cell
A device made of a light-sensitive substance, such as selenium, which detects or measures light by triggering a flow of current, a deflection, or a change in voltage.

Pickup
A device used in record players and electric guitars that detects vibration and transmits it to an amplifier as an electric signal. The detecting mechanism may be a small electromagnet or a piezoelectric crystal that changes voltage in response to mechanical pressure.

Planar technique
Technique whereby a circuit is built up on a thin slice of a semiconductor, particularly silicon, in layers. The silicon wafer is repeatedly coated with a photoresist (a substance that is photosensitive), etched, and cleaned to create complex circuits.

Polyphase
Having phases of two or more alternating voltages of equal frequency that are displaced by a fraction every cycle.

Primary cell
A type of battery in which a one-way chemical reaction is converted into electrical energy. Most dry-cell batteries are primary cells.

Printed circuit
An insulated board on which the circuit track is "printed" by applying a conducting material. The circuit components are then soldered, wired, or pinned in place.

Radiation
The release of electromagnetic energy in the form of waves.

Random access memory (RAM)
The computer's working memory that is used to store the programs and data that are currently in use. The random access means that contiguous bytes of data do not have to be accessed sequentially.

Reactor
A device that produces nuclear energy. All types of nuclear reactor have the same two basic components: rods of fissile material, such as uranium, which spontaneously decay and release heat, are surrounded by a "moderator," such as carbon or heavy water, which controls the rate of reaction by preventing overheating.

Read-only memory (ROM)
A memory chip that permanently stores the most vital instructions that the computer needs to operate. The data in read-only memory cannot be altered or deleted.

Rectifier
A device, such as a diode, that converts alternating current into direct current.

Resistance
The property of a conductor that restricts the flow of electric current through it and is expressed by the release of energy as heat. The standard unit of measurement of resistance is the ohm.

Resistor
A device, such as a rheostat, placed in an electric circuit to inhibit the flow of current.

Rheostat
A rheostat is described as a variable resistor because its sliding contact means that it is not continuously in contact with the circuit. One form of the rheostat is the lighting dimmer switch.

Secondary cell
A battery in which the potential for the chemical reaction that is converted to electrical energy can be renewed by recharging. When the secondary cell is exhausted, it is recharged by receiving mains electricity, thereby reversing the original reaction. Wet-cell batteries, such as car batteries, are secondary cells.

Semiconductor
Substances, such as silicon and germanium, that are neither conductors nor insulators. Their conducting properties depend on the number of free electrons that have escaped their molecular bonds. Their conductivity can be selectively increased by adding tiny impurities, a process known as "doping."

Silicon chip
An integrated circuit or microprocessor.

Solid-state
A term used to describe electronic appliances that use solid transistors rather than hollow thermionic valves.

Thermionic valve
The earliest type of electronic component that consists of a hollow glass or metal tube with a minimum of two electrodes. Electric current is passed through the evacuated tube. Depending on the number, type, and arrangement of the electrodes, the thermionic valve can perform a number of functions in an electric circuit, including rectifying and amplifying current. Diodes, triodes, and pentodes are types of thermionic valve, defined by the number of electrodes. Thermionic valves are also known as vacuum tubes.

Thin film transistor (TFT)
See Active matrix.

Transformer
A device whereby the voltage of alternating current is changed by electromagnetic induction. The incoming current passes to the primary coil of the transformer and induces current in the secondary coil, which shares the same magnetic iron core. The change in voltage is determined by the ratio of the respective number of turns in the coils.

Transistor
A miniature solid electronic component made of one or more semiconducting substances. It performs the same range of functions as its predecessor, the thermionic valve. The advantage of the transistor over the thermionic valve, aside from compactness, was that it consumed less electricity.

Triode
A thermionic valve containing three electrodes, an anode, a cathode, and a negative control grid. Varying the voltage on the grid by small amounts produces large variations in current, hence the triode's use as an amplifier.

Vacuum tube
See Thermionic valve.

Voltage
The potential difference between two points in an electric circuit, expressed as a number of volts, the standard unit of measurement. The potential difference refers to the amount of electrical energy converted by the flow of a unit of electric charge from one point to the other.

INDEX

Page numbers in boldface denote main entries.